Guide to Electronics in the Home

GUIDE TO ELECTRONICS IN THE HOME

The Editors of Consumer Reports Books
with Monte Florman

CONSUMERS UNION
Mount Vernon, New York

Special thanks to A. Larry Seligson
for his review of the contents of this book.

Library of Congress Catalog Card Number: 87-73412
ISBN: 0-89043-215-5
Second printing, April 1989

Design by Joy Taylor
Manufactured in the United States of America

Guide to Electronics in the Home is a Consumer Reports Book published by Consumers Union, the nonprofit organization that publishes *Consumer Reports,* the monthly magazine of test reports, product Ratings, and buying guidance. Established in 1936, Consumers Union is chartered under the Not-For-Profit Corporation Law of the State of New York.

The purposes of Consumer Union, as stated in its charter, are to provide consumers with information and counsel on consumer goods and services, to give information on all matters relating to the expenditure of the family income, and to initiate and to cooperate with individual and group efforts seeking to create and maintain decent living standards.

Consumers Union derives its income solely from the sale of *Consumer Reports* and other publications. In addition, expenses of occasional public service efforts may be met, in part, by nonrestrictive, noncommercial contributions, grants, and fees. Consumers Union accepts no advertising or product samples and is not beholden in any way to any commercial interest. Its Ratings and reports are solely for the use of the readers of its publications. Neither the Ratings nor the reports nor any Consumers Union publications, including this book, may be used in advertising or for any commercial purpose. Consumers Union will take all steps open to it to prevent such uses of its material, its name, or the name of *Consumer Reports.*

CONTENTS

Guide to Electronics in the Home

■ INTRODUCTION

In 1984, Consumer Reports Books published the first edition of *Guide to Electronics in the Home.* Prepared with the assistance of Consumers Union's emeritus Technical Director Monte Florman, the book featured test reports and brand-name Ratings and recommendations that were originally prepared for publication in *Consumer Reports* magazine and carefully reviewed and revised for the book. Two years later, a revised edition provided additional information about products in the burgeoning home electronics marketplace.

Now, this 1988 edition of *Guide to Electronics in the Home* updates the prior revision and presents a great deal of new product data.

Today, more than ever before, the overwhelming number and variety of models available to consumers may make it extremely difficult to know what electronic product is the best choice for you. *Guide to Electronics in the Home* is filled with useful and objective information to help you make an intelligent and informed decision about what product best suits your needs *before you buy.* It combines the virtues of *Consumer Reports'* concise and expert presentations with the convenience of a book—a ready reference, easy to use, with the information you want to compare all in one place. Each product section gives general information about what you should expect in the way of performance, convenience, and features, as well as maintenance tips.

The Ratings charts

Ratings of individual brands and models are based on CU's laboratory tests, controlled-use tests, and/or expert judgments. Although the Ratings are not an infallible guide, they do offer comparative buying information that can greatly increase the likelihood that you will receive value for your money.

You may be tempted to look first at whatever brand appears at the top of the Ratings order. Resist the impulse if you can. Instead, begin with the introductory section that precedes each set of Ratings, and then read the notes and footnotes. In those sections you will find the features, qualities, or deficiencies shared by the products in the test group. You may find out, in a sentence that starts with "Except as noted," what qualities most of these electronic products have in common.

The first sentence in the introduction to the Ratings tells you the basis on which the order of the Ratings was decided. Sometimes the products may be listed alphabetically. Usually, Ratings are "Listed in order of estimated quality." That means CU's electronics engineers judged the brand (or product type) listed first to be the best, the one listed next to be second best, and so on. If the introduction says, "Listed, except as noted, in order of estimated overall quality," it means that somewhere in the Ratings one or more groups of products are about equal to one another and are therefore listed in a special fashion—for example, "models judged approximately equal in overall quality bracketed and listed alphabetically."

Prices. Each Ratings chart includes the month and year in which the test group appeared in *Consumer Reports.* Usually the prices are the ones published at the time of the original report.

Model changes. Manufacturers commonly change models once a year or so to match their competition, to stimulate sales by offering something "new," or to incorporate improvements in technology and design. Quite often, however, retailers will carry over older models when a manufacturer introduces new ones. Where the dealer carry-over is heavy, older models may remain available for months, even years, after they have been discontinued.

There is a good chance that the particular brand and model of electronic equipment that you select from a Ratings chart will either be out of stock when you go to buy it or it will have been superseded by a later version. One way in which manufacturers try to outdistance their competitors is by introducing more and more "convenience" features. Some contribute to the general utility of the product. Others are simply frivolous, may actually impede the product's usefulness, or cause confusion. Our evaluation helps guide you through the maze of features and the performance Ratings.

To keep as up to date as possible on the rapidly changing electronics marketplace, you should refer to current issues of *Consumer Reports* as additional information about products covered in this book becomes available. That way you can also find out about new products introduced since this book was revised and updated.

■ Component TV Systems

TELEVISION

Manufacturers hope to persuade consumers to buy separate TV tuner, monitor, and speakers as part of a video system that could expand to include a VCR, home computer, a stereo sound system, access to interactive videotex data bases and services, and more. In time, the traditional TV receiver with a cabinet housing built-in tuner, speakers, and picture tube could give way to the more expensive component concept. Before that happens, consumers will have to be convinced to accept higher prices for the promise of much better sound and picture, greater flexibility, and improved convenience.

Figure on spending at least $800 for a component-TV package. For about the same money, you can buy a deluxe monitor/receiver, a one-piece set with its own stereo decoder and amplifier, inputs for accessories, and output jacks for separate speakers.

Running counter to the component trend is the TV entertainment "center" containing the sound system and (sometimes) the loudspeakers. The quality of these units is high. Your choice of a particular brand and model of either video product will often be governed by the features you want, and how much you're willing to pay for them.

Not so long ago, buying a new TV set posed only a few basic choices: color or black and white, a big screen or a small one, a console or a table model. There are a lot more entries on the TV menu these days.

Because stereo and high-fidelity sound are becoming commonplace, in broadcasts as well as in prerecorded tapes, TV sets can now hold their own with other audio and video gear. That's a welcome and long-overdue development. Of course, while you can still park the set at one end of the living room and never mind the home-entertainment system, to take full advantage of a TV set's capabilities, you may soon be connecting it to one or two other components.

Sets with a 19- or 20-inch screen are the most popular choice—large enough for the whole family to watch yet small enough that they won't dominate the room. Many models with this screen size are available with a remote control. Some are relatively expensive sets, with stereo sound and, typically, a welter of input and output jacks for such peripheral equipment as a stereo VCR, external speakers, and a computer. Others are simpler and cheaper; they have monophonic sound, fewer accessory jacks, and, in some cases, a more rudimentary remote control.

Picture quality

The most important aspect of any TV set is still picture quality.

The smallest sets around are pocket-sized. Black-and-white versions list for around $160, color sets for around $250. Tiny TV is more a gimmick than anything else; the sets deliver a picture of only moderate quality.

Color sets cost a good deal more than black-and-white models. For as little as $45, you can get a basic 12-inch black-and-white set as a handy second set for the kitchen, den, or a child's room. Color TV prices start at about $120 for a set with a 13-inch screen. The 19-inch color sets start at about $170.

With table models and portables, you may be able to save some money by tailoring the TV set to the size of your room. You can view a 13-inch set comfortably from 4½ feet or so, a 19-inch set from about 6½ feet, and a 25-inch set from about 9 feet.

You may want to consider a set with a 35-inch "direct-view" screen or a projection TV set if you have a particularly big viewing room or if you want to approach the big-screen impact of a movie theater. But be prepared for a major outlay—$1,800 to $4,000.

Fortunately, just about any set will produce an acceptable TV picture. But differences do exist, with some sets offering exceptionally clear picture quality.

Picture clarity. Good picture clarity means that images are rendered in fine detail, without fuzzy or unnatural outlines. Individual strands of hair or textures of fabric should be clearly discernible.

Contrast. The factors that affect contrast the most are *black-level retention* and *screen reflectance.* They give a TV picture "punch," regardless of the program material or the lighting level in the room. On a set with good black-level retention, black areas remain black when they should—in night scenes, for example. On sets with poor black-level retention, large black areas will wash out to gray. A TV set with the desirable low-screen reflectance appears dark when the set is off. A dark picture tube inhibits screen reflectance by absorbing room light, which means that contrast doesn't suffer as much in bright room lighting as with a lighter-colored screen.

Color fidelity. Good color fidelity means that the grass on a golf course, for example, will be nearly the same green as you are likely to see outdoors, that an egg yolk will be yellow-orange, and that flesh tones will be those of living human beings. Colors will remain consistent as a scene progresses.

Cable TV reduces reception problems. Without it your television set may have to overcome such hurdles as weak reception and interference from electrical gadgets, passing vehicles, and the fluttering TV signals that bounce off airplanes—an especially important consideration if you live near an airport.

Geometric distortion. Images on the screen should maintain their shape and size, without warping near the sides or corners of the screen.

Interlace. On a set with good interlace, the horizontal lines that compose a TV picture will be evenly spaced and indiscernible at normal viewing range.

Automatic color control. Circuitry in most sets will do a good job of keeping color levels consistent from channel to channel or in the face of such adversities as airplane flutter.

Adjacent-channel rejection. If you subscribe to a cable-TV system that fills all the available channels, a set with good adjacent-channel rejection will ignore strong signals from channels next to the one being watched.

VHF fringe reception. A set should be able to pull in signals and produce a good picture on channels 2 through 13 even if you live on the periphery of a viewing area.

UHF fringe reception. Channels 14 through 83 are generally more difficult to receive than VHF stations. Nevertheless, any set should be able to give you a usable picture on any UHF station that's within reach.

Spark rejection. Electrical interference from an automobile ignition or from a sparking appliance such as an oil burner, a vacuum cleaner, or a hair dryer shouldn't distort the TV picture with streaks or dots.

Sound

Over the years, TV viewers have had to put up with second-rate sound while manufacturers concentrated on picture quality. All that has changed.

Producers and broadcasters are devoting more attention to the sound tracks of their programs. Manufacturers have improved the tuners and speakers in TV sets. Most important, stereo broadcasts reproduce the aural spaciousness of programs broadcast in stereo.

A stereo TV set may be advertised merely as "stereo-equipped" (or "stereo-ready") or as having a "built-in MTS decoder with SAP." Either way, the ads mean that a TV set can receive and reproduce "multichannel TV sound" broadcasts. The set can also receive any "separate audio program" transmitted along with the TV picture. SAP could give viewers a choice of watching a foreign film in its own language or a dubbed version. A stereo-equipped set has an MTS adapter jack that lets you add a decoder for stereo later on. With most such sets, you must feed the decoder's sound output into a hi-fi system to get stereo sound.

Stereo programming is gradually becoming more common. Most cable companies pass along the stereo signal from the networks. Even if you're not eager to listen to stereo, a stereo set may be worth considering. Stereo sets deliver markedly better sound than their monophonic counterparts.

Tone quality. The ideal frequency response is "flat," meaning that the set neither overemphasizes nor underemphasizes any frequency in the audible spectrum. Sets with the best tone quality sound good, but not as good as high-quality low-priced audio components.

The built-in speakers on most stereo sets are positioned far enough apart to provide a discernible stereo effect. Surprisingly, even such unlikely candidates as talk shows benefit from stereo. And concert performances are genuinely improved, as adherents of simulcast stereo broadcasts are already aware.

Sets with speakers mounted below the screen tend to produce the most meager stereo effect. Speakers that flank the screen are generally better.

To take full advantage of stereo sound, you can move toward the home-entertainment-system approach and connect the TV set to other audio components. Many models have jacks for external speakers driven by the TV set's own amplifier. Those jacks typically deliver power levels of five watts or less, which isn't really enough to drive a decent pair of stereo speakers at reasonably loud volume. The alternative: Feed the audio signal to a hi-fi receiver, which will pipe it through the speakers. Hooking up a TV set that way isn't difficult, but it can create some pretty messy wire spaghetti, depending on the placement of components. It's best to put the speakers a couple of feet away from each side of the TV set.

Remote controls

Walking over to the TV set to change channels, fine-tune the picture, or adjust the sound is almost a thing of the past. You can now do all those things and more with the remote control. Indeed, many sets have only a rudimentary set of controls on the cabinet itself, so you must use the remote control.

The basic infrared remote control lets you change channels (or scan quickly from station to station), adjust the volume, mute the sound, and turn the set on and off. Most remotes let you do much more. Among the handier features are these:

Picture controls let you adjust brightness, contrast, color intensity, and tint to your liking.

Dual antenna inputs let you switch from a regular broadcast TV channel to a cable channel.

Dual video sources let you switch easily between the TV set and the VCR.

Last-channel recall lets you shuttle quickly between two stations. It's a handy feature when you want to watch two sporting events.

Sleep timer shuts off the set automatically after a predetermined interval (typically, 30, 60, or 90 minutes). That way, if you fall asleep in front of the late movie, the set won't drone on until dawn.

Main sound and SAP (separate audio program), on stereo sets only, lets you change from the main stereo program to SAP, broadcast by a few stations; it might, for example, let you listen to a foreign film in its original language or switch to a dubbed version.

Tuner/external video refers to a remote control that can switch from a TV broadcast to a taped program from, say, a VCR.

Tone/balance on a remote control allows you to adjust the sound on a stereo set.

Channel block-out allows you to use the control to restrict the viewing of certain channels.

Stereo/mono allows listening to stereo programs in mono if the stereo effect becomes intrusive.

Other features

Screen size. The size of the screen is measured diagonally. A number of sets sport a flat-faced, squared-off tube—a design notable mainly for its promotional value. The square corners add an advertisable extra inch or so to the screen—a 19-inch set with a rounded-corner tube becomes a 20-inch set when the tube is square. A flatter tube is somewhat better at avoiding reflections from lights in the room. Otherwise, the gain in what you see is pretty small.

Channel presets. This term describes how many channels you can receive without resetting the tuner. If you want to receive a channel that isn't among the preset ones, you have to fiddle with tiny knobs, dials, or switches. Where no number is given in the chart, it means the set will tune in any channel it can receive at the touch of one or two buttons.

Direct access. This is a feature that permits using a keypad to tune from one channel directly to any other, without scanning or skipping through the intervening stations.

Numeric keypad. Most remote controls have a numeric keypad. Among other things, a keypad lets you tune from one channel directly to any other.

Last-channel recall. With this feature, you can switch quickly between the channel you're watching and the last channel viewed. This is handy for, say, sports fans who want to keep up with two games at once.

Dual antenna. TV sets with this feature have buttons that let you switch from, say, a cable-TV channel to an off-the-air channel.

TV/video. This refers to buttons that let you switch from the antenna input to input from a video source, such as a videocassette recorder or a video camera. It's a handy feature if you decide you want to watch a program you're taping.

MTS decoder. Sets with the decoder are stereo sets, with all the necessary circuitry to allow programs broadcast in stereo to be heard in stereo. Stereo sets also give you access to the separate audio program (SAP), an additional audio channel that can provide sound in a second language.

MTS adapter jack. Sets with this feature can easily be upgraded by adding a separate stereo decoder.

Audio input/output. For stereo TVs, you should have a pair of output jacks; they let you pipe the sound from the TV set to separate speakers.

Video input/outputs. These allow you to plug accessories, such as a VCR or videodisc player, into the TV set.

Headphone jack. Sets with this feature let you plug in a pair of headphones so you can listen without disturbing others.

RGB input. This is a feature of interest if you own video games or a computer and don't have a separate color monitor. The RGB (red-green-blue) input bypasses some of the set's circuitry to deliver a sharper, crisper computer display.

On-screen displays. These indicate that the set will display the time of day, the channel number, or other data at the touch of a button. This is generally a handy feature.

Channel block-out is a control that allows you to prevent the viewing of certain channels. Useful for parents who want to control what the kids watch and when they watch it.

Sharpness control. This is useful if you live near the edge of a reception area, for example, and need to adjust the picture to "soften" it as a way of reducing picture "snow."

Future enhancements. Sets with digital circuitry crop up here and there in some brand lines. Digital technology could allow zoom or freeze-frame viewing and improve the picture's clarity. Manufacturers have been singing the praises of digital TV for several years but have yet to market sets that take full advantage of the technology.

High-definition TV, another promising development, can deliver a picture that's as clear and crisp as a high-quality photograph. Manufacturers are arguing over the proper standards for high-definition TV. But the biggest obstacle is likely to be governmental rather than technical. High-definition broadcasts require a bigger chunk of the broadcast spectrum than stations and networks currently command. It's up to the Federal Communications Commission to decide whether to allocate additional frequencies.

Recommendations

If you're in the market for a new TV set, you should make a stereo set your first choice. You'll get a set that should deliver a very good picture. (Good picture quality is more the rule than the exception with 19- and 20-inch TV sets these days.) You're also more likely to get a set that delivers good sound in both stereo and mono. And you'll have a set that will remain up to date for some time to come.

There's nothing wrong with mono sets. If all you want is a fairly basic set that delivers a good picture, you have plenty to choose from, at list prices about $200 to $300 less than comparable stereo sets.

If you think you may want stereo someday soon, look for a set that has an MTS adapter jack. It allows a nonstereo set to be upgraded with the addition of a separate stereo decoder (see page 26).

Ratings of 19- and 20-inch color TV sets

Listed in order of estimated quality. Prices are as published in a **March 1987** report.

◉ ◓ ○ ◒ ●

Better ⟵ ⟶ *Worse*

Brand and model	Price	Dimensions, in.	Picture clarity	Black-level retention	Screen reflectance	Color fidelity	Geometric distortion	Interlace	Airplane flutter	Automatic color control	Adjacent-channel rejection	Fringe reception, VHF	Fringe reception, UHF	Spark rejection	Tone quality
Stereo TV sets															
✓ Sony KV-20XBR	$700	18 × 20 × 22	◉	◉	◓	◉	○	◉	◓	◉	◓	○	○	●	◉
Quasar TT6298YW	799	20 × 31 × 20	◉	◓	○	◓	◉	◓	◉	◉	◓	◓	◓	◒	○
Magnavox RG4314WR	630	18 × 27 × 19	◉	◉	○	◓	○	◓	◒	◉	○	◓	◓	◓	◓
Sears 42751	590	18 × 30 × 20	◉	◓	○	◓	○	◉	◉	◉	◓	◒	◒	◒	◉
Sylvania RKF198	657	19 × 20 × 19	◉	◉	○	◓	○	◉	○	◉	◓	◓	◓	◓	○
Toshiba CX2066	425	19 × 20 × 19	◉	◓	◒	◓	◓	◓	◉	◉	◉	○	○	○	◓
Mitsubishi CS-2053R	599	19 × 21 × 20	◉	◉	○	◓	○	◉	○	◓	◉	○	◓	○	◓
Fisher PC226W	529	19 × 21 × 21	◉	○	○	◓	◓	○	◓	◓	○	○	◓	◓	◓
General Electric 8-2090	482	38 × 27 × 31	◉	○	○	◓	◓	◓	◓	◉	◓	○	○	○	◉
Hitachi CT-1958	487	17 × 28 × 19	◉	◓	◒	◓	◓	◉	◓	◉	◓	◓	◓	◓	○
Panasonic CTG-2067R	615	20 × 27 × 21	◉	◓	○	◓	◓	◉	◓	◉	◓	◓	○	◒	◒
Sharp 20LP86	500	21 × 27 × 20	◉	○	◓	◓	◉	◒	◉	◉	○	◒	◒	◒	◓
Philco R3971AWA	468	17 × 26 × 19	◉	◉	○	◓	◓	○	○	◉	○	○	○	○	◓
Sanyo 12C700	500	20 × 26 × 21	◉	◓	○	◓	○	◓	◉	◓	○	◓	◒	○	◓

RCA FMR570ER	499	18 × 25 × 16	◉	◓	○	○	◒	◓	◓	◉	◓	◓	○	◓	◉
Zenith SC1935W	549	19 × 21 × 19	◉	◓	○	◓	○	◉	○	◓	○	○	◒	○	◓
J.C. Penney 2220	700	20 × 28 × 20	◉	◒	◒	◓	◓	◉	○	◉	◓	○	○	●	◓
Curtis Mathes M2075RW	800	20 × 21 × 20	○	◒	○	◓	◉	◉	○	◉	◓	◓	◓	◒	○
Montgomery Ward 12696	450	19 × 25 × 20	◉	◒	◓	○	◓	◉	◉	●	○	◒	○	○	◓
Emerson MS30R	460	19 × 27 × 20	◓	◓	○	◓	◉	◓	◓	◓	○	●	●	○	◓
Nonstereo TV sets															
Magnavox RG 4250WA	480	18 × 20 × 19	◉	◉	○	◉	◓	◒	○	◉	◓	◓	◓	○	◒
Hitachi CT-1965	399	19 × 23 × 19	◉	◓	○	◓	◓	◉	◓	◓	◉	◓	◒	◓	◒
Mitsubishi CS2001R	449	18 × 20 × 20	◉	◉	◒	◉	○	◉	◓	◓	◓	◓	◓	○	○
Sears 42002	290	17 × 23 × 20	◉	◓	○	◓	○	◓	◉	◉	◓	◓	◒	○	◒
Panasonic CTH-1942R	330	17 × 22 × 19	◉	◓	○	◓	◉	◒	◓	◉	◓	○	○	○	●
Sylvania RXF169	430	18 × 20 × 19	◉	◉	○	◓	◓	◉	○	◉	◒	◓	◓	○	◒
Curtis Mathes M1959RW	550	17 × 25 × 18	◉	◓	◒	◓	◓	◉	◒	◉	◓	◒	◒	◓	◓
General Electric 8-1930	330	18 × 23 × 19	◉	○	○	◓	◓	◓	◉	◉	◓	○	○	◒	●
J.C. Penney 2115	420	17 × 24 × 19	◉	○	◓	◓	○	◉	◓	◓	◉	◓	○	◓	●
Quasar TT5947AW	375	17 × 22 × 19	◉	◓	○	◓	◓	○	◓	◉	○	◒	○	◒	●
RCA FMR468WR	299	17 × 25 × 19	◉	◓	○	◓	○	◓	◓	◉	○	○	○	○	○
Sharp 19LP56	275	17 × 23 × 20	◉	◉	◓	◓	◒	◓	◉	◉	◓	○	○	○	◒
Sony KV-1926R	412	19 × 24 × 19	◓	◉	○	◓	○	◉	◓	◉	◓	○	○	○	○
Zenith SC1911W	338	17 × 24 × 19	◉	◓	◓	◓	○	◉	◓	◉	◓	○	○	◓	●
Philco R3932AWA	339	18 × 20 × 20	◉	◉	○	◉	◓	○	○	◓	○	○	○	●	◒
Gold Star CMT-9325	288	19 × 20 × 19	◉	◓	○	○	◓	◓	◓	◓	◓	○	◒	○	◒
Montgomery Ward 12636	300	17 × 19 × 19	◓	○	○	◓	○	◉	◓	◓	○	○	◓	◓	●
Sanyo 91C625	400	18 × 25 × 19	◉	○	◒	◓	○	◓	◉	◓	◓	○	○	○	◓
Emerson ECR212	250	16 × 24 × 19	◉	◒	◒	◒	◉	◒	◓	◓	○	◒	◒	○	◒

Ratings of 19- and 20-inch color TV sets (continued)

Brand and model	Remote control capabilities: Tint	Brightness/contrast	Color intensity	Main sound and SAP	Tuner/external video	Tone/balance	Channel blockout	Stereo/mono	Advantages	Disadvantages	Comments
Stereo TV sets											
✓ Sony KV-20XBR	✓	✓	✓	✓	✓	✓	—	—	B,D,E,J	f	*B,D,F,H,J,K,L*
Quasar TT6298YW	—	—	—	✓	✓	✓	—	✓	B,C,D,E	g,h,j	*A,B,C,D,F,G,H,N*
Magnavox RG4314WR	—	—	—	—	—	—	—	—	E,L,M	i,j	*E,G,J,L,N*
Sears 42751	✓	✓	✓	✓	✓	✓	—	✓	C,D,F	f,l	*B,F,G,H,J,L*
Sylvania RKF198	—	—	—	✓	✓	✓	✓	✓	C,D,E,G,H,I,M	c,h,j,l	*B,E,F,H,J,L*
Toshiba CX2066	✓	✓	✓	✓	✓	✓	—	✓	B,C,D,E,L	c,f	*B,F,L*
Mitsubishi CS-2053R	✓	✓	✓	✓	✓	—	—	—	A,C,E,G	c,f,l	*B,D,E,F,H*
Fisher PC226W	✓	✓	✓	✓	✓	✓	—	✓	C,D,F	f,g,h	*A,B,D,F,G,H,J,L*
General Electric 8-2090	—	—	—	✓	✓	—	✓	✓	B,C,D,E	b,e,j,l	*C,D,H,J,K*
Hitachi CT-1958	—	—	—	✓	✓	—	—	—	C,D,E	f,h,l	*C,D,G*
Panasonic CTG-2067R	—	—	—	✓	✓	—	—	✓	C,D,E	—	*A,C,D,G,J*
Sharp 20LP86	—	—	—	✓	✓	—	—	—	B,E,L	g	*A,G,H*
Philco R3971AWA	—	—	—	—	—	—	—	—	K,L,M	—	*E,G,J,L,N*
Sanyo 12C700	—	—	—	✓	✓	—	—	✓	F	i	*A,C,D,E,G,H*
RCA FMR570ER	—	—	—	✓	✓	—	—	—	C,D,E,K,L	b	*G,N*
Zenith SC1935W	—	—	—	—	—	—	—	—	A,E	b,c,e,f,k,l	*E,H,L,N*

J.C. Penney 2220	—	—	—	—	✓	—	—	—	B,C,D,E	h	*A,C,D,G,H,I*
Curtis Mathes M2075RW	—	—	—	✓	✓	—	—	—	B,C,D,E	c,h,i	*A,D,E,H,I*
Montgomery Ward 12696	—	—	—	—	—	—	—	—	C,D,E,L	f	*G*
Emerson MS30R	—	—	—	—	—	—	—	—	C,D,F,L	f	*A,D,G,H*
Nonstereo TV sets											
Magnavox RG 4250WA	—	—	—	—	—	—	—	—	—	—	*J,L*
Hitachi CT-1965		—	—	—	—	—	—	—	D	l	*C,D,J*
Mitsubishi CS2001R	—	—	—	—	✓	—	—	—	F	l	*D*
Sears 42002	—	—	—	—	—	—	—	—	E	l,m	*C*
Panasonic CTH-1942R	—	—	—	—	—	—	—	—	E	—	*C,D*
Sylvania RXF169	—	—	—	—	—	—	—	—	—	—	*J,L*
Curtis Mathes M1959RW	—	—	—	—	—	—	—	—	A	f,g,l	*I*
General Electric 8-1930	—	—	—	—	—	—	—	—	E	b,l	*C,D,J*
J.C. Penney 2115	—	—	—	—	—	—	—	—	E	b,d	*C,D*
Quasar TT5947AW	—	—	—	—	—	—	—	—	E	—	*C,D*
RCA FMR468WR	—	—	—	—	—	—	—	—	D,E	b	—
Sharp 19LP56	—	—	—	—	—	—	—	—	A	—	—
Sony KV-1926R	—	✓	—	—	—	—	—	—	D,E	l	*F,L*
Zenith SC1911W	—	—	—	—	—	—	—	—	E	k	*L*
Philco R3932AWA	—	—	—	—	—	—	—	—	—	a	*J*
Gold Star CMT-9325	—	—	—	—	—	—	—	—	A	e,m	*A,C,D,H,N*
Montgomery Ward 12636	—	—	—	—	—	—	—	—	E	l	*C*
Sanyo 91C625	—	—	—	—	—	—	—	—	A,F	l	*C,D,M*
Emerson ECR212	—	—	—	—	—	—	—	—	E	f,m	—

Ratings of 19- and 20-inch color TV sets (continued)

Specifications and Features

Except as noted, all have: • Picture shrinkage at low line voltage and horizontal overscan (a loss of some part of the picture) that are acceptably small. • Coaxial-input connector for VHF broadcast channels and cable TV. • Ability to receive VHF channels 2 through 13 and UHF channels 14 through 83. • Channel numbers that are easy to read at normal distances. *Unless otherwise indicated, all sets with audio-line outputs have:* • Satisfactory signal-to-noise ratio and total harmonic distortion, excellent frequency response, satisfactory stereo separation.

Key to Advantages

A–Receiver has numeric keypad for channel selection.
B–Has dual antenna connectors; allows full use of remote control if set is to receive one scrambled cable channel.
C–Displays time of day on screen.
D–Displays channel number on screen.
E–Can scan through channels; scan feature can be programmed to bypass unused channels.
F–Can scan through channels; scan feature stops automatically at next active channel.
G–Has multiple antenna connectors; external signal splitter required to allow full use of remote control if set is used with cable-TV converter box.
H–Can be programmed to retain and recall 1 set of audio settings; useful if you frequently switch between different sources.
I–Has extensive on-screen graphics to aid in operating set.
J–Has white-balance-compensation switch, useful for correcting small errors in color when you make your own videotapes.
K–Signal-to-noise ratio (a measure of background noise) better than most.
L–Total harmonic distortion (a muddying of the sound) lower than most.
M–Can be programmed to control VCRs of other brands.

Key to Disadvantages

a–Lacks VHF coaxial input; separate "balun" transformer required if this set is hooked to some outdoor antennas or to cable TV.
b–Picture or sharpness controls inconveniently located on side or back of set.
c–Speakers located below screen; severely limits stereo effect.
d–To reinstate a channel bypassed for scan feature, you must use the remote control.
e–One-button control can't be used without affecting room-light sensor.
f–One-button control switches picture settings to nonadjustable factory settings.
g–Signal-to-noise ratio (a measure of background noise) worse than most
h–Total harmonic distortion (a muddying of the sound) worse than most.
i–Stereo separation worse than most.
j–Frequency response worse than most.
k–Has excessive horizontal picture loss (12 to 14 percent).
l–Picture shrinkage worse than most.
m–Channel numbers harder to read at normal viewing distance than most.

Key to Comments

A–Has removable pane of glass over screen.
B–Status of several controls—not just time and channel—can be displayed on screen.
C–Has one-button control that restricts the range of other picture controls.
D–Has vertical hold control, a feature needed primarily for older video games.
E–Can produce "pseudostereo" sound from monophonic broadcasts. Pseudostereo isn't a substitute for true stereo, but it can enhance the viewing experience somewhat.
F–Push-button audio and/or picture controls.
G–Has built-in speakers flanking screen.
H–Can drive external speakers without additional amplifier.
I–Has "vacation" power switch that shuts down standby power; allows you to disconnect the set without unplugging it.
J–Cannot receive UHF channels beyond channel 69. (Channels above 69 are very rare.)
K–Has detachable speakers.
L–Has combination VHF/UHF coaxial input.
M–Lacks stereo but provides separate audio program (SAP).
N–Has room-light sensor, which automatically adjusts picture brightness somewhat to compensate for changes in room light.

A guide to features

Brand and model			Tuning			Remote control					Sound		Auxiliary inputs				Other	
	Price	Screen size, in.	Channel presets	Total channels/ cable channels	Direct access	Numeric keypad	Last-channel recall	Dual antenna	TV/video	Sleep timer	MTS decoder	MTS adapter jack	Audio input/output	Headphone jack	Video input/output	RGB input	On-screen displays	Channel block-out
Curtis Mathes																		
M2075RW	—	20	—	142/72	✓	✓	✓	✓	✓	✓	✓	—	2/3	2/2	✓	—	✓	—
M1959RW	—	19	—	142/72	✓	✓	✓	—	—	—	—	✓	—	—	—	—	—	—
A1940MW	—	19	12	82/0	—	—	—	—	—	—	—	—	—	—	—	—	—	—
A2060R	—	20	—	142/[1]	—	✓	—	—	✓	—	✓	—	1/1	1/1	—	—	✓	—
Emerson																		
EC194	$390	19	—	82/0	—	—	—	—	—	—	—	—	—	—	✓	—	—	—
ECT1900	430	19	12	82/0	—	—	—	—	—	—	—	—	—	—	✓	—	—	—
ECR212	460	19	12	82/0	✓	✓	—	—	—	—	—	—	—	—	✓	—	—	—
ECR215	—	19	12	82/0	✓	✓	—	—	—	—	—	—	—	—	✓	—	—	—
ECR222	500	19	—	139/69	✓	✓	—	—	—	✓	—	✓	0/1	—	✓	—	—	—
M195R	550	19	—	139/69	—	✓	—	—	—	—	—	—	1/1	1/1	✓	—	✓	—
M20R	600	20	—	139/69	—	✓	✓	—	—	✓	—	—	1/1	1/1	✓	—	✓	—
MS198R	700	19	—	139/69	—	✓	—	—	✓	✓	✓	—	1/1	1/1	✓	—	✓	—
MS30R	750	20	—	139/69	—	✓	✓	—	—	✓	✓	—	1/1	1/1	✓	—	✓	—
Fisher																		
PC202W	550	19	—	112/42	✓	✓	✓	—	—	—	✓	—	—	1/0	—	—	—	—
PC203W	550	19	—	112/42	—	✓	✓	—	—	—	✓	—	2/2	1/0	—	—	—	—
PC205W	600	20	—	140/58	—	✓	✓	—	—	✓	✓	—	2/2	1/0	—	—	—	—
PC206W	650	20	—	181/113	—	✓	✓	—	✓	✓	✓	—	4/4	2/2	—	—	✓	—
PC226W	650	20	—	181/113	—	✓	✓	—	✓	✓	✓	—	2/2	2/1	✓	—	✓	—

[1] *Information not available.*

General Electric																		
8-1900	249	19	—	68/0	—	—	—	—	—	—	—	—	—	—	—	—	—	—
8-1910	279	19	—	94/38	✓	—	—	—	—	—	—	—	—	—	—	—	—	—
8-1930	349	19	—	94/38	—	—	—	—	—	—	—	—	—	—	—	—	—	—
8-1931	349	19	—	94/38	—	—	—	—	—	—	—	—	—	—	—	—	—	—
8-2050	449	20	—	155/99	—	✓	✓	—	—	✓	—	—	—	—	—	—	✓	✓
8-2060	549	20	—	155/99	—	✓	✓	—	✓	✓	✓	—	2/2	1/1	—	—	✓	✓
8-2090	649	20	—	155/99	—	✓	✓	✓	✓	✓	✓	—	2/2	2/1	—	—	✓	✓
Gold Star																		
CMR-2030	320	19	—	82/0	—	—	—	—	—	—	—	—	—	—	✓	—	—	—
CMR-9080	320	19	—	82/0	—	—	—	—	—	—	—	—	—	—	✓	—	—	—
CMR-9163	320	19	—	82/0	—	—	—	—	—	—	—	—	—	—	✓	—	—	—
CMT-9165	400	19	12	82/0	✓	✓	—	✓	—	—	—	—	—	—	✓	—	—	—
CMT-9322	420	19	12	105/23	✓	✓	—	✓	—	—	—	—	—	—	✓	—	—	—
CMT-9325	420	19	12	105/23	✓	✓	—	—	—	—	—	—	1/1	1/1	✓	—	—	—
CMT-9408	500	19	—	139/69	—	✓	—	✓	—	—	—	—	—	—	✓	—	✓	—
CMT-9428	500	19	—	139/69	—	✓	—	✓	—	—	—	—	—	—	✓	—	✓	—
CMT-2135	500	20	—	139/69	—	✓	—	✓	✓	✓	✓	—	—	—	✓	—	—	—
CMT-2132	600	20	—	139/69	—	✓	—	✓	✓	✓	✓	—	1/2	1/2	✓	—	✓	—
KMV-9002	700	19	12	105/23	✓	—	—	—	✓	—	—	—	—	—	✓	—	✓	—
Hitachi																		
CT-1958	—	19	—	139/69	✓	✓	—	—	✓	—	✓	—	1/1	1/0	—	—	✓	—
CT-1951	420	19	—	82/0	—	—	—	—	—	—	—	—	0/1	—	✓	—	—	—
CT-1953	520	19	—	106/36	—	✓	—	—	—	—	—	—	0/1	—	✓	—	—	—
CT-1955	520	19	—	106/36	—	✓	✓	—	—	✓	—	—	—	—	✓	—	✓	—
CT-1965	550	19	—	119/49	—	✓	✓	—	—	✓	—	—	0/1	—	✓	—	✓	—
CT-1966	570	19	—	119/49	—	✓	✓	—	✓	✓	—	—	1/1	1/0	✓	—	✓	—
CT-2065W	600	20	—	119/49	—	✓	✓	—	✓	✓	—	—	—	—	✓	—	✓	—
CT-2066/B	620	20	—	119/49	—	✓	✓	—	✓	✓	—	—	1/1	1/0	✓	—	✓	—
J.C. Penney																		
2106	350	19	—	83/0	—	—	—	✓	—	—	—	—	—	—	—	—	✓	—
2115	420	19	—	113/45	—	✓	✓	—	—	—	—	—	—	—	—	—	—	—
2220	—	20	—	142/70	✓	✓	✓	✓	✓	✓	✓	—	2/2	2/2	✓	✓	✓	—

A guide to features (continued)

Brand and model	Price	Screen size, in.	Channel presets	Total channels/ cable channels	Direct access	Numeric keypad	Last-channel recall	Dual antenna	TV/video	Sleep timer	MTS decoder	MTS adapter jack	Audio input/output	Headphone jack	Video input/output	RGB input	On-screen displays	Channel block-out
			Tuning						Remote control			Sound		Auxiliary inputs			Other	
Magnavox																		
CG4137WA	—	19	—	68/0	[1]	—	—	—	—	—	—	—	—	—	—	—	—	—
CG4147WA	—	19	—	68/0	[1]	—	—	—	—	—	—	—	—	—	—	—	✓	—
CG4150WA	—	19	—	152/96	✓	—	—	—	—	—	—	—	—	—	—	—	✓	—
RF4254WA	—	19	—	68/84	✓	✓	—	—	—	—	✓	—	2/2	[1]	—	✓	✓	✓
RG4250WA	—	19	—	152/96	✓	✓	✓	—	—	—	—	—	—	—	—	—	—	—
RF4378SL	—	20	—	68/110	✓	✓	✓	✓	—	✓	✓	—	2/2	[1]	✓	—	✓	✓
RG4314WR	—	20	—	152/96	✓	✓	✓	—	✓	—	✓	—	1/1	1/1	✓	—	—	—
RG4332WA	—	20	—	152/96	[1]	✓	✓	—	—	✓	—	✓	—	—	✓	—	✓	✓
RG4352WA	—	20	—	152/96	✓	✓	✓	—	—	—	✓	—	2/2	—	—	—	✓	✓
RG4378BK	—	20	—	178/122	[1]	✓	✓	✓	✓	✓	✓	—	4/4	2/1	✓	✓	✓	✓
Mitsubishi																		
CS-1945R	—	19	—	139/69	[1]	✓	✓	—	—	✓	—	—	—	—	—	—	✓	—
CS-2000	—	20	—	139/69	[1]	—	✓	—	—	—	—	—	1/0	—	—	—	✓	—
CS-2001R	—	20	—	139/69	✓	✓	✓	—	✓	✓	—	—	1/0	1/0	—	—	—	—
CS-2011R	—	20	—	139/69	[1]	✓	✓	—	—	✓	✓	—	2/0	—	—	—	✓	—
CS-2051R	—	20	—	139/69	[1]	✓	✓	✓	✓	✓	✓	—	4/2	2/1	✓	—	✓	—
CS-2053R	—	20	—	139/69	✓	✓	✓	✓	✓	✓	✓	—	2/1	2/1	—	—	✓	—

Montgomery Ward																	
12636	—	19	—	139/69	✓	✓	—	—	—	—	—	✓	1/0	1/0	—	—	—	—
12801	$280	19	—	82/[1]	—	—	—	—	—	—	—	✓	—	—	—	—	—	—
12516	320	19	—	105/[1]	✓	—	—	—	—	—	—	—	—	—	—	—	—	—
12646	400	19	—	111/[1]	—	✓	✓	—	—	—	—	✓	—	—	—	—	✓	—
12637	400	19	—	110/[1]	—	✓	—	—	✓	—	—	✓	1/1	1/1	—	—	—	—
12656	450	19	—	111/[1]	—	✓	✓	—	—	—	✓	—	1/1	1/0	✓	—	✓	—
12676	470	20	—	140/[1]	—	✓	✓	—	—	✓	—	—	1/1	1/1	✓	—	✓	—
12696	—	20	—	139/69	—	✓	✓	—	—	✓	✓	—	1/1	1/0	✓	—	✓	—
12686	—	20	—	139/[1]	—	✓	✓	—	—	✓	✓	—	1/1	1/0	✓	—	✓	—
Panasonic																		
CTH-1900	—	19	12	68/0	—	—	—	—	—	—	—	—	—	—	✓	—	—	✓
CTH-1942R	—	19	—	155/99	—	✓	✓	—	—	✓	—	—	—	—	—	—	—	—
CTH-1911	400	19	14	80/24	—	—	—	—	—	—	—	—	—	—	✓	—	—	—
CTH-1922	430	19	—	155/99	—	—	—	—	—	—	—	✓	—	—	✓	—	—	—
CTH-1952R	480	19	—	155/99	—	✓	✓	—	—	✓	—	—	—	—	✓	—	—	—
CTH-2049R	530	20	—	155/99	—	✓	✓	—	—	✓	—	✓	—	—	✓	—	—	—
CTH-2053R	550	20	—	155/99	—	✓	✓	—	—	✓	—	✓	1/0	1/0	—	—	—	—
CTH-2063R	580	20	—	155/99	—	✓	✓	✓	—	✓	—	✓	1/0	1/0	—	—	✓	✓
CTG-2067R	700	20	—	155/99	✓	✓	✓	—	✓	✓	✓	—	1/0	1/0	✓	—	✓	—
Philco																		
R3932AW	—	19	—	82/0	✓	✓	✓	—	—	—	—	—	—	—	—	—	—	—
C3924AW	—	19	—	82/0	✓	—	—	—	—	—	—	—	—	—	—	—	—	—
R3971AW	—	19	—	166/96	✓	✓	✓	—	—	✓	✓	—	1/1	—	—	—	—	—
R3940AW	—	19	—	166/96	✓	✓	—	—	—	—	—	—	—	—	—	—	—	—
C3901AW	—	19	—	82/0	✓	—	—	✓	—	—	—	—	—	—	—	—	—	—
Quasar																		
WT5965YW	350	19	14	94/24	✓	✓	—	—	—	—	—	—	—	—	—	—	—	—
WT5966YW	370	19	—	155/99	✓	✓	—	—	—	—	—	—	—	—	—	—	—	—
TT5957AW	430	19	—	155/99	✓	✓	✓	—	—	✓	—	✓	—	—	—	—	—	—
TT5947AW	—	19	—	155/99	✓	✓	✓	—	—	✓	—	—	—	—	—	—	—	—
TT5948AW	—	19	—	155/99	✓	✓	✓	—	—	✓	—	—	—	—	—	—	—	—
TT5958YW	—	19	—	155/99	✓	✓	✓	—	—	✓	—	✓	—	—	—	—	—	—
WT5946AW	—	19	—	155/99	✓	✓	—	—	—	—	—	—	—	—	—	—	—	—
WT5951AW	—	19	—	82/0	✓	✓	—	—	—	—	—	—	—	—	—	—	—	—
WT6245YW	370	20	14	94/24	✓	✓	—	—	—	—		✓	—	—	—	—	—	—
TT6277YW	470	20	—	155/99	✓	✓	✓	—	—	✓	—	✓	—	—	—	—	—	✓
TT6278YW	480	20	—	155/99	✓	✓	✓	—	—	✓	—	✓	1/0	1/0	—	—	—	✓
TT6248YM	540	20	—	155/99	✓	✓	✓	—	—	✓	✓	—	0/2	—	—	—	—	✓
TT6290XE	650	20	—	[1]	✓	✓	—	—	—	—	—	—	3/2	3/2	—	✓	—	—
TT6298YW	990	20	—	137/69	✓	✓	✓	✓	✓	✓	✓	—	3/2	3/2	✓	✓	✓	—

A guide to features (continued)

Brand and model	Price	Screen size, in.	Channel presets	Total channels/ cable channels	Tuning: Direct access	Tuning: Numeric keypad	Tuning: Last-channel recall	Tuning: Dual antenna	Remote control: TV/video	Remote control: Sleep timer	Remote control: MTS decoder	Sound: MTS adapter jack	Sound: Audio input/output	Sound: Headphone jack	Auxiliary inputs: Video input/output	Auxiliary inputs: RGB input	Other: On-screen displays	Other: Channel block-out
RCA																		
FMR425E	309	19	—	68/0	—	—	—	—	—	—	—	—	—	—	—	—	—	—
FMR425W	309	19	—	68/0	—	—	—	—	—	—	—	—	—	—	—	—	—	—
FMR455W	329	19	—	113/57	✓	—	—	—	—	—	—	—	—	—	—	—	—	—
FMR505W	349	20	—	150/94	✓	—	✓	—	—	—	—	—	—	—	—	—	—	—
FMR468HR	389	19	—	150/94	—	✓	✓	—	—	—	—	—	—	—	—	—	—	—
FMR520WR	—	20	—	150/94	✓	✓	✓	—	✓	—	—	✓	—	—	—	—	✓	—
FMR510WR	399	20	—	150/94	—	✓	✓	—	—	—	—	—	—	—	—	—	✓	—
FMR514TR	439	20	—	150/94	—	✓	✓	—	—	—	—	—	—	—	—	—	✓	—
FMR555WR	569	20	—	150/94	—	✓	✓	—	✓	—	✓	—	1/0	1/0	—	—	✓	—
FMR560ER	649	20	—	150/94	✓	✓	✓	—	✓	—	✓	—	4/4	2/0	—	—	✓	—
FMR560WR	649	20	—	150/94	✓	✓	✓	—	✓	—	✓	—	4/4	2/0	—	—	✓	—
FMR570ER	649	20	—	150/94	—	✓	✓	—	✓	—	✓	—	2/2	2/0	—	—	✓	—
FMR570WR	649	20	—	150/94	—	✓	✓	—	✓	—	✓	—	4/4	2/0	—	—	✓	—
Sanyo																		
91C511	300	19	—	82/0	—	—	—	—	—	—	—	—	—	—	—	—	—	—
91C601	300	19	—	140/70	✓	—	—	—	—	—	—	✓	—	—	✓	—	—	—
91C621	430	19	—	140/70	✓	✓	✓	—	—	—	—	✓	—	—	✓	—	—	—
91C625	440	19	—	140/70	✓	✓	✓	—	—	—	—	—	—	—	✓	—	—	—
91C900	500	19	—	112/42	✓	✓	✓	—	—	—	✓	—	—	—	✓	—	—	—
12C910	550	20	—	140/70	✓	✓	✓	—	✓	✓	✓	—	2/1	2/1	✓	—	—	—
AVM210	550	20	—	140/70	—	✓	✓	—	✓	✓	✓	—	4/4	2/1	✓	✓	✓	—
AVM215W	580	20	—	181/125	✓	✓	✓	—	✓	✓	✓	—	4/4	2/2	✓	—	✓	—
12C700	630	20	—	140/70	✓	✓	—	—	✓	✓	✓	—	4/2	4/2	✓	✓	—	—

Sears																		
41002	199	19	—	82/0	—	—	—	—	—	—	—	—	—	—	—	—	—	—
41301	240	19	—	82/0	—	—	—	—	—	—	—	—	—	—	—	—	—	—
41401	290	19	—	119/[1]	—	—	—	—	—	—	—	—	—	—	—	—	—	—
42002	290	19	12	82/0	—	—	—	—	—	—	—	—	—	—	—	—	—	—
42091	320	19	12	105/[1]	—	—	—	—	—	—	—	—	—	—	—	—	—	—
42442	340	19	—	111/[1]	—	✓	—	✓	—	—	—	—	—	—	—	—	—	—
42151	390	19	—	119/[1]	—	✓	—	—	—	✓	—	✓	1/0	1/0	—	—	✓	—
42351	440	19	—	119/[1]	—	✓	—	✓	—	✓	✓	—	1/1	1/0	—	—	✓	—
42451	440	19	—	119/[1]	—	✓	—	✓	—	✓	✓	—	1/1	1/0	—	—	✓	—
42551	440	19	—	119/[1]	—	✓	—	✓	—	✓	✓	—	1/1	1/1	✓	—	✓	✓
42701	490	20	—	140/[1]	—	✓	—	✓	—	✓	✓	—	3/1	3/1	—	✓	✓	—
42751	590	20	—	181/125	—	✓	✓	—	✓	✓	✓	—	2/0	2/0	✓	—	✓	—
Sharp																		
19LP16	380	19	—	82/0	—	—	—	—	—	—	—	—	—	—	—	—	—	—
19LP36	460	19	—	110/40	—	—	—	—	—	—	—	—	—	—	—	—	—	—
19LP56	510	19	—	110/40	—	✓	✓	—	—	—	—	—	—	—	—	—	—	—
20LV56	600	20	—	110/40	—	✓	✓	✓	✓	—	—	✓	1/0	1/0	—	—	—	—
20LV76	690	20	—	110/40	—	✓	✓	✓	✓	—	✓	—	2/2	1/0	—	—	—	—
20LP86	760	20	—	126/70	—	✓	✓	✓	✓	—	✓	—	1/2	1/0	—	—	—	—
Sony																		
KV-1967	500	19	—	155/99	—	—	—	—	—	—	—	—	—	—	—	—	—	—
KV-1926R	610	19	—	181/125	—	✓	—	—	—	✓	—	—	—	—	✓	—	✓	—
KV-2071R	640	20	—	181/125	—	✓	✓	—	✓	✓	—	✓	1/0	1/0	—	—	✓	✓
KV-2075R	700	20	—	181/125	—	✓	✓	—	✓	✓	✓	—	2/2	1/0	—	—	✓	✓
KV-1981R	740	19	—	181/125	—	✓	—	—	✓	✓	✓	—	2/0	1/0	—	—	✓	—
KV-2080R	740	20	—	181/125	—	✓	✓	✓	✓	✓	✓	—	2/2	1/1	—	—	✓	✓
KV-2084R	840	20	—	181/125	—	✓	—	—	✓	✓	✓	—	6/2	3/1	✓	—	✓	—
KV-20XBR	940	20	—	181/125	—	✓	—	✓	✓	✓	✓	—	2/1	3/2	✓	✓	✓	—
Sylvania																		
CXF165	—	19	—	[1]	✓	—	—	✓	—	—	—	—	—	—	—	—	✓	—
RXF155	—	19	—	[1]	✓	✓	✓	—	—	—	—	—	—	—	—	—	✓	—
RXF169	—	19	—	152/96	✓	✓	✓	—	—	—	—	—	—	—	—	—	—	—
CXF176	—	20	—	68/84	✓	—	—	—	✓	—	—	✓	—	—	—	—	✓	—
RKF178	—	20	—	68/84	✓	✓	✓	—	✓	—	—	✓	—	—	—	—	✓	—
RKF192	—	20	—	24/152	—	✓	✓	—	—	—	✓	—	2/2	1/1	—	—	✓	—
RKF195	—	20	—	68/84	✓	✓	✓	—	—	—	—	—	2/2	1/1	—	—	✓	—
RKF198	—	20	—	178/122	✓	✓	✓	✓	✓	✓	✓	—	2/3	2/2	✓	—	✓	✓

A guide to features (continued)

Brand and model	Price	Screen size, in.	Channel presets	Total channels/cable channels	Direct access	Numeric keypad	Last-channel recall	Dual antenna	TV/video	Sleep timer	MTS decoder	MTS adapter jack	Audio input/output	Headphone jack	Video input/output	RGB input	On-screen displays	Channel block-out
			Tuning				Remote control				Sound		Auxiliary inputs				Other	
Toshiba																		
CF2026	490	20	—	117/[1]	✓	—	—	—	—	—	—	—	1/0	1/0	✓	—	—	—
CX2056	600	20	—	141/[1]	✓	—	—	—	—	—	✓	—	2/0	1/0	—	—	✓	—
CF2036	620	20	—	117/[1]	✓	✓	✓	—	✓	✓	—	—	2/0	1/0	✓	—	✓	—
CX2016	620	20	—	117/[1]	✓	✓	✓	—	✓	✓	—	—	1/0	1/0	✓	—	✓	—
CF2046	720	20	—	117/[1]	✓	✓	✓	—	✓	✓	✓	—	2/2	1/0	—	—	✓	—
CX2066	750	20	—	141/71	✓	✓	✓	✓	✓	✓	✓	—	1/1	1/1	✓	—	✓	—
CS2076	760	20	—	141/[1]	✓	✓	✓	✓	✓	✓	✓	—	2/2	1/1	✓	—	✓	—
Zenith																		
C1910B	—	19	12	82/0	—	—	—	—	—	—	—	✓	—	—	—	—	—	—
C1908W	350	19	12	82/0	—	—	—	—	—	—	—	✓	—	—	—	—	—	—
C1012W	370	19	—	178/122	—	—	—	—	—	—	—	✓	—	—	—	—	—	—
C1920W	390	19	—	178/122	✓	—	—	—	—	—	—	✓	—	—	—	—	—	—
C2020H	420	20	—	178/122	✓	—	—	—	—	—	—	✓	—	—	—	—	—	—
SC1911W	430	19	—	178/122	✓	✓	—	—	—	—	—	—	—	—	—	—	—	—
SC1923W	470	19	—	178/122	✓	✓	✓	—	✓	—	—	✓	—	—	—	—	—	—
SC2023H	510	20	—	178/122	✓	✓	✓	—	✓	—	—	✓	—	—	—	—	—	—
SC2027S	530	20	—	178/122	✓	✓	✓	—	✓	—	—	✓	2/1	2/1	✓	—	—	—
SC1935S	611	19	—	178/122	✓	✓	✓	—	✓	—	✓	—	4/2	2/1	✓	—	—	—
SC1935W	611	19	—	178/122	✓	✓	✓	—	✓	—	✓	—	2/2	2/0	✓	—	—	—

■ Helping the hearing impaired ■

People with impaired hearing may hear little or no television sound. To remedy this, increasing numbers of programs are being offered with closed captioning—that is, subtitles that appear only on the screen of sets equipped with a decoder. (TV listings designate "CC" for closed-captioned programs.) TV stations transmit the captions using the ordinarily invisible "vertical interval" of a TV picture (the black bar you see when the picture rolls).

The closed-captioning system works remarkably well. It delivers crisp white letters on a black band. If you live in a fringe reception area, however, neither the picture nor the closed captions will be letter-perfect. With a signal equivalent to fringe reception, an average of one letter per line may be missing from the caption. Nevertheless, that's still a reasonably good performance.

■ The most reliable brands ■

The repair index shown here is based on the reports of readers who told about their experience with some 120,000 19-inch color TVs. The data were derived from Consumers Union's 1986 Annual Questionnaire.

The longer the bar in the graph, the more often sets of that brand needed repair.

Differences between closely ranked brands, such as the four at the top, are not significant, but differences between brands at the top and those at the bottom are dramatic. For example, the index indicates that sets from Philco broke down nearly four times as often as sets made by Toshiba. As a rule, a spread of five points or more is worth noting.

Some caveats: The repair index applies only to 19-inch monophonic sets from the brands listed. Larger or smaller sets from the same manufacturer may well have a different repair history. Because stereo is such a recent addition, the respondents didn't have enough experience with stereo sets to allow judging how stereo affects reliability. Finally, the data are historical: The sets produced by a company four or five years ago could be either better or worse than those the company is turning out this year. But this kind of data has been fairly consistent over the years, so your chances of getting a reliable set will be increased if you choose a brand that has been trouble-free in the past.

TV Service

Any TV set should be designed so that repairs, when necessary, are simple, speedy, and inexpensive. Features such as a readily accessible

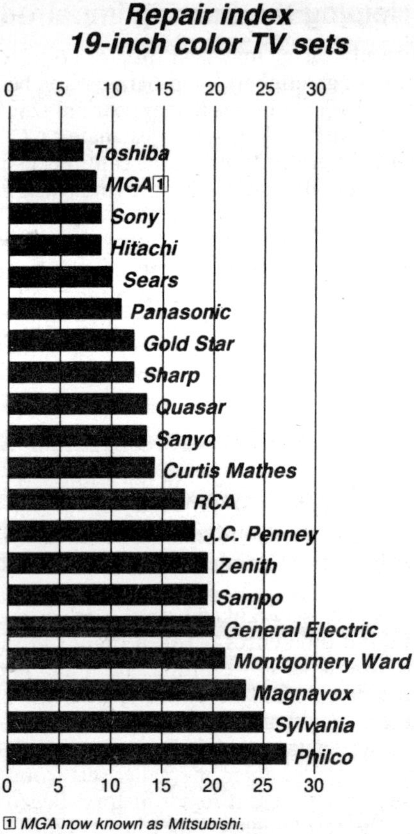

printed circuit board help make a television set easy to service. It's comforting to note the continuing advances in solid-state technology, resulting in sets with fewer and fewer parts. Large numbers of transistors are giving way to small numbers of integrated circuits.

Integrated circuits may be designed to plug into sockets in the circuit board rather than soldered in place. Removable plug-in integrated circuits and modules are an asset if repairs are needed because a qualified technician can usually remove and replace a defective part quickly.

Color consoles. Nowadays, these models are simple to service—easier than earlier ones because there aren't as many circuit boards, and the boards are smaller and easier to reach. The advantage is that a technician may be able to make a quick repair in the home rather than being forced to take the chassis to the shop.

Smaller color sets. Not all small-screen color sets are easy to service. It's harder to repair a set where every integrated circuit and every transistor is solidly soldered in or where the boards are interconnected with a formidable array of cables.

Small black-and-white sets. Their relatively simple circuitry makes black-and-white TV sets less likely to need repair than a color model. If repairs are needed, the lightweights should be easy to tote to a service shop, especially those with a carrying handle or handhold and those sets weighing no more than about 15 pounds. Considering the discounted price at which many of these small black-and-white sets are sold, they might just as well be discarded rather than repaired: They would probably not be worth the expenditure of $50 to $60 to repair.

■ What more money can buy ■

Just as spending more generally gets you a bigger screen, spending more also gets you a fancier, more versatile TV set. The differences show up in these features:

Channel selector. The old-fashioned rotary knob tuner, now nearly extinct, turns up only on the cheapest sets. Higher-priced models generally have some sort of electronic selector that gives you "direct access": You can go directly from, say, channel 2 to channel 9.

One sort of electronic tuner has to be adjusted before you can use it. The adjustment involves tedious fiddling with tiny buttons, dials, or switches to set up the desired channels. Thereafter, you change channels with a button or knob. Some electronic tuners limit the number of channels you can receive at one time—an annoyance to cable-TV users who have two dozen or more channels at their disposal.

A quartz-locked tuner that works with a push-button keypad is a much better setup. It automatically locks onto a station's proper frequency. In general, sets with a quartz-locked tuner can receive many more of the available channels.

Often, you can use a quartz-locked tuner to program out the channels you don't want to watch. Press the appropriate button and the set scans up or down the TV band, stopping only at the stations you want.

Remote control. Practically a standard feature these days, the remote lets you change channels, adjust sound, and, with some designs, adjust the picture or switch program sources without getting up from your armchair. On some sets, the remote control is a necessity; some tuning, sound, and picture controls are only on the remote, not on the TV set.

Displays and doodads. A number of sets will display the channel you're watching—or the time—on screen at the press of a button. The channel readout might be handy if your set lacks an illuminated tuner. The clock could be a nuisance if power interruptions are common in your area; they may wipe out the TV's time-keeping memory, which you'll then have to reprogram.

Auxiliary inputs and outputs. Sets designed specifically to serve as the command center of an entire home-entertainment system are apt to be more satisfactory than basic models. That's because TV picture and sound signals go through a good deal of technological massage in their passage from studio to screen. Many sets offer input and output jacks that let you bypass some of the set's electronic hardware and so reduce the deterioration.

The cable connection. A set advertised as "cable ready" will be able to receive at least 23 channels reserved for nonscrambled cable programs, with no need for the cable company's converter box. But a cable-ready set may not be all you need to watch cable shows. You'll need a converter box if the signal is delivered scrambled, often the case with the so-called premium sports and movie channels. You'll need a special "addressable decoder" if you want to watch pay-per-view programs. The decoder lets the company keep track of the extra-fee programs you've watched and bill you accordingly.

If you tie into cable, you also may not be able to make full use of a VCR without acquiring some additional hardware. The way you interconnect the VCR, the cable-converter box, and the TV set may prevent you from watching one program while recording another.

■ Stereo sound for older TV sets ■

You should make a stereo set your first choice. But what if you want to listen to stereo TV and don't want to buy a new stereo set? Consider buying a *Recoton FRED II.*

FRED (for Friendly Recoton Entertainment Decoder) is a book-sized box that gives just about any TV set stereo sound capability. The unit is installed near a television set, where it takes audio signals from the channel the set is tuned to and routes them to your hi-fi system. The *FRED II* lists for $170.

The *FRED II* can tap into the TV signal in several ways. If the TV set has an MTS or MPX output jack, the *FRED* plugs in there. If the set is connected to a cable-TV converter, the *FRED* plugs into the box. The *FRED* has a pickup probe that can be mounted somewhere on the TV cabinet, where it can sense magnetic fields produced by the set's sound circuitry.

Using the pickup to eavesdrop on the TV is somewhat akin to finding a heartbeat with a stethoscope, and placing the pickup can be a little tricky. With most sets, the pickup should be placed somewhere underneath the cabinet. But sets with a thick wood or metal cabinet may provide such a weak magnetic field to the outside of the set that a special pickup—one that's mounted inside—has to be used.

When it is properly hooked up, the *FRED* works very well. It produces a stereo signal about on a level with the best stereo decoders in top-of-the-line TV sets. The *FRED* can also add a "pseudostereo" effect to monophonic broadcasts. The device has a couple of useful additional features, including controls that reduce background hiss and adjust the stereo separation effect.

In sum, the *FRED II* is a good way to gain the benefits of stereo TV without incurring the cost of a new set.

Rent-to-own TV set deals

"Rent-to-own" is a payment method that allows you to rent a product by the week or month and to gain ownership once you've made a specified number of payments. Renting may sound like a good way to buy a television set: no money down, free delivery, and free repairs (often at your home) during the period you're making your payments. If you find yourself strapped for funds, you just give the TV set back and no one will ask you for additional payments.

The convenience of rent-to-own, however, can come at a stiff price. The effective interest rate you will have to pay can be very high. In Baltimore, for example, one appliance store was selling a particular TV set for $608 in cash, or $698 in total payments if financed. Another store was offering two rent-to-own plans for the same brand of TV. Under those plans, monthly payments brought the total price to $1,301 over a year and a half, while weekly payments added up to $1,552. The implicit interest rate on those transactions ranges from 115 to 152 percent a year.

Unsophisticated consumers may not realize the high cost involved because rent-to-own transactions aren't covered by the disclosure requirements of the Federal Truth in Lending Act. In some states, usury laws don't apply either, because renting a product is technically considered different from financing a purchase. In some states, a rent-to-own dealer can legally deliver a used product instead of a new one, without disclosing that fact.

Ratings of large-screen color TVs

Listed in order of estimated quality. Prices are as published in a **January 1986** report. + indicates shipping is extra.

⊙	◓	○	◒	●
Better ←				→ *Worse*

Brand and model	Mfr. suggested price	Type: Console or table model	Color fidelity	Picture clarity	Black-level retention	Screen reflectance	Freedom from geometric distortion	Interlace	Automatic color control	Adjacent-channel rejection	Airplane-flutter rejection	Fringe reception, VHF	Fringe reception, UHF	Spark rejection	Resolution [1]
Wards 17737	$ 850	C	⊙	◓	⊙	◓	◓	◓	⊙	○	●	○	○	⊙	—
Philco R5883WAK	750	C	◓	⊙	⊙	◓	◓	◓	⊙	○	◒	○	◓	⊙	—
Magnavox RF4950AK	899	C	⊙	◓	⊙	◓	○	⊙	⊙	○	●	○	◓	⊙	—
RCA GLR2538P	800	C	◓	○	⊙	⊙	○	⊙	⊙	○	◓	◓	◓	⊙	—
Sony KV-2680R	1250	C	◓	◓	⊙	○	◓	◓	⊙	◒	◓	○	◓	●	310
Panasonic CTG-2587R	1099	C	◓	○	○	◓	○	⊙	⊙	○	○	○	○	⊙	350
JVC AV-2590	1100	T	⊙	⊙	◒	○	○	⊙	⊙	○	◒	○	○	◒	350
Quasar TU9950YP	975	C	◓	◓	◒	◓	◓	⊙	⊙	○	◓	○	○	⊙	—

Fisher PC-340W	850	T	◓	◓	◓	○	◓	◒	◉	○	○	◒	○	○	330
Mitsubishi CK-2587R	1100	C	◓	◓	○	◓	◉	◉	◉	○	○	○	○	●	350
Toshiba CZ2685	1490	T	◉	◓	○	○	◓	◉	◉	○	◓	○	◒	◒	350
GE 26CP6869	899	T	◓	◓	◓	◉	○	○	◓	○	○	○	○	◒	330
Zenith SB2527P	960	C	○	◓	◒	◉	◓	◉	◉	○	◒	○	○	◉	290
Zenith SB2729N	1080	C	○	◓	◒	◉	○	◉	◉	○	○	○	◒	◉	300
Sanyo AVM260	800	T	○	◓	◓	○	◓	○	◉	○	◒	●	○	○	330
Sears Cat. No. 4870	680+	C	◓	○	●	◓	◉	◉	◉	◒	○	◒	○	●	240
Curtis Mathes M2658RL	[2]	C	○	◓	●	◉	○	◉	◓	◒	◒	○	○	◒	370
Hitachi CT-2559	1080	C	◓	○	○	○	○	◉	◓	◒	◒	○	◒	◉	350

[1] *As measured at set's video input. Via antenna (RF) inputs, sets with a comb filter provided approx. 300 lines of resolution; the* **Sears,** *which lacks a comb filter, provided approx. 250 lines.*

[2] *Mfr. does not provide suggested retail price.*

Specifications and Features

All have: • Built-in stereo/SAP decoder and stereo speakers. • Quartz (frequency-synthesized) tuner. • Automatic fine-tuning to pull in off-set channels. • Ability to receive channels 2 through 13 (VHF) and 14 through 69 (UHF). • Sharpness control. • Coaxial input connector for VHF and cable TV. • Ability to retain channel and volume settings when turned off but not unplugged. • No horizontal-hold control (none needed).

Except as noted, all have: • Comb filter. • Tone and stereo balance controls. • Illuminated channel numbers on control panel. • Room-light sensor. • Vertical hold control. • 25-in screen.

Except as noted, all lack: Built-in numeric keypad on set.

A guide to features for large-screen TVs

Model		Screen size; Table or Console	Tuning method [1]	No. channels	Remote control	Stereo decoder	Stereo adapter jacks	No. speakers	Audio inputs	Audio outputs	Video inputs	Video outputs	Cabinet style(s) [2]	Additional features
Curtis Mathes	M2582RG	25T	Q	134	✓	✓	—	2	✓	✓	✓	✓	Mo, Mo(**M2582RW**)	*A,C,E,G,H,I*
	K2504MW	25C	Q	82	—	—	—	1	—	—	—	—	Ts, Cl(**K2506MD**), Ur(**K2507ML**)	*E,I*
	K2510R0	25C	Q	134	✓	—	✓	1	—	—	—	—	Mo	*E,I*
	K2530MK	25C	Q	134	—	—	✓	1	—	—	—	—	Cl, Cl(**K2530MD**)	*C,E,I*
	K2530RK	25C	Q	134	✓	—	✓	1	—	—	—	—	Cl, Cl(**K2530RD**)	*C,E,I*
	K2536ME	25C	Q	134	—	—	✓	1	—	—	—	—	Mo	*C,E,I*
	K2536RE	25C	Q	134	✓	—	✓	1	—	—	—	—	Mo, Ea(**K2532RM**), Cy(**K2534RP**), Ur(**D2537RL**)	*C,E,I*
	M2520RM	25C	Q	134	✓	✓	—	4	—	—	—	—	Ea, Cy(**K2522RP**), Mo(**K2524RK**), Cl(**K2526RC**)	*A,C,E,H,I*
	M2600RW	26T	Q	142	✓	✓	—	4	✓	✓	✓	✓	Mo	*A,C,E,G,H,I*
	M2647RL	26C	Q	142	✓	✓	—	4	✓	✓	✓	✓	Ur	*A,C,E,H,I*
	M2658RL	26C	Q	142	✓	✓	—	4	✓	✓	✓	✓	Mo, Cl(**M2650RD**), Cy(**M2653RP**), Cl(**M2650RC**)	*A,C,E,G,H,I*
	M2660RC	26C	Q	142	✓	✓	—	4	✓	✓	✓	✓	Cl, Cy(**M2662RH**), Mo(**M2664RE**)	*A,C,E,H,I*

Fisher	PC-340W	25T	Q	112	✓	✓	—	2	✓	✓	✓	✓	Ct	*A,E,H,I*
	HT-770	25C	Q	112	✓	✓	—	2	✓	✓	✓	✓	Ct	*E,H,I*
	PC-350W	26T	Q	140	✓	✓	—	2	✓	✓	✓	✓	Ct	*E,H,I*
	HT-790	26C	Q	140	✓	✓	—	4	✓	✓	✓	✓	Ct	*E,H,I*
	HT-880	26C	Q	140	✓	✓	—	4	✓	✓	✓	✓	Ct	*E,H,I*
General Electric	25PF6802	25C	K	82	—	—	—	1	—	—	—	—	Ts	*I*
	25PF6812	25C	Q	112	—	—	—	1	—	—	—	—	Tr, Ct**(25PF6813)**	*I*
	25PF6830	25C	Q	112	—	—	—	1	—	—	—	—	Ct, Ea**(25PF6831),** Tr **(25PF6832)**	*I*
	25PF6841	25T	Q	112	✓	—	—	2	—	—	—	—	Ct Tr, Ct**(25PF6843)**	*I*
	25PF6842	25C	Q	112	✓	—	—	1	—	—	—	—	EA**(25PF6846)**	*I*
	25PF6818	25C	Q	112	—	✓	—	2	—	✓	—	—	Tr	*A,I*
	25PF6850	25C	Q	112	✓	—	—	1	—	—	—	—	Ct, Ea**(25PF6851),** Tr**(25PF6852)**	*I*
	25PF6849	25T	Q	112	✓	✓	—	2	—	✓	—	—	Ct	*A,F,H,I*
	25PP5859K	25T	Q	130	✓	—	—	2	✓	✓	✓	✓	Ct	*A,E,F,I*
	25PF6853	25C	Q	112	✓	✓	—	2	—	✓	—	—	Tr, Ea**(25PF6854),** Ct**(25PF6855)**	*A,F,H,I*
	25PF6856	25C	Q	112	✓	✓	—	4	—	✓	—	—	Ct, Tr**(25PF6857)**	*A,F,H,I*
	26CP6869	26T	Q	130	✓	✓	—	4	✓	✓	✓	✓	Mo	*A,C,D,E,F,G,H,I*
	26CP6863	26C	Q	130	✓	✓	—	4	✓	✓	✓	✓	Ct, Ea**(26CP6864),** Tr**(26CP6865)**	*A,C,E,F,G,H,I*
	26CP6870	26C	Q	130	✓	✓	—	4	✓	✓	✓	✓	Ct, Ea**(26CP6871),** Tr**(26CP6872)**	*A,C,E,F,G,H,I*
	25PM5882	25T	Q	130	✓	✓	—	2	✓	✓	✓	✓	Mo	*A,C,D,E,F,G,H,I*
Gold Star	CMT-2515	25C	Q	105	✓	—	—	1	—	—	—	—	Ct	*C*
	CMT-2525	25T	Q	139	✓	✓	—	2	✓	✓	✓	✓	Mr	*A,C,E,G,H*

A guide to features for large-screen TVs (continued)

Model	Screen size; Table or Console	Tuning method [1]	No. channels	Remote control	Stereo decoder	Stereo adapter jacks	No. speakers	Audio inputs	Audio outputs	Video inputs	Video outputs	Cabinet style(s) [2]	Additional features
Hitachi CT2550	25T	Q	106	✓	—	✓	2	—	—	—	—	—	*A*
CT2551	25C	Q	106	✓	—	✓	2	—	—	—	—	—**(CT2552)**	*A*
CT2555	25C	Q	106	✓	—	✓	2	—	—	—	—	—**(CT2556)**	*A,E,I*
CT2553	25C	Q	139	✓	✓	—	4	—	—	—	—	—	*A,H*
CT2557	25C	Q	139	✓	✓	—	4	—	—	—	—	—	*A,E,H,I*
CT2559	25C	Q	139	✓	✓	—	4	✓	✓	✓	—	—	*A,E,H,I*
CT2647	26T	Q	139	✓	✓	—	2	✓	✓	✓	✓	—	*A,E,H,I*
CT2648	26C	Q	139	✓	✓	—	4	✓	✓	✓	✓	—	*A,E,H,I*
CT2649	26C	Q	139	✓	✓	—	4	✓	✓	✓	✓	—**(CT2652), (CT2653)**	*A,E,H,I*
J.C. Penney 685-2500	25T	V	127	✓	—	—	3	—	—	—	—	Ct	—
685-4207	25C	V	127	✓	—	✓	1	—	—	—	—	Ct	—
685-2506	25T	V	127	✓	—	✓	3	—	—	—	—	Ct	—
685-2502	25T	Q	127	✓	✓	—	4	✓	✓	✓	✓	Ct	*A,B,C,E,H*
685-4018	25C	V	127	✓	—	—	1	✓	—	✓	—	Ct, Tr**(685-4018)**, Cy**(685-4020)**	—
685-2504	26T	Q	130	✓	✓	—	2	✓	✓	✓	✓	Ct	*A,C,F,G,H*
685-4217	26C	Q	130	✓	✓	—	2	✓	✓	✓	✓	Ct, Tr**(685-4218)**	*A,C,F,G,H*

JVC C-2550	25T	Q	108	✓	—	—	1	—	—	—	—	—	*G,I*
C-2565	25C	Q	108	✓	—	—	1	—	—	✓	—	Ct	*G,I*
AV-2590	25T	Q	134	✓	✓	—	2	✓	✓	✓	✓	Sa	*A,D,E,H,I*
AV-2690	26T	Q	142	✓	✓	—	2	✓	✓	✓	✓	Sa	*A,B,C,D,E,G,H,I*
C-2685	26C	Q	142	✓	✓	—	6	✓	✓	✓	✓	Ct	*A,C,E,G,H,I*
Magnavox CF4700AK	25C	K	82	—	—	—	1	—	—	—	—	Ct	*I*
CF4712AK	25C	Q	82	—	—	—	1	—	—	—	—	Ct, Co,**(CF471HP)** Co**(CF4716PE)**	*I*
RF4800AK	25C	Q	82	✓	—	—	1	—	—	—	—	Ct, Ts**(RF4802PE)**, Tr**(RF4806PE)**	*I*
CF4740AK	25C	Q	152	—	—	✓	1	—	—	—	—	Ct, Tr**(CF4748PE)**	*I*
R4940AV	25C	Q	152	✓	—	✓	1	—	—	—	—	Ct, Tr**(R4948PE)**	*I*
CF4750AK	25C	Q	152	—	✓	—	2	✓	✓	—	—	Ct	*A,E,H,I*
RF4950AK	25C	Q	152	✓	✓	—	2	✓	✓	—	—	Ct, Co**(RF4954HP)**, Tr**(RF4956PE)**	*A,E,H,I*
RF4552SL	26T	Q	152	✓	✓	—	2	✓	✓	✓	✓	Sa	*A,C,D,E,H,I*
RF5950AK	26C	Q	152	✓	✓	—	2	✓	✓	—	—	Ct	*A,E,H,I*
RF6020PE	26C	Q	178	✓	✓	—	4	✓	✓	✓	✓	Ts	*A,C,E,H,I*
RF6140PE	26C	Q	178	✓	✓	—	4	✓	✓	✓	✓	Ct, Co**(RF6144HP)**, Tr**(RF6146PE)**	*A,C,E,H,I*
Mitsubishi CS-2667R	26T	Q	139	✓	✓	—	3	✓	✓	✓	✓	Ct	*A,B,C,E,G,H,I*
CK-2587R	25C	Q	139	✓	✓	—	4	✓	✓	✓	✓	Tr	*A,C,E,G,H,I*
CK-2662R	26C	Q	139	✓	✓	—	5	✓	✓	✓	✓	Ur	*A,C,E,G,H,I*
CK-2688R	26C	Q	139	✓	✓	—	4	✓	✓	✓	✓	Ct	*A,C,E,H,I*

A guide to features for large-screen TVs (continued)

Model	Screen size; Table or Console	Tuning method [1]	No. channels	Remote control	Stereo decoder	Stereo adapter jacks	No. speakers	Audio inputs	Audio outputs	Video inputs	Video outputs	Cabinet style(s) [2]	Additional features
Montgomery Ward 12826	25T	Q	105	✓	—	✓	1	✓	✓	✓	✓	Mr	*H*
17116	25C	Q	130	—	—	—	1	—	—	—	—	Tr, Ct**(1726)**	—
17316	25C	Q	125	✓	—	✓	2	—	—	—	—	Tr	*H,I*
17216	25C	Q	125	✓	—	✓	1	—	—	—	—	Tr, Ct**(17226)**	*H,I*
17416	25C	Q	130	✓	—	—	2	—	—	—	—	Tr, Ct**(17426)**	*I*
17916	25C	Q	125	✓	—	✓	2	✓	✓	—	—	Tr	*H*
17737	25C	Q	152	✓	✓	—	2	✓	✓	—	—	Ea, Ct**(17727)**, Tr**(17717)**	*A,E,H,I*
17927	26C	Q	152	✓	✓	—	2	✓	✓	✓	✓	Tr	*A,C,E,H*
NEC CT-2510A	25T	Q	142	✓	✓	—	2	✓	✓	✓	✓	Sa	*A,C,D,E,G,H,I*
CT-2610A	25T	Q	142	✓	✓	—	2	✓	✓	✓	✓	Sa	*A,C,D,E,G,H,I*
Panasonic CTG-2501	25C	K	68	—	—	—	1	—	—	—	—	Ct, Ts**(CTG2502)**	*A,I*
CTG-2530	25T	Q	155	—	—	✓	2	✓	—	✓	—	Mo	*A,E,I*
CTG-2521	25C	Q	155	—	—	—	1	—	—	—	—	Ct	*A,I*
CTG-2524	25C	Q	155	—	—	—	1	—	—	—	—	Md	*A,I*
CTG-2552R	25C	Q	155	✓	—	✓	1	—	—	—	—	Ts	*A,E,G,I*

	CTG-2551R	25C	Q	155	✓	—	✓	1	—	—	—	—	Ct	*A,E,G,I*
	CTG-2554R	25C	Q	155	✓	—	✓	1	—	—	—	—	Md	*A,E,G,I*
	CTG-2556R	25C	Q	155	✓	—	✓	1	—	—	—	—	Ea	*A,E,G,I*
	CTG-2557R	25C	Q	155	✓	—	✓	1	—	—	—	—	Mo	*A,E,G,I*
	CTG-2560R	25T	Q	155	✓	—	✓	2	✓	—	✓	—	Mo	*A,C,E,F,I*
	CTG-2561R	25C	Q	155	✓	—	✓	1	✓	—	✓	—	Ct	*A,C,E,F,G,I*
	CTG-2562R	25C	Q	155	✓	—	✓	1	✓	—	✓	—	Ct	*A,C,E,F,G,I*
	CTF-2670R	26T	Q	155	✓	—	✓	2	✓	—	✓	—	Ct	*A,C,E,G,I*
	CTG-2568R	25C	Q	155	✓	—	✓	2	✓	—	✓	—	Mo	*A,C,E,F,G,I*
	CTG-2567R	25C	Q	155	✓	—	✓	2	✓	—	✓	—	Mo	*A,C,E,F,G,I*
	CTG-2587R	25C	Q	155	✓	✓	—	2	✓	✓	✓	✓	Mo	*A,C,E,F,G,H,I*
	CTG-2690R	26T	Q	125	✓	—	✓	2	✓	✓	✓	✓	Ur	*A,B,C,D,E,G,I*
	CTG-2588R	25C	Q	155	✓	✓	—	4	✓	—	✓	—	Mo	*A,C,E,F,G,H,I*
	CTF-2684R	26C	Q	125	✓	—	✓	4	✓	✓	✓	✓	Md	*A,C,E,G,I*
	CTF-2688R	26C	Q	125	✓	—	✓	4	✓	✓	✓	✓	Mo	*A,C,E,G,I*
Philco	R4930TWA	25T	Q	82	✓	—	—	1	—	—	—	—	Ct	*I*
	C5800WPE	25C	K	82	✓	—	✓	1	—	—	—	—	Ct	*I*
	C5813WPN	25C	Q	82	✓	—	✓	1	—	—	—	—	Co, Tr(**C5814WPE**)	*I*
	R4950WWA	25T	Q	139	✓	—	—	1	—	—	—	—	Ct	*E,G*
	C6814WPE	25C	Q	82	—	—	✓	2	—	—	—	—	Tr	*I*
	R5852WPE	25C	Q	82	✓	—	✓	1	—	—	—	—	Tr	*I*
	R5860WAK	25C	Q	152	✓	—	✓	1	—	—	—	—	Ct, Tr(**R5862WPE**)	*I*
	R6860WAK	25C	Q	152	✓	—	✓	2	—	—	—	—	Ct	*I*
	R5883WAK	25C	Q	152	✓	✓	—	2	✓	✓	—	—	Ct, Tr(**R5885WPE**)	*A,E,H,I*
	R6980WAK	26C	Q	152	✓	✓	—	2	✓	✓	—	—	Ct, Tr(**R6982WPE**)	*A,E,H,I*

A guide to features for large-screen TVs (continued)

Model	Screen size; Table or Console	Tuning method [1]	No. channels	Remote control	Stereo decoder	Stereo adapter jacks	No. speakers	Audio inputs	Audio outputs	Video inputs	Video outputs	Cabinet style(s) [2]	Additional features
Quasar WU9410YU	25C	K	68	—	—	✓	1	—	—	—	—	Ct	*I*
WU9420YK	25C	K	68	—	—	✓	1	—	—	—	—	Ct, Tr**(WU9428YP)**, Ea**(WU9425YD)**	*I*
WU9512YU	25C	Q	82	—	—	✓	1	—	—	—	—	Ct	*I*
WU9249YP	25C	K	82	—	—	✓	1	—	—	—	—	Tr	*H,I*
WU9610YU	25C	Q	155	—	—	✓	1	—	—	—	—	Ct	*I*
WL9439YP	25C	K	68	—	—	✓	2	—	—	—	—	Md	*I*
WU9625YD	25C	Q	155	—	—	✓	1	—	—	—	—	Ea, Tr**(WU9628YP)**	*I*
TU9810YU	25C	Q	155	—	—	✓	1	—	—	—	—	Ct	*E,G,I*
TS9713YK	25C	Q	155	—	—	✓	1	—	—	—	—	Ct	*G,I*
TU9820YK	25C	Q	155	—	—	✓	1	—	—	—	—	Ct, Ea**(TU9823YD)**, Tr**(TU9828YP)**	*E,G,I*
TT9803YU	25T	Q	155	✓	—	✓	2	—	—	—	—	Ct	*I*
TT9804YK	25T	Q	155	✓	—	✓	2	✓	✓	✓	✓	Ct	*I*
TU9833YU	25C	Q	155	—	✓	✓	2	—	—	—	—	Ct, Md**(TL9839YP)**	*E,G,I*
WL9458YP	25C	K	68	—	—	✓	2	—	—	—	—	Md	*A,H,I*
TU9840YK	26C	Q	155	✓	—	✓	2	—	—	—	—	Ct	*C,E,F,G,I*

	TU9950YP	25C	Q	155	✓	✓	✓	2	—	—	—	—	Ct	*A,E,G,H,I*
	TT9906YK	26T	Q	155	✓	✓	—	2	✓	✓	✓	✓	Ct	*E,G,I*
	TL9958YP	26C	Q	155	—	✓	—	2	—	—	—	—	Md	*A,E,G,H,I*
	TU9962YK	26C	Q	155	✓	✓	—	2	✓	✓	✓	✓	Ct	*A,C,E,F,G,H,I*
	TL9960YP	26C	Q	155	✓	✓	—	2	✓	✓	✓	✓	Ct	*A,C,E,F,G,H,I*
RCA	GLR640T	25C	K	68	—	—	—	1	—	—	—	—	Ct, Cy(**GLR644H**), Tr(**GLR648P**)	*I*
	GLR680T	25C	Q	125	—	—	✓	1	—	—	—	—	Ct, Mo(**GLR682E**), Ct(**GLR682T**), Cy(**GLR684H**)	*I*
	FLR620TR	25T	Q	125	✓	—	✓	3	—	—	—	—	Ct	*I*
	GLR641TR	25C	Q	125	✓	—	—	1	—	—	—	—	Ct, Cy(**GLR645HR**), Tr(**GLR649PR**)	*I*
	FLR622TR	25T	Q	125	✓	—	✓	3	✓	—	✓	—	Ct(**FLR624ER**)	*I*
	GLR681TR	25C	Q	125	✓	—	✓	1	—	—	—	—	Ct, Mo(**GLR683ER**), Ct(**GLR683TR**), Cy(**GLR685HR**)	*I*
	GLR841TR	26C	Q	137	✓	—	✓	1	—	—	—	—	Ct, Co(**GLR845FR**), Co(**GLR845HR**), Tr(**GLR849PR**)	*I*
	FLR2522	25T	Q	125	✓	✓	—	4	✓	✓	✓	—	Ct(**FLR2523**)	*A,E,H*
	GLR891	26C	Q	162	✓	✓	—	2	✓	✓	✓	—	Ct, Ea(**GLR895TR**), TR(**GLR899PR**)	*A,H*
	GLR2538P	25C	Q	113	✓	✓	—	2	—	—	—	—	Tr, Mo(**GLR2531E**), Ct(**GLR2530T**)	*A,E,H,I*
	FLR2622T	26T	Q	125	✓	✓	—	4	✓	✓	✓	✓	Ct(**FLR2623E**)	*A,E,H,I*
	GLR2640T	26C	Q	125	✓	✓	—	2	✓	✓	✓	—	Ct, Cy(**GLR2645H**), Tr(**GLR2648P**)	*A,E,H,I*
	GLR2650P	26C	Q	125	✓	✓	—	2	✓	✓	✓	—	Ct, Co(**GLR2655C**), Tr(**GLR2658P**)	*A,E,H,I*

A guide to features for large-screen TVs (continued)

Model	Screen size; Table or Console	Tuning method [1]	No. channels	Remote control	Stereo decoder	Stereo adapter jacks	No. speakers	Audio inputs	Audio outputs	Video inputs	Video outputs	Cabinet style(s) [2]	Additional features
GLR2750N	27C	Q	137	✓	✓	—	4	✓	✓	✓	—	Ct, Cf(**GLR2760T**), Ct(**GLR2780N**)	*A,E,H,I*
GLR2788A	27C	Q	137	✓	✓	—	4	✓	✓	✓	—	Ct	*A,E,H,I*
GLR2790B	27C	Q	137	✓	✓	—	4	✓	✓	✓	—	Sa	*A,E,H,I*
Sanyo 52C100	25C	Q	112	✓	—	—	1	—	—	—	—	Ct	—
52C200	25C	Q	140	✓	✓	—	2	—	—	—	—	Mo	*A,H*
AVM260	25T	Q	112	✓	✓	—	2	✓	✓	✓	✓	Sa	*D,H*
AVM270	26T	Q	140	✓	✓	—	2	✓	✓	✓	✓	SA	*A,B,D,E,G,H*
Sears 4303	25C	K	82	—	—	—	1	—	—	—	—	Ct	*I*
4905	25C	V	82	✓	—	—	1	—	—	—	—	Ct	*I*
49076	25C	V	111	✓	—	—	1	—	—	—	—	Ct	*I*
4810	25C	V	111	✓	—	✓	1	—	—	—	—	Ct	*I*
4820	25C	V	111	✓	—	✓	1	—	—	—	—	Ct	*I*
4806	25C	Q	111	✓	—	✓	1	—	—	—	—	Ea	*I*
4816	25C	Q	111	✓	—	✓	1	—	—	—	—	Ts	*I*
4870	25C	Q	111	✓	✓	—	2	✓	—	✓	—	Ts	*A,G,H,I*
4876	26C	Q	140	✓	✓	—	4	✓	—	✓	—	Ea	*E,I*
4886	26C	Q	140	✓	✓	—	4	✓	—	✓	—	Ts	*E,I*

Sharp 25KT15	25T	K	82	—	—	—	1	—	—	—	—	—	—
25KT35	25T	V	105	—	—	—	1	—	—	—	—	—	—
25KT55	25T	V	105	✓	—	—	1	—	—	—	—	—	—
25KT65	25T	V	105	✓	—	—	1	—	—	—	—	—	—
25KV775	25T	V	105	✓	—	✓	2	✓	—	✓	—	—	*C,E*
25KC165	25C	V	105	✓	—	—	1	—	—	—	—	Ct	—
25KV785	25T	V	140	✓	✓	—	2	✓	✓	✓	—	—	*C,E,G,H,I*
25KC265	25C	V	140	✓	✓	—	2	✓	✓	✓	—	Ct	*A,C,G,H,I*
26KC485	26C	V	140	✓	✓	—	2	✓	✓	✓	—	Ct	*A,C,E,G,H,I*
Sony KV-2670R	26T	Q	181	✓	—	✓	1	—	—	✓	—	Ct(**KV-2690R, 2697R)**	*E,G,I*
KV-2781R	27T	Q	181	✓	✓	—	4	✓	—	✓	—	Ct	*A,E,F,G,H,I*
KV-25XBR	25T	Q	181	✓	✓	—	2	✓	✓	✓	✓	Um	*A,B,C,D,E,G,H,I*
KV-2782R	27C	Q	181	✓	✓	—	4	✓	✓	✓	✓	Ct	*A,C,E,F,G,H,I*
KV-2680R	26C	Q	181	✓	✓	—	2	✓	—	✓	—	Ct	*A,C,E,F,G,H,I*
KV-25DXR	25T	Q	181	✓	✓	—	2	✓	✓	✓	✓	Um	*A,B,C,D,E F,G,H,I*
KV-2783R	27C	Q	181	✓	✓	—	4	✓	✓	✓	✓	CL	*A,C,E,F,G,H,I*
Sylvania CLE235PE	25C	K	82	—	—	✓	1	—	—	—	—	Tr	*I*
RLE308WA	25T	Q	152	✓	—	✓	1	—	✓	—	—	Ct	*I*
CLE242AK	25C	Q	82	—	—	✓	1	—	—	—	—	Ct, Cy(**CLE244HP**), Tr(**CLE246PE**)	*I*
RLE335PE	25C	Q	82	✓	—	✓	1	—	—	—	—	Tr	*I*
RLE342AK	25C	Q	152	✓	—	✓	1	—	—	—	—	Ct, Tr(**RLE346PE**), Cy(**RLE344HP**)	*I*
CLE253PN	25C	Q	152	—	✓	—	2	✓	✓	—	—	Cy, Ct(**CLE251AK**), Tr(**CLE255PE**)	*A,E,H,I*

A guide to features for large-screen TVs (continued)

Model	Screen size; Table or Console	Tuning method [1]	No. channels	Remote control	Stereo decoder	Stereo adapter jacks	No. speakers	Audio inputs	Audio outputs	Video inputs	Video outputs	Cabinet style(s) [2]	Additional features
RLE353PN	25C	Q	152	✓	✓	—	2	✓	✓	—	—	Cy, Ct**(RLE351AK)**, Tr**(RLE355PE)**	*A,E,H,I*
RLE368PE	25C	Q	152	✓	✓	—	2	✓	✓	—	—	Tr, Ct**(RLE362AK)**	*A,E,H,I*
RNE602SL	26T	Q	152	✓	✓	—	2	✓	✓	✓	✓	Sa	*A,C,D,E,H,I*
RNE372AK	26C	Q	152	✓	✓	—	2	✓	✓	✓	✓	Ct, Cy**(RNE375PN)**, Tr**(RNE378PE)**	*A,C,E,H,I*
RNE590SL	26C	Q	178	✓	✓	—	4	✓	✓	✓	✓	Ct	*A,C,E,G,H,I*
RNE595PN	26C	Q	178	✓	✓	—	4	✓	✓	✓	✓	Cy, Ct**(RNE592BK)**	*A,C,E,G,H,I*
Toshiba CX2675	26T	Q	117	✓	—	—	1	✓	—	✓	—	Mo	*A,D,E,G,I*
CF2514	25C	Q	133	✓	✓	—	4	✓	—	✓	—	Mo	*A,D,E,H,I*
CZ2685	26T	Q	139	✓	✓	—	4	✓	✓	✓	✓	Mr	*A,B,D,E,G,H,I*
Zenith B2500W	25T	Q	157	—	—	—	1	—	—	—	—	Mo	*I*
B2504G	25C	Q	157	—	—	—	1	—	—	—	—	Ct, Cl**(B2508P)**, Cy**(B2510N)**	*I*
SB2501W	25T	Q	157	✓	—	—	1	—	—	—	—	Mo	*I*
SB2505G	25C	Q	157	✓	—	—	1	—	—	—	—	Ct, Cl**(SB2509P)**, Cy**(SB2511N)**	*I*
SB2569K	25T	Q	157	✓	—	—	2	—	—	—	—	Mo**(SB2569S)**	*I*

B2712G	27C	Q	157	—	—	—	1	—	—	—	—	Ct, Cl**(B2716P)**	*I*
SB2591P	25T	Q	178	✓	—	✓	2	✓	✓	✓	✓	Mo	*A,E,F,I*
SB2713G	27C	Q	157	✓	—	—	1	—	—	—	—	Ct, Cy**(SB2715N)**, Cl**(SB2717P)**	*I*
SB2595P	25T	Q	178	✓	✓	—	2	✓	✓	✓	✓	Mo, Mo**(SB2595A)**, Mo**(SB2597Y)**	*A,E,F,I*
SB2527P	25C	Q	178	✓	✓	—	2	✓	✓	✓	✓	Cl, Ct**(SB2523K)**, **(SB2523G)**, Ea**(SB2531N)**	*A,E,F,H,I*
SB2795P	27T	Q	178	✓	✓	—	2	✓	✓	✓	✓	Mo	*A,E,F,I*
SB2541X	25C	1	178	✓	—	✓	4	✓	—	—	—	Um	*A,E,F,I*
SB2729N	27C	Q	178	✓	✓	—	2	✓	✓	✓	✓	Ea, Md**(SB2727P)**	*A,E,F,H,I*
SB2797Y	27T	Q	178	✓	✓	—	2	✓	✓	✓	✓	Mo	*A,E,F,I*
SB2737Y	27C	Q	178	✓	✓	—	4	✓	✓	✓	✓	Um	*E,F,H,I*
SB2741X	27C	Q	178	✓	✓	—	4	✓	✓	✓	✓	Um, Tr**(SB2771R)**, Fp**(SB2777P)**	*E,F,H,I*
SB2731G	27C	Q	178	✓	✓	—	2	✓	✓	✓	✓	Ct	*E,F,H,I*

[1] *✓ = Electronic varactor; channels may need to be programmed prior to selection. K = Mechanical knob. Q = Quartz-synthesized tuner.*

[2] *Styles, where given are as designated by mfr. Ct = Contemporary. Tr = Traditional. Ea = Early American. Co = Colonial. Cy = Country. Fp = French Provincial. Md = Mediterranean. Ts = Transitional. Mo = Modern. Cf = Country French. Sa = State-of-the-art. Cl = Classical. Ur = Upright. Um = Ultramodern. Mr = Monitor/receiver. Models in parentheses are similar to listed model except for cabinet style.*

Key to Additional Features
A–Tone controls.
B–RGB input, used with computer.
C–Dual VHF antenna input.
D–Earphone or headphone jack.
E–Comb filter, which provides improved picture resolution.
F–Channel censor.
G–Sleep timer.
H–Can receive separate audio program (SAP) signal provided by some stations.
I–Sharpness control.

Ratings of small-screen black-and-white TVs

Listed in order of overall quality. Prices are as published in a **March 1982** report, and updated for 1984 publication. + indicates shipping extra.

Better ←——→ *Worse*

◉ ◓ ○ ◒ ●

Brand and model	Price	Overall picture quality: Observed picture quality	Freedom from geometric distortion	Black-level retention	Interlace	Airplane-flutter rejection	Adjacent-channel rejection	Spark rejection	Citizens Band rejection	VHF at antenna terminals	UHF at antenna terminals	Fringe reception: VHF reception with built-in antenna	UHF reception with built-in antenna	Comments
RCA AFR120S	[1]	◓	◒	◓	◓	◒	○	◓	◒	○	○	○	◒	A,B,C
Toshiba T288	$130	◓	○	◒	◓	●	○	○	●	○	○	◓	◒	B
Philco B422MWH	[2]	◓	●	◒	◓	●	◓	○	●	○	◒	○	◒	E
Sylvania MW0130WH	[3]	◓	●	◒	◓	●	◓	○	●	○	◒	○	◒	E
Zenith N121S	[4]	○	◒	◉	◓	◒	◉	◓	◒	◒	◒	○	◒	A,D,H,J
General Electric 12XE2104T	[5]	○	○	○	○	●	●	○	◒	○	○	○	◒	A,C,D,G,I
Magnavox BB3732	[6]	○	◒	○	○	◒	◓	○	○	○	◒	◒	○	A,J

Quasar AP3230TH	7	○	○	○	○	◓	○	◉	◒	○	○	○	○	B,C,F
Panasonic TR-1215T	8	○	○	○	◒	◓	◓	◉	◒	○	○	○	◒	C,D,F
Sampo B-1201BK	9	○	○	●	○	●	●	◒	◒	◒	○	◒	○	C
Sharp 3K82B	10	○	○	●	○	●	◒	◒	◒	○	○	◒	○	C
Sanyo 21T66A	11	○	◒	◒	◒	◒	◓	○	●	○	◒	○	○	C

1 **RCA AFR 120S** *superseded by* **AJR120W; AJR120Y** *essentially similar; price not available.*

2 **Philco B422MWH** *superseded by* **B424SWH** *($85);* **B438SWA** *($86) essentially similar.*

3 **Sylvania MW0130WH** *superseded by* **MWB132AL** *($100);* **MWB134WA** *($110) essentially similar.*

4 **Zenith N121S** *superseded by* **BT121** *($95).*

5 **General Electric 12XE2104T** *superseded by* **12XF3104S; 12XF3114W** *and* **12XG3122W** *essentially similar; price not available.*

6 **Magnavox BB3732** *superseded by* **BD3733** *($90).*

7 **Quasar AP3230TH** *superseded by* **AP3230WH;** *price not available.*

8 **Panasonic TR-1215T** *superseded by* **TR-1241T** *($95);* **TR-1240T** *($90) essentially similar.*

9 **Sampo B-1201BK** *essentially similar to* **B-1201BW** *($100) and* **B-1211BW** *($100).*

10 **Sharp 3K82B** *superseded by* **3K89** *($110).*

11 **Sanyo 21T66A** *designated* **21T66-1** *($80).*

SPECIFICATIONS AND FEATURES

All have: • 12-in. screen measured diagonally. • Built-in VHF antenna. • Unlighted channel numbers. • Handle or well for carrying. *Except as noted, all have:* • Separate UHF antenna. • 70-channel click-stop UHF tuner. • Overscan of 12 percent or less. • Tone quality judged fair. • Serviceability judged average. *Except as noted, all lack:* • Preset fine-tuning feature.

KEY TO COMMENTS

A–Preset fine-tuning.
B–At least one frequently used control at set's rear; judged inconvenient.
C–Combined VHF/UHF built-in antenna.
D–Built-in antenna replaceable by user.
E–Earphone jack; earphone supplied.
F–Earphone jack; earphone not supplied.
G–Serviceability judged worse than average.
H–Overscan was 15 percent.
I–Tone quality judged poor.
J–Worse than other sets in receiving some nonstandard signals.

Industry spokesmen insist that the high price of renting-to-own is justified by conveniences like free delivery, free repairs, and a replacement if the item has to go into the shop.

It seems unlikely that the convenience of the rent-to-own arrangement is really worth more than the TV set itself.

HOME SATELLITE DISHES

Most people still rely on roof antennas for TV reception (see page 50), but a fair number of consumers have taken a giant step forward, or upward. Home satellite dishes for receiving TV broadcasts directly from satellites in space have become a minor craze among people who can't receive ordinary over-the-air TV broadcasts, aren't served by cable, or who revel in getting "free" TV instead of paying a monthly cable fee, as well as people who just really like watching TV. In 1986, roughly 1.5 million people owned satellite earth stations—called TVRO systems, for "TV receive only."

As the home dishes spread, prices dropped. Back in 1979, a common price was $30,000 or more. The full-featured outfit installed on the roof of Consumers Union's Mount Vernon headquarters cost $2,650 in 1986. There are no-frills units advertised for under $1,000. The slide in prices became even steeper when *Home Box Office* and other cable companies started to scramble their signals. (See page 47.)

Tuning in

A home dish system lets you latch onto the signals from one of the two dozen or so communications satellites that carry domestic television programs, along with computer data, radio traffic, and telephone calls. These satellites are strung along an arc above the equator, high enough so that their orbital speed matches the rotation of the earth. They have a "geosynchronous," or "geostationary," orbit and thus appear to stay still in the sky.

Each satellite has what amounts to as many as 24 transponders that recieve signals from earth and transmit them back again. So each "bird" can accommodate as many as 24 different TV programs. The signals are in the microwave portion of the radio band, where the waves are "short" enough to be beamed at the satellite's antenna. Most of the video traffic travels in what's known as the C-band, a portion of the microwave spectrum corresponding to 4,000 to 8,000 megahertz. That was the band assigned years ago to the cable-TV industry.

The signals coming back from the satellites are very weak—the equivalent of a CB radio broadcasting from 22,000 miles away. The signals are weak partly to conserve power. But the signals are also kept weak so as not to interfere with ground-based microwave traffic such as long-distance telephone relays, which use the same frequencies.

As a consequence, the owner of a home earth station needs a large dish to gather the signals. How large depends on the location. Domestic satellite signals are strongest in the Middle West, so a dish that's 5 or 6 feet across might be sufficient for residents of Kansas City or Denver. Those living near the coasts need a dish that's 8 to 12 feet in diameter.

What you need

Setting up a dish isn't an easy do-it-yourself project. The dish itself is but one component of a satellite system. Indoor components include the receiver, which acts as a tuner and power source for the dish's electronics. Hooking up the cable from the dish to the satellite receiver is a little tricky, but the average do-it-yourselfer should be able to handle it. Connecting the receiver to the TV set is like hooking up a VCR. When you add a descrambler, however, the setup starts to get complicated, with more than a dozen separate wires and cables connecting the various parts.

Watching satellite television is a little more complicated than just turning on the set and flipping the dial. Altogether, Consumers Union engineers, who set up a dish in Mount Vernon, New York, could choose from about 150 different channels. To get a particular channel, you have to aim the dish at a satellite, which may take some "tweaking." Upscale receivers let you program the setting—store it in memory—so you can then tune in that satellite at the press of a button. Buttons on the remote-control unit are usually labeled with letters and numerals. The satellites have names like *Galaxy 1* and *Westar 4* and go by shorthand designations, such as *G1* or *W4*. You tap those two-character codes into the remote-control unit. The dish then homes in on the satellite so you can start receiving one of its channels. Using the remote control, you flip through the transponders to find the program you want.

If you have more than one TV set in the house and want to watch satellite programming on the second set as well, bear in mind that the dish can only "look" at one satellite at a time. Changing satellites for the first set will automatically change satellites for the other one. The second set can be wired to the satellite receiver and can watch other stations on the same satellite as long as the dish is equipped with a block downconverter. But each extra TV set may need its own descrambler, a considerable extra expense.

What about programming?

At first, you may be intrigued by just about every program. But the novelty of watching the proceedings of the Canadian Parliament, for instance, may tend to wear off. You will quickly learn to use a dish owner's program guide to find out which satellites offer the programs you want.

There are several monthly program guides that list all the scheduled satellite broadcasts. The guides are about the size of a small city's telephone book. At any given time, the "bird-watcher" can find sports, news, public affairs, series reruns, and movies. Especially movies. Consumers Union estimated that more than 1,200 different films were shown in one month alone in 1986.

As a dish owner you get to eavesdrop on various not-ready-for-primetime morsels known as "feeds" and "backhauls." Those include the programs that the national networks feed to local stations throughout the day as well as unedited news or sports footage being hauled back via satellite to the network from TV crews in the field. While "raw" television may be diverting, broadcasters and pay-cable programmers alike regard such viewing as aerial trespassing.

What next?

What troubles a lot of people about dishes is that they're so obtrusive. Not only do dishes have to be large enough to collect the weak satellite signal, they also need a clear view of the sky, unobstructed by buildings, trees, fences, or foliage.

A few years ago there was talk about powerful "direct broadcast" satellite TV that would use dishes only one foot in diameter, but the companies promoting that system never really got it off the ground. Instead, the future is said to be in Ku-band (12,000 to 18,000 MHz), a band assigned for commercial satellite use. Ku-band signals can be made much stronger than C-band signals, so you can pick them up with a smaller dish. On the East Coast, for instance, you'd need only a 2- or 3-foot dish.

A handful of Ku-band satellites are now in orbit, but so far they carry little programming that might interest the average viewer. If and when Ku-band programming takes off, you'll be able to retrofit many C-band dishes.

Buying a dish

You need a fair amount of space for a satellite dish. And even if you have the land, the aesthetic cost of a huge, ungainly object near your house may still be too high.

Be sure to check local zoning restrictions. While a federal ruling makes dishes legal to own and prohibits discriminatory zoning, local ordinances can still ban dishes if they also ban other types of antennae.

You also need a site with a clear view of the satellite belt in the southern sky. Watch out for local sources of interference, such as a nearby microwave transmitter. Such terrestrial interference makes some sites, particularly in urban areas, unsuitable for dishes.

If you're content with the 100 or so stations that remain unscrambled (and you want to gamble that more programmers won't hop on the scrambling bandwagon), you may want to buy a dish.

You should be able to get a reasonably well-equipped system for about $1,600, including installation. That price includes a 10-foot aluminum-mesh dish, a motorized mount, the dish electronics and cables, and a remote-controlled receiver. Site preparation, stereo sound provisions, a descrambler box, and the television set are all extra. Buy from a reputable local dealer who can survey your site as well as install and service the unit. Don't try to install it yourself unless you really know what you're doing.

■ Scrambled signals ■

Home Box Office, the biggest and richest of the pay-cable programmers, began to scramble its signal early in 1986. Several other cable programmers soon followed *HBO*'s lead with other pay-cable programmers following along.

At issue is the question of whether the satellite signals that fill the air are justifiably the private property of the companies that produce and distribute them. The cable-TV industry claims they are. The cable companies purchase nonbroadcast programming to sell to subscribers, they say, not to give away to poaching dish owners. Dish owners say they are not "pirates" or "thieves" stealing services. The pay-cable signals rain down on their property anyway, like any other sort of radio transmission. Merely tuning in is not the same as tapping into electricity from a utility pole.

Dish manufacturers and dealers claim to be the victims of a monopoly. In their view, cable franchises and cable programmers are effectively stifling competition by making access to programs prohibitively expensive for dish owners. They point out that two-thirds of the dish owners are located outside of areas that cable serves.

Home Box Office isn't trying to prevent dish owners from seeing its programs, just from seeing them without paying *HBO*. Dish owners are free to descramble its signal—for a price.

The first expense is the descrambler, a black box of electronics that puts the scrambled signal back together. Once *HBO* chose its scrambling format, most other pay-cable programmers chose the same format. Dish owners are thus obliged to buy only one descrambler, which decodes 90 percent of scrambled programming. (Other formats were chosen by CBS, some Canadian programmers, and private cable services such as those that supply X-rated movies to hotels.)

The descrambler, called *Videocipher II*, is made by a company called M/A-Com and is sold under the brand names *M/A-Com* or *Channel Master* for $395.

While one *Videocipher II* descrambler is able to decode all those signals, the dish owner must still pay each programmer a monthly fee in most cases. The programmer makes sure of that by, in essence, switching on each user's descrambler by satellite. Programmers turn on a descrambler by encoding a special number in the signal they send up to their satellite. If you fall behind in your payments, they stop sending up your account number.

The monthly payments are not inconsiderable. To get all the channels that were formerly free, a dish owner may have to pay something like $80 a month. Why should they have to pay the same price as cable subscribers when no cable franchise has had to spend money laying cable to their homes? Besides, since their dishes eliminate the middleman—the cable franchise—shouldn't the price be lower, not higher? Cable franchises pay *HBO* as little as $5 a month in some areas.

A *Home Box Office* brochure explains that cable-franchise operators are "volume buyers," and that cable companies spend "millions of dollars to market our service to consumers." Since *HBO* has to shoulder those "promotion costs" when it sells direct to the consumer, the price has to be higher.

Scrambling, now that it's here, is probably here to stay. The real question is how much dish owners will have to pay to decipher scrambled programming.

The pieces of a satellite TV receiving system

Descrambler. This is necessary to decode scrambled signals.

Dish. A dish may be solid, perforated, or metal mesh. There is little functional difference among types. Mesh types may shed wind loads better and blend more easily into the landscape, especially if painted black. Solid dishes (steel, aluminum, or fiberglass-covered metal mesh) are said to be sturdier, but they may be harder to keep steady because of their weight and wind- or snow-loading. Black-painted aluminum mesh is a popular choice.

Downconverter. This converts signals from the microwave range down to a more usable radio frequency—typically 70 megahertz—for transmission to the house. A simple downconverter sends one transponder signal at a time. A more sophisticated block downconverter sends all 24 transponder signals at once, thus allowing you to view more than one satellite channel at a time, say, by a second TV.

Feedhorn. This receives the focused signal and feeds it through the waveguide to the antenna.

Low-noise amplifier. The LNA boosts the strength of the signal picked up by the antenna. Signal "noise" is measured in degrees Kelvin. LNA noise rating of 100° K or less is considered desirable.

Motor drives. Motorized mounts allow aiming the dish from indoors by remote control. The linear type uses a screw rod to move the dish assembly. The length of the rod limits a sweep to perhaps half the sky, which is good enough for most satellite viewing. A costlier rotary type can readily sweep the dish from horizon to horizon.

Mounting hardware. The mounting stability is often the weak link of a home satellite setup. Any mount must remain steady in severe weather. Options of varying complexity—and cost—are available. The simplest is a patio mount, which must be moved by hand. Most mounts are polar type, oriented so that simple side-to-side movement sweeps the sky along the satellite arc. The altazimuth type moves up, down, and sideways to aim at satellites.

Multiconductor cable. This includes conductors for TV signals, power for dish electronics, and a mount motor.

Receiver. The receiver tunes in satellite signals. It is worth looking for this component on its own rather than as part of a package. Simple units provide manual tuning and monophonic sound. Upscale versions provide stereo sound and remote control.

Site preparation. Concrete footing or a slab is often necessary for the dish to be mounted stably on the ground. The dish may also be mounted on a rooftop or patio. An important consideration is steadiness and a clear sight line to the satellite belt in the southern sky.

TV ROOF ANTENNAS

An indoor antenna, whether separate rabbit ears or one built into a set, has very limited power to remove picture snow or get rid of multiple images, or "ghosts." What's more, an indoor antenna that provides adequate black-and-white reception may not be able to render colors properly.

A TV roof antenna, however, has the potential to make the best of an inferior signal. The improvement that can result in a television picture is often remarkable.

The type of roof antenna you need for good reception depends on the quality of the TV signal in your location. The farther you are from the transmitter, the weaker the signal. A roof antenna becomes almost a necessity in weak-signal areas—places roughly 30 to 50 miles from the transmitter. But to describe antenna performance solely in terms of distance from the transmitter doesn't tell the whole story. Terrain, transmitter height and power, and other variables also affect signal strength.

When installed, a typical antenna has the look of a capital T, with its crossbar—the antenna boom—holding the signal-sensitive elements. Some booms measure more than 10 feet in length, but are reasonably easy to handle because they're made of lightweight metal.

If you want to watch UHF channels 14 through 83, as well as VHF channels 2 through 13, a roof antenna could be a necessity because UHF signals don't travel as well as VHF. You can have separate VHF and UHF antennas, but a combination model not only makes for a neater installation than two separate antennas, it's also less expensive to install. It requires only a single lead-in wire running from the roof to the set instead of two separate leads. You need a signal splitter to route the VHF and UHF signals to the appropriate terminals on the TV set.

Antenna performance can be thought of as consisting of six factors.

Gain. The main job of an antenna is to capture signals from a TV transmitter and feed them to the lead-in wire of a TV set. To pull in relatively weak signals from a distant station, an antenna needs high gain, which is one way of describing good signal-gathering ability.

Other things being equal, the higher a TV station's broadcast frequency—that is, the higher its VHF or UHF channel number—the weaker the signal relayed by the antenna to the TV set. Therefore, it's also critical that gain be adequate from channel to channel across the VHF and UHF bands.

Front-to-back ratio. A TV antenna should be more sensitive to signals arriving at its front than to signals that reach it from other directions.

This is especially important if you live between two cities, each of which has one or more TV stations (see the section on antenna rotators, page 57). A high front-to-back ratio helps keep the picture ghost-free by ignoring or rejecting signals that buildings and other obstacles reflect to the antenna's rear.

Beam width. If the angle of reception, or beam width, is too great, the antenna may have trouble rejecting signals reaching it from either side. But if the beam is too narrow (rarely a problem with TV roof antennas), you might find it hard to set the antenna to the precise position that gives the best reception.

Flatness. A TV channel occupies a specific bandwidth of frequencies. Unless an antenna's response to all the signals in the bandwidth is smooth and uniform—or "flat," as engineers say—the colors in a TV picture may leave something to be desired.

FM band. A TV roof antenna can also help to pull in the FM radio band. This can be a benefit if you live in an area where FM signals are not sufficiently strong. In such locations, a TV antenna is likely to provide better FM reception than the line-cord antenna of an FM radio. For demanding applications, however, a separate roof-mounted FM antenna is necessary.

While most antennas are usually ready to receive FM when you take them from the box, some need a little preparation before hookup. You may find a few stublike removable elements that are supposed to *reject* the FM band. With stubs in place, such an antenna suppresses possible interference from nearby strong FM stations. If you want to pull in FM, however, you cut or break off the removable stubs from the antenna, as the instructions direct.

Impedance match. An antenna's impedance—a measurable electrical characteristic—should come reasonably close on all channels to matching the impedance of the cable that feeds signals to the TV set. Otherwise, you may get unpleasant multiple reflections in the TV picture. An antenna is usually designed for connection to ribbon cable (also called "twin-lead"), which has a 300-ohm impedance.

If oil burner motors, car ignition systems, and other forms of electrical interference make a TV picture streaky, you might try coaxial cable rather than ribbon cable as a lead-in to the set. Often, coaxial cable is able to reduce such interference. To use coaxial cable with most antennas, you'll need one or more balun transformers. (Baluns are available at electronics supply stores.) However, you should think twice about using coaxial cable on the high VHF and UHF channels in areas where signals are extremely weak: Losses in signal strength at those high frequencies are much greater with coaxial cable than with ribbon cable.

Assembly and installation

The typical antenna is sturdily constructed. It's designed to assume its working shape when you unfold the snap-out elements. Manufacturers generally keep to a minimum the number of screws needed for assembly, though some models have subassemblies that must be fastened to the main boom.

An antenna comes with the right hardware for fastening the unit to a mast—the vertical part of the T assembly. But you'll have to buy the mast itself, plus hardware for attaching it to the roof or strapping it to the chimney, as well as enough cable and standoff cable supports (to be spaced at about 3-foot intervals).

Read the installation instructions at least once before beginning assembly and then again as you work. Don't tackle the job, though, unless you're reasonably handy and willing to clamber about on the roof. And be sure to work as far away as possible from any power lines in the vicinity of the roof. Reports of people being electrocuted while installing antennas are not uncommon.

If you plan to install your own antenna, here are some suggestions.

- If you aren't installing a rotator, you can find the best permanent orientation for the antenna by having one person turn it slowly while another watches the picture on the TV set.
- Use cadmium-plated or other rustproof hardware at all points.
- Strap the mast firmly to a strong upright structure, such as a chimney. Use guy wires for additional support on all masts more than 8 feet high.
- Avoid long lengths of free-hanging lead-in. Fasten the lead-in to the mast and to the building with standoffs. Don't squeeze the standoff insulator hooks tightly over the lead-in. Where the lead-in passes over a sharp corner, protect it from abrasion with plastic tape or a split section of rubber or plastic hose.
- Use a feed-through insulator at the wall or window frame. Fasten the lead-in against window frames or baseboards with staples designed for the purpose which can be obtained at electronic parts supply stores. Make sure there is no wiring or other metal in the wall behind the tacking surface.
- Give some thought to protecting your TV set from antenna-drawn lightning. There are small install-it-yourself devices available to protect against voltage overload by near misses. Defense against direct hits of lightning require professional installation of elaborate lightning arresters.

Recommendations

Before buying a new TV roof antenna to improve your picture, consider first whether you really need a new one. Don't assume that you live in a weak-signal area because you get a barely acceptable picture or a picture that varies from good to bad, particularly on windy days. It's easier and cheaper to fix your existing antenna than to buy and install a new one. The problem with your picture might be an antenna whose connections have become weathered or corroded. Try cleaning and refurbishing the connections first. The trouble might even be a lead-in wire that's in poor condition, that's become partially or totally disconnected from the antenna terminals, or whose insulation has deteriorated.

If your problems persist, ask your neighbors what kind of reception they are getting with what kind of antenna. Or check with TV dealers and television repair people. Repair people especially usually know about local conditions and may be able to suggest a relatively low-priced antenna that performs well enough in your area.

If investigation confirms that you need a better-performing antenna, use the Ratings as a preliminary guide, but note that the Ratings order is based only on VHF capabilities. Surprisingly, Consumer Union's tests showed that some combination VHF/UHF units outperformed antennas designed for VHF reception alone.

Ratings of TV roof antennas

Listed by groups in order of estimated quality; within groups, listed alphabetically except as noted. All judged to have good VHF beam width and excellent VHF flatness. Where applicable, all judged to have excellent UHF flatness and, except as noted, good UHF beam width. Prices are as published in a **March 1979** report, and updated for **1984** publication.

Excellent ● **Very good** ◓ **Good** ○ **Fair** ◒ **Poor** ●

Brand and model	Price	Boom length (in.)	Band(s)	VHF gain	VHF front-to-back ratio	UHF gain	UHF front-to-back ratio	Comments
■ *The following four models listed in order of estimated overall quality.*								
Jerrold VU935S	$ 169	148	VHF/UHF	Excellent	Excellent	Good	Good	A
Channel Master 3612	[1]	111	VHF	Excellent	Very good	—	—	
Jerrold VIP304	113	110	VHF	Very good	Excellent	—	—	A
Winegard CH8100	170	142	VHF/UHF	Very good	Very good	Very good	Good	A,B
Channel Master 3672A	[2]	154	VHF/UHF	Very good	Good	Fair	Fair	
Finco CSV10	98[3]	119	VHF	Very good	Good	—	—	
Finco F88C	155[3]	141	VHF/UHF	Very good	Good	Very good	Good	
Jerrold VU933S	105	103	VHF/UHF	Very good	Good	Very good	Very good	A
Sears Cat. No. 7933	[4]	157	VHF/UHF	Very good	Good	Very good	Very good	

Antennacraft CDX1050	94	160	VHF/UHF	○	○	◓	◓	
Winegard CH4053	90	109	VHF	○	○	—	—	A,B
Antennacraft CCS1025	55	100	VHF/UHF	○	◒	◓	○	
Antennacraft CS900	70	120	VHF	○	◒	—	—	
Channel Master 3675A	74	96	VHF/UHF	○	◒	◒	◒	C
Finco 94	[3]	111	VHF/UHF	○	◒	◒	○	
Sears Cat. No. 7931	[4]	97	VHF/UHF	○	◒	○	◓	
Ultron 321204	60	111	VHF/UHF	○	◒	○	○	D
Winegard CH7082	100	108	VHF/UHF	◒	○	◓	◓	A,B
Archer VU110	44	100	VHF/UHF	◒	◒	◓	○	
Archer VU160	70	140	VHF/UHF	◒	◒	●	○	
Channel Master 1165A	[5]	114	VHF/UHF	◒	◒	◓	◓	E

[1] **Channel Master 3612** *superseded by* **3612B** *($108).*

[2] **Channel Master 3672A** *superseded by* **3672B** *($161);* **3671B** *($193) and* **3678B** *($144) essentially similar to* **3672B.**

[3] **Finco** *brand name changed to* **Finco-Sonim;** *price unavailable.*

[4] **Sears Cat. No. 7933** *and* **7931** *superseded by essentially similar* **79322** *($60) and* **79302** *($40), respectively.*

[5] **Channel Master 1165A** *superseded by* **1165B** *($97).*

KEY TO COMMENTS
A–As received, designed to reject the FM band, but tabs could be removed to improve FM reception.
B–Could be connected either to 300-ohm ribbon (twin-lead) or to coaxial cable; has built-in balun.
C–Beam width judged poor in UHF reception.
D–No splitter for VHF/UHF separation.
E–Optional FM rejection trap available.

Ratings of VHF/UHF antennas

For those whose viewing options include the UHF band, here is a separate ranking of combination antennas. Antennas are listed below in order of estimated quality, based on their overall quality for *both* VHF and UHF reception. (Within groups, models listed alphabetically.)

Jerrold VU935S

Jerrold VU9333S
Sears Cat. No. 7933
Winegard CH8100

Finco F88C

Antennacraft CDX 1050

Antennacraft CCS1025
Winegard CH7082

Archer VU160
Channel Master 1165A
Sears Cat. No. 7931

Archer VU110
Channel Master 3672A
Ultron 321204

Finco 94

Channel Master 3675A

OTHER TV RECEPTION AIDS

You may need a TV coupler, TV/FM signal splitter, and/or rotator if you have more than one TV set, live in a weak-signal area, or are concerned about the quality of FM reception.

Splitters

If you live in an area with strong signals, an inexpensive indoor antenna atop or alongside a second TV set may provide a satisfactory picture. But in a weak-signal area, an indoor antenna is apt to bring in a great deal of snow. And no matter what the local signal strength, an indoor antenna often can't get rid of ghosts from reflected TV signals.

A good way to improve the reception on your second set without putting up another roof antenna is to install a TV coupler. It's a little box, about the size of a small bar of soap, to which you attach the lead-in from the roof antenna and from which you run separate lines to each of your sets. You can buy a two-set coupler for about $5.

Although you may find claims that a coupler is suitable for mounting outdoors on an antenna mast, think twice about doing so. Not every coupler is adequately sealed against the weather, so it should be installed indoors—perhaps on the back of one of the television sets or on any convenient baseboard.

Most couplers are likely to work well for VHF reception (channels 2 through 13), but performance in UHF reception (channels 14 through 83) can be marginal.

A coupler can also be used to improve FM reception by connecting the receiver to a TV roof antenna. (However, to get excellent FM reception, you'll need a specially designed FM roof antenna.)

VHF/UHF signal splitters

A combination VHF/UHF antenna usually does well enough where you want to receive both VHF and UHF bands. Without a combination antenna, each TV band requires its own antenna and lead-in wire.

Even if you use a combination antenna, you'll still need a splitter at the set to separate VHF and UHF signals at the terminals on your TV set. There are also slightly more elaborate splitters designed to let one lead-in carry the two TV bands plus the FM band as well.

You can't ordinarily expect the very best possible FM reception when you use the TV antenna for FM, but it may be adequate in many locations.

Two-way (VHF/UHF) splitters are marketed for either outdoor or indoor use. The outdoor splitter is designed to prevent the terminals from getting wet and comes with a clamp or other arrangement for fastening it to the antenna mast. The indoor type needs neither weather protection nor mast-connecting hardware, so it costs less. You usually attach this type of splitter to the back of the TV set. A splitter on the roof must obviously be the outdoor type. The roof antenna splitters and the splitters on the set don't have to be the same brand or model.

Three-way splitters (VFH/UHF/FM) may not work as well as two-way models. The three-way models could compromise your reception at some points on TV or FM bands. You would be better off using two antennas—one for TV and one for FM—with one lead to your TV set and a separate lead for FM.

Installation of a splitter seldom presents any serious problems. One suggestion: Mount the outdoor splitter with the connecting terminals on the down side to minimize weather damage to the terminals.

Antenna rotators

If you live where television and FM radio signals come from different directions, or if you live near hills or high buildings that create ghost images on TV or multipath reflections with FM, you can probably improve your reception by using an antenna rotator. The reason: Optimum reception is likely to depend on careful—and different—positioning of the antenna for each source of transmission. This is especially true with high-gain antennas, which must be aimed precisely in the right direction to pinpoint distant or weak signals.

An antenna rotator commonly has a control box that plugs into a wall outlet. You can keep it near your TV set or hi-fi. The control box connects (by way of a cable you purchase separately) to an electric motor mounted at the base of the antenna mast. There may be a built-in provision for guy wires to steady the rotator, if necessary. An automatic antenna rotator lets you set an indicator to a point on the dial, and the motor rotates the antenna to the desired point and stops. With a manual control box, you hold down a button until the antenna reaches the desired point. The rotator has a series of markings around the dial. The smaller the increment of rotation, the better the rotator will pinpoint the reception.

An antenna should point precisely in the direction indicated by the pointer on the control-box dial—that's good accuracy. If the antenna comes back to a particular place consistently each time the pointer is set there, that's good repeatability.

It's nice for the position of the control-box pointer and the antenna to be the same, but a high degree of accuracy is not all that important. A few degrees off won't matter much. Just mark or remember the points on the control-box dial where each station gets best reception.

If a rotator's repeatability isn't good, the unit is still useful; it's just less convenient to use.

Rotators often have a quirk in their repeatability. Because of gear play in the rotator motor, repeatability holds true only for one direction of rotation, be it clockwise or counterclockwise. (The antenna rotates in the same direction that you rotate the pointer.) In other words, when changing stations, if you should reverse the direction of antenna rotation, the motor would have to run a bit to take up the slack in the gears before the antenna moved again, and the antenna would stop a few degrees short. There are two things you can do to minimize the problem. You can note two separate control-box settings for each station—one for each direction of antenna rotation. Or you can overshoot the setting when approaching from the opposite direction and then dial back to it.

Some control boxes have a knob with a pointer to be rotated over a compasslike face. In addition to the dial, there may be push buttons that work like those on a car radio. Some models use two pointers in the control box: one that moves as you turn the dial, and a second that shows the position of the antenna as it rotates. Others don't show the antenna's position while it's rotating. Instead, they have lights that go on when the motor is running—maybe one to show if the rotator is moving clockwise and another if it is going counterclockwise.

VIDEOCASSETTE RECORDERS

A VCR frees you from TV schedules; you can time-shift—record your favorite programs and watch them at your leisure. If you want to watch a show without commercials, promotions, or credits, you can "zap" them away. You can enjoy movies not only at a theater but also in your own home, by renting tapes.

The selection of VCRs has never been better, but that's a mixed blessing. You'll find hundreds of models offering every conceivable combination of features. Prices range from a rock-bottom $270 (or less) to an intimidating $1,900. Unfamiliar brands are mixed in with the more familiar names and with names previously associated only with hi-fi equipment. Salespeople seem to be speaking an unknown language as they talk about time-shifting, six- and seven-headed machines, hi-fi sound, HQ picture, and MTS.

You have a better chance of getting the unit that's right for you if you familiarize yourself with what's available and the jargon that goes with it.

Source basics

If you're buying your first VCR, there are some elementary questions you need to answer before you head for the local audio-video discount house.

How will you use the VCR? Some people use their VCRs primarily for time-shifting—recording a TV program for viewing later, at a more convenient time. If that's how you plan to use a VCR, it's important to get one that's easy to program and that can record all the programs you want within a certain time period. Every VCR on the market lets you do some time-shifting; as you move up in price, you generally get greater time-shifting capability. The VCR should also have a tuner that will let you bring in the broadcast and cable channels in your area without a lot of dial-turning.

A lot of VCR owners use their machine primarily to play prerecorded tapes. For that kind of use, the most basic VCR will suffice. You can even buy a playback-only videocassette player for about $150. However, if you want to hear taped musical programs at their best, you should consider a higher-priced VCR that has hi-fi sound.

Growing numbers of people are creating their own video entertainment—home movies and such shot on a camcorder (camera with built-in video recorder) and played back through the VCR. If you can afford $1,000 or so for a camcorder, then you should consider buying a fairly full-featured VCR; it should have the editing and sound-dubbing capabilities you'll need for your videos.

Camcorders have been a popular choice for home videos. You can also buy a camera and a portable VCR, but a camcorder is lighter and easier to use.

What format should you choose? Three formats exist—VHS, Beta, and 8mm. They are incompatible with one another. The format you select determines the variety of rental movies available to you, how easily you can swap tapes with friends, and possibly the price of the VCR. Most machines use the VHS format, which makes it the safest choice.

The Beta format is dwindling in usage. Beta is fine if all you want to do is time-shift, but you may have trouble finding Beta tapes to rent.

The 8mm format is relatively new. It's fine for time-shifting, but until it becomes more widely used, you'll find very little in the way of tape rental in this format. Because 8mm tape cassettes are so small, the format lends itself to use in camcorders. The 8mm machines use two audio recording methods, one of which is a digital method similar to the one used to make compact discs. That gives 8mm machines inherently fine sound quality.

Features and characteristics

A VCR's picture quality doesn't necessarily get better as you move up in price. The better models are designed so that frequently used controls such as Fast-Forward and Rewind are logically arranged and easy to use; they have a tuner that lets you bring in any available channel without retuning or reprogramming; and they have a complement of useful features that many other machines lack. Here's a rundown.

Number of heads. It's not necessarily true that "the more heads, the better." At least two heads are required for basic video recording. The VHS models with hi-fi have two additional heads devoted exclusively to sound. Special visual effects such as slow motion (a feature used mainly by sports fans) requires additional heads. Playback at the slowest speeds (of interest to people who want to conserve tape) often uses more heads.

Picture enhancement. This is a desirable feature. Look for an "HQ" label. That denotes models that can deliver a higher-quality picture than other machines.

A VCR's slowest tape speed makes the greatest demands on the recorder's ability to reproduce a good picture, but is appealing if you time-shift and want to conserve tape. For the best picture, however, run your machine at its standard speed. In addition to getting optimum picture quality, it will permit better reproduction if you lend or borrow tapes. VCR tapes recorded at slower-than-standard speed may not do very well when played back on a machine other than the one used to record them in the first place.

The best VCRs deliver a picture that is acceptably crisp and clear, with little visual "noise"—graininess, intrusive streaks of color, or such other defects as a jittery picture. But the pictures produced by signals from even the best VCR are no match for those from a good TV set.

Programming. Because time-shifting is so popular, manufacturers and retailers emphasize a VCR's ability to record automatically. The typical VCR can record four events (programs) over a fourteen-day span. A few, generally lower-priced models, can handle only two events or only a seven-day span. Some can be set to record over 365 days. Note, however, that you'll have to reload the VCR after every five to eight hours of recording.

It's one thing to know that a VCR can record eight events in fourteen days. It's quite another to make the machine behave. The first thing to do is master the instruction book. Take your time. Then, when you want to program, it won't seem so formidable to tell the VCR when to start recording, what channel to record, how long to run the tape, and so on.

Some VCRs are much easier to program than others. The best come with clear instructions in the owner's manual and on the machine itself. They also have controls that let you back up if you overshoot a time or channel-number setting. Best of all is a feature called on-screen programming. Commands that appear on the TV screen help you enter the time, date, and channel of the program you want to tape. If you do a lot of time-shifting, you should give on-screen programming a great deal of weight.

Cable-TV. Most VCRs are "cable-ready" in the sense that you can attach the cable to a connector on the back of the machine. True cable readiness means that the VCR can receive nonscrambled cable-TV channels directly, with no need for a cable-converter box or other hardware. The cable won't impair the VCR's time-shifting capability. If the number of channels a VCR can receive is greater than eighty-two, that's a sign the machine is cable-ready.

You may, however, need a converter box to receive channels such as HBO and Showtime. In that case, the converter becomes the channel selector and so limits the time-shifting you can do, even on a cable-ready VCR. You can restore the time-shifting capability in several ways—using one converter for the TV and one for the VCR, for example, or buying a cable-TV adapter for those VCRs that accept one.

Fine editing. The pause control should allow connecting recorded segments smoothly and unobtrusively. A common use of fine editing is to eliminate commercials when you're recording the program you're watching. If you have a video camera, or plan to buy one, a VCR with good fine-editing ability will be an asset.

Selectivity. This is a measure of adjacent-channel rejection; that is, how well a VCR blocks out signals from the channels next to the one being watched. It is particularly important if you use a cable-TV system that fills all the available channels.

Sensitivity. This is the ability to receive a weak signal and deliver a viewable, if snowy, picture. Sensitivity is important if you live in a fringe reception area.

Flutter. Flutter is an audio defect in linear-track recording. It makes music sound sour or watery. Compared with a good low-priced cassette deck, many VCRs have more flutter. They will, however, do an adequate job of reproducing speech. Flutter is likely to be negligible with a VCR operating in its hi-fi mode.

S/N (signal-to-noise ratio), tape. The rushing or hissing that can be heard during quiet passages, particularly at the slowest tape speed, is evidence of a low ratio of signal-to-noise. Here, too, VCRs may not be as good as a low-priced tape deck. However, background noise will be almost inaudible in the hi-fi modes.

Recording noise. In the high-fidelity mode, unwelcome background noise sometimes intrudes when music undergoes fluctuations in loudness.

S/N, tuner. This indicates how much background noise the VCR's tuner will introduce on a videotape. You'll hear the noise if you pipe sound from the VCR tuner through a stereo system.

Distortion, tuner. This is a measure of how muddy or gritty musical or vocal peaks may sound.

Sharpness control. This is useful for adjusting the picture to produce softer or harder edges. It lets you adjust the picture for the best balance between sharp images and video "noise."

One-touch recording. This feature allows you to start taping immediately, without having to use the programming controls, and means you won't miss any of the program you're watching if the phone or doorbell rings. One press of the button gives you thirty minutes of taping; two, sixty minutes; up to eight hours on some units.

Memory backup. Memory saves your programming settings for a while in case of a power outage. Some VCRs will retain settings for only a few minutes; others, for several hours.

Noise reduction. This feature reduces audio background hiss.

Auto index. An auto-index feature electronically marks the tape every time you start to record so that you can find the beginning of each recording automatically during rewind or fast-forward.

Time-left indicator. This tells you the recording time left on a tape.

Slow motion. A much-heralded feature of VCRs for years, slow motion is used mostly by sports fans.

Single frame advance. Another traditional but possibly little-used feature, this allows you to look at a tape segment one frame at a time.

Video dub. This allows you to insert video segments smoothly without disturbing the linear sound track.

Audio dub. This lets you replace the linear sound track with a new one. You might use audio dubs to jazz up a home video. Most hi-fi VCRs let you mix sound from the hi-fi and linear tracks.

Tuning in. "Quartz synthesized," "electronic," "quartz-locked," and "voltage synthesized" are among the terms that describe a VCR's channel tuner. The terms themselves don't tell you much about a VCR's operation. What matters most are two factors: How many channels you can choose from at any one time, and how easy it is to change the channel settings.

Some VCRs have only twelve channel presets. That means the machine can receive, say, only channels 2 through 13; if you want to bring in UHF channel 21 or a cable-TV channel, you have to adjust tiny knobs, dials, or buttons to change one of the presets to the desired channel. Some VCRs can bring in any channel at the touch of one or two buttons. That's desirable.

VCR sound

Amplifiers and loudspeakers in most TV sets haven't been designed to provide quality sound. But that has changed with the advent of stereo broadcasts, stereo TV sets, and VCRs with true high-fidelity audio circuitry.

Now, with the right VCR, you can record a program of music, rent a movie, and get sound that does justice to the picture. You can even use a VCR in place of a cassette tape deck to record only sound. Of course, you need a TV set or a hi-fi system with good speakers. A VCR with hi-fi won't be of much use if you play the sound through an old TV set with a single little speaker.

The audio features to consider are:

Linear stereo. This is the smallest and least significant step up from the conventional monophonic recording. Stereo enhances the sound image. But because stereo VCRs record sound on a relatively narrow strip of the tape (the so-called linear track) and record at a relatively slow speed, a linear stereo sound track may be marred by background tape hiss, wavery musical notes, and other defects. Adding noise-reduction circuitry helps, but only a little. Linear stereo generally shows up on mid-priced VCRs.

Hi-fi. This is the top of the line in sound quality. Hi-fi VCRs record stereo sound in a form very similar to an FM broadcast, so you get a tape with a wide frequency response and low background noise. The sound from a hi-fi VCR is a match for the sound quality of the best movie sound tracks. Although hi-fi is primarily a feature of higher-priced VCRs, it may also show up on mid-priced units. Many prerecorded movie tapes now have the sound track on both the linear and hi-fi tracks.

A hi-fi VCR can record up to eight hours of sound on a single tape, and the sound will be free of the flutter and inaccuracies in frequency response that can plague a conventional tape deck. A hi-fi VCR also has a wider dynamic range and a higher signal-to-noise ratio than a tape deck.

MTS decoder. This feature, which is coupled with the hi-fi circuitry, has become more important as increasing numbers of TV programs are broadcast in stereo. The signal used for stereo broadcasts is technically known as an MTS, or multichannel TV sound, signal. The MTS decoder allows you to record stereo off the air. A VCR that has stereo without MTS can only play recorded tapes in stereo. It will record stereo broadcasts in mono.

PCM recording. This is what 8mm VCRs use for "premium" audio recording. It's a digital system similar to the one used to make compact discs, and so can deliver sound that's even better than that from a hi-fi VCR. PCM recording doesn't suffer as much from modulation noise. The 8mm models have a second sound-recording method, which is similar to the one used in hi-fi VCRs but doesn't provide quite the same level of quality.

The hi-fi circuitry should deliver very good sound quality, with higher-priced VCRs slightly better than low-priced models overall. The higher-priced models have less flutter, and their tuners do a better job of receiving sound without unwanted hiss or distortion.

Recommendations

The simplest, cheapest VCR in the store may well deliver a decent picture and have the basic features you need. But if you buy a VCR just because it's low-priced, you could be making a mistake.

For one thing, you could wind up with a brand that's destined to spend a lot of time in the repair shop (see page 80). For another, you could wind up with a VCR that frustrates you whenever you try to change the channel presets or set it up to record a program when you're away.

No matter which brand you choose, look for a model with the extras that can make the VCR convenient to use.

In particular, consider these:

- a tuner that lets you bring in any available channel easily, and that can receive all the channels in your area
- on-screen programming, which lets you set the VCR to record automatically with a minimum of fuss and frustration
- one-touch recording, which lets you start recording instantly
- auto-index, which makes it quite easy to find programs on a tape

You don't have to pay a premium to get a VCR with those features.

Don't assume that you'll get a better picture by spending more money. What you do get by moving up in price are VCRs with greater versatility, high-quality audio recording, and the ability to receive stereo TV sound. You have to go toward the middle of a line to get a VCR with the tape-editing and dubbing controls some people desire, or a unit that can record a lot of programs.

Ratings of videocassette recorders

Listed in groups; within groups, listed in order of estimated quality. Most use the VHS format (exceptions are noted). Prices are as published in a **January 1987** report.

Brand and model	Price	Picture quality	Features	Programming	Fine editing	Selectivity	Sensitivity	Flutter	Sound quality: S/N, tape	Sound quality: Recording noise	Sound quality: S/N, tuner	Sound quality: Distortion, tuner	Advantages	Disadvantages	Comments
High-priced, full-featured models															
Sony SL-HF900 (Beta)	$1500	◉	◓	◉	◓	○	○	◓	●	◓	◓	○	B,G,J,K, L,M,O,P	g,i,l,m	*B,C,D,E,G, H,J,K,M*
Magnavox VR9565AT	855	◉	◉	◉	○	○	○	◓	●	○	○	◓	A,D,J,K, L,M,N	m	*A,B,C,D, J,K,L,M*
Quasar VH-5865	1290	◉	◉	◉	○	○	○	◓	●	○	○	◓	A,D,J,L, M,N	m	*A,B,C,D,F, J,K,L,M*
Panasonic PV-1742	1250	◉	◓	◉	○	○	○	◓	●	○	○	◓	A,D,J,L, M,N	m	*A,B,C,D,F, J,K,L,M*
GE 9-7400	849	◉	◓	◉	○	○	○	◓	●	○	○	◓	A,J,K,L, M,N	c,m	*A,B,C,F, J,K,L,M*
Zenith VR4100	1099	◓	◓	◓	◉	○	○	○	◒	◓	○	◓	A,I,J,L,M	b,i,m	*B,C,D,E,H, I,K,L,M*
Sony SL-HF450 (Beta)	750	◉	○	○	◉	○	○	○	●	○	○	○	F,M,O,P	e,i,j	*C,G,J,M*
JVC HR-D566U	899	◓	○	◓	◓	○	○	○	◒	◓	○	◓	A,J,M	i,m	*B,C,D,E,I, K,L,M*
Kenwood KV-917HF	1000	◓	◓	◓	◉	○	○	○	◒	◓	○	◓	A,J,L,M	i,m	*B,C,D,I,K,L*

Mitsubishi HS-430UR	1000	◓	◓	◒	◉	◓	○	○	◒	○	◓	◒	B,K	i,k,m	*C,D,E,G, J,K*
Realistic Model 40	700	◉	◒	◓	◉	◒	○	○	◒	◓	○	○	K,L	b,h,m	*C,G,K*
Kodak MVS-5380 (8mm)	1600	○	◓	◓	◉	◓	○	◉	—	○	○	◒	D,E,H,K,N	g,m	*A,C,D,L,M*
Fisher FVH-960	850	◓	○	○	◉	◓	○	◒	●	○	◓	○	N	e,i,m	*E,K*
RCA VMT670HF	899	○	◓	◉	◒	○	◓	○	◒	○	○	○	A,J,N,O	d,e,m	*A,B,C,D,E, J,M*
Sony EV-S700U (8mm)	1395	◓	◓	◒	◉	○	○	◉	—	◓	○	○	B,E,H,J,K	e,g,i,m	*B,C,D,E,G, M*
Pioneer VE-D70 (8mm)	1450	◓	○	◒	◉	○	○	◉	—	◓	○	○	B,E,H,J,K	e,g,i,m	*B,C,D,E,G, M*
Toshiba M-5900	1000	○	○	○	◉	◓	○	○	●	○	○	◓	A,M	m	*B,C,D,E,G, J,M*
Akai VS-626U	1050	◓	◒	◒	◉	◒	◒	◓	●	◓	○	○	D,J,K,M	a,b,h,j,m	*C,G,L*
Emerson VCS966	730	◒	◒	◒	◓	○	○	◒	●	◓	○	○	F,L	b,e,i,m	*C,D,K*
Low-priced, basic models															
GE 9-7145	—	◉	○	◓	◒	○	○	◓	●	—	◒	○	F,J,L,M,N	a,c	*K*
Magnavox VR9530AT	355	◉	○	◓	◒	○	○	◓	●	—	◒	○	A,F,J,L, M,N	a,c	*K*
Panasonic PV-1364	500	◉	◒	◓	◒	○	○	◓	●	—	◒	○	A,F,J,L, M,N	a,c	*F,K*
JVC HR-D180U	499	◓	○	◓	◉	○	○	●	●	—	◒	○	A,C,F,M,P	e,i	*I,K*
Quasar VH-5163	450	◉	◒	○	◒	○	○	◓	●	—	○	◒	A,J,L,M,N	a,c,h	*K*
Zenith VR1810	459	◓	◒	◓	◉	○	○	○	●	—	◒	◒	L,M,P	e	*H*
Mitsubishi HS-339UR	480	◓	◒	◓	○	○	○	◓	◒	—	◒	○	C,N	a,e,k	*G,M*
Sanyo VCR7250 (Beta)	380	◓	◒	◓	◉	●	○	○	●	○	◒	◒	P	e,h,l,m	*G,K*

Ratings of videocassette recorders (continued)

Listed in groups; within groups, listed in order of estimated quality. Most use the VHS format (exceptions are noted). Prices are as published in a **January 1987** report.

Brand and model	Price	Picture quality	Features	Programming	Fine editing	Selectivity	Sensitivity	Flutter	Sound quality: S/N, tape	Sound quality: Recording noise	Sound quality: S/N, tuner	Sound quality: Distortion, tuner	Advantages	Disadvantages	Comments
NEC N-915	449	◉	◒	◒	○	◒	○	○	●	—	◒	○	K,L,M	e,i	*E,G,H*
Sears 53350	370	○	◒	◓	◉	◒	◓	○	◒	—	○	○	M,N,O	e,f,h,i,m	*G,K,M*
Fisher FVH-905	500	◓	◒	○	◉	○	○	○	◒	—	○	◓	—	a,e,i	*K*
Toshiba M-2430	680	○	◒	○	○	○	○	◒	◒	—	○	◓	C,M,P	e,h	*E,G*
RCA VMT295	399	○	◒	○	◓	○	○	◓	◒	—	○	◓	C,N,O	e,h,i,m	*M*
Realistic Model 14	330	◓	●	○	◓	○	○	○	●	—	○	○	—	e,h,i	*K*
Sharp VC-6846UB	550	◓	●	○	○	○	○	◓	◒	—	◒	○	M	a,e,h,i	*G,K*
Hitachi VT-1100A	495	○	◒	◓	◉	○	○	○	●	—	○	◒	M,N,O	e,f,h,i	*G,M*
Emerson VCR870[1]	—	◒	●	○	●	○	○	○	●	—	◒	○	L	a,h,i	*K*

[1] *Superseded by HQ-equipped* **VCR872,** *which delivered a notably better picture than the VCR* **870;** *did better in fine editing; records eight events over 21 days; has 16 channel presets and one-touch recording.*

Specifications and Features

All have: • Adequate MTS stereo channel separation. • Excellent MTS and hi-fi frequency response. *Except as noted, all:* • Have poor audio frequency response on linear ("normal") tracks at slowest tape speed. • Hi-fi units can record TV/FM simulcasts, an advantage. • Lack V-lock control to stabilize vertical jitter in freeze-frame picture (unless otherwise indicated, that control must be adjusted with a screwdriver).

Key to Advantages

A–Programming instructions clearer than most.
B–Dual RF inputs let unit retain time-shifting capability with one scrambled cable channel.
C–Tape-transport controls (Record, Play, etc.) are easier to operate than most.
D–Can be programmed using either on-screen or console display.
E–*Kodak* can record up to 12 hr. of very high quality audio; *Pioneer* and *Sony*, up to 24 hr.
F–Audio high-frequency response on linear tracks at slowest speed better than most.
G–Jog/shuttle control makes editing especially easy.
H–Significantly smaller than most.
I–Comes with videotape explaining VCR use.
J–Freeze-frame picture quality at slowest speed judged better than most.
K–Tape compartment has clear window; lets you see tape movement at a glance.
L–Can eject tape with power off.
M–Inserting a tape turns on the machine.
N–One-touch recording has "standby" mode. It lets you delay recording by up to 24 hr.
O–V-lock control can be adjusted without a screwdriver.
P–Can be set to play back automatically after rewinding.

Key to Disadvantages

a–Plugging accessory such as video camera into line input disconnects tuner.
b–Tape-transport controls more difficult to operate than most.
c–Closely spaced digits make console display difficult to understand; lacks on-screen programming.
d–Programmed only from remote control.
e–Tape counter and clock use the same digits on the display; you have to press a button if you want to switch from counter to clock and vice versa.
f–Lacks line/tuner switch; you must cycle through numbers on the channel selector to get to the line input.
g–ALC (automatic audio-level control) cannot be disabled; that can limit the dynamic range of recorded music.
h–Remote control lacks numeric keypad.
i–Lacks both a quick programming guide on the VCR and on-screen programming.
j–Lacks clock.
k–Has 24-hr clock; more confusing to use than a conventional 12-hr clock.
l–High frequency audio response in hi-fi or PCM mode slightly worse than most.
m–Cannot record TV/FM simulcasts.

Key to Comments

A–Has provision for cable-TV adapter to retain VCR's channel-changing capability.
B–Can mix hi-fi with linear-track sound.
C–Has headphone jack or jacks. Volume level of headphones can be adjusted on all but *Akai* and *Emerson*.
D–Has microphone jack or jacks.
E–Has ac outlet; switched on *Mitsubishi-HS-430UR*, unswitched on others.
F–Ac outlet can be switched or unswitched.
G–Has one-button record. Other units will record only when you press Record and Play.
H–Has "unified" remote control that can control some TV sets from same mfr.
I–Console has numeric keypad.
J–Has variable-speed slow motion.
K–Has remote jack for camera to run VCR.
L–Memory backup time greater than 60 min.
M–Has V-lock control.

A guide to features

Brand and model				Picture		Programming						Tuning		Sound				Special features			
	Price	HQ (High quality)	Heads	Sharpness control	Capability (programs/days)	On-screen programming	One-touch recording	Memory backup	Channels	Presets	Hi-fi	Linear stereo	Noise reduction	MTS decoder	Auto index	Time-left indicator	Slow motion	Frame advance	Video dub	Fine editing	Audio dub
Akai VS-220U	$ 390	✓	2	—	4/14	✓	✓	—	107	14	—	—	—	—	—	—	—	✓	—	✓	—
VS-515U	750	✓	2	✓	6/14	✓	✓	✓	107	32	✓	—	—	✓	✓	—	✓	✓	—	✓	✓
VS-525U	780	✓	2	✓	6/14	✓	✓	✓	107	32	✓	—	—	✓	✓	—	✓	✓	—	✓	✓
VS-555U	850	✓	4	✓	6/14	✓	✓	✓	107	32	✓	—	—	✓	✓	—	✓	✓	—	✓	✓
VS-565U	880	✓	4	✓	6/14	✓	✓	✓	107	32	✓	—	—	✓	✓	—	✓	✓	—	✓	✓
VS-626U	1050	✓	4	✓	8/28	✓	✓	✓	142	16	✓	—	—	✓	✓	—	✓	✓	—	✓	✓
Emerson VCR872	500	✓	2	✓	8/21	—	✓	✓	110	16	—	—	—	—	—	—	—	—	—	—	—
VCS966	730	—	2	✓	4/14	—	✓	—	139	16	✓	—	—	✓	—	—	✓	✓	—	—	✓
VCS966H	800	✓	2	✓	4/14	—	✓	✓	139	16	✓	—	—	✓	—	—	✓	—	—	—	✓
VCR870	[3]	—	2	✓	4/14	—	—	—	105	12	—	—	—	—	—	—	—	—	—	—	—
Fisher FVH-906	450	✓	2	—	4/14	—	✓	✓	111	All	—	—	—	—	—	—	—	—	—	✓	—
FVH-905	500	✓	2	—	4/14	—	✓	—	111	All	—	—	—	—	—	—	—	—	—	—	—
FVH-919	550	✓	2	—	7/14	—	✓	✓	111	All	—	✓	✓	✓	—	—	—	✓	—	✓	—
FVH-920	600	✓	2	—	6/385	✓	✓	✓	111	All	—	✓	✓	✓	—	—	—	✓	—	✓	—

	FVH-940	700	—	4	—	8/365	✓	✓	✓	111	All	—	✓	✓	✓	—	—	—	✓	—	✓	—
	FVH-950	800	✓	4	—	8/365	✓	✓	✓	111	All	✓	—	—	✓	—	—	—	✓	—	✓	—
	FVH-960	850	✓	4	—	9/14	—	✓	✓	140	All	✓	✓	—	✓	✓	✓	—	✓	—	✓	✓
	FVH-980	900	✓	4	✓	8/365	✓	✓	✓	140	All	✓	✓	✓	✓	✓	—	—	✓	—	✓	—
	FVH-990	1000	✓	4	✓	8/365	✓	✓	✓	181	All	✓	✓	✓	✓	✓	✓	✓	✓	—	✓	✓
General Electric	9-7100	349	✓	2	✓	4/14	—	✓	✓	93	14	—	—	—	—	—	—	✓	✓	—	✓	—
	9-7115	369	✓	2	✓	4/14	—	✓	✓	93	14	—	—	—	—	—	—	✓	✓	—	✓	—
	9-7135	369	✓	2	✓	4/14	—	✓	—	93	All	—	—	—	—	—	—	✓	✓	—	✓	—
	9-7120	419	✓	2	✓	4/14	—	✓	✓	93	14	—	—	—	—	—	—	✓	✓	—	✓	—
	9-7140	419	✓	2	✓	4/14	—	✓	—	93	All	—	—	—	—	—	—	✓	✓	—	✓	—
	9-7150	429	✓	2	✓	4/14	—	✓	✓	93	All	—	—	—	—	—	—	✓	✓	—	✓	—
	9-7156	449	✓	2	✓	4/14	✓	✓	✓	93	All	—	—	—	—	—	—	✓	✓	—	✓	—
	9-7215	449	✓	4	✓	4/14	—	✓	✓	93	14	—	—	—	—	—	—	✓	✓	—	✓	—
	9-7175	499	✓	2	✓	4/14	—	✓	✓	93	All	—	✓	✓	✓	—	—	✓	✓	—	✓	—
	9-7245	499	✓	4	✓	4/14	—	✓	✓	93	All	—	—	—	—	—	—	✓	✓	—	✓	—
	9-7250	519	✓	4	✓	4/14	—	✓	✓	93	All	—	—	—	—	—	—	✓	✓	—	✓	—
	9-7176	529	✓	2	✓	4/14	✓	✓	✓	93	All	—	✓	✓	✓	—	—	✓	✓	—	✓	—
	9-7256	549	✓	4	✓	4/14	✓	✓	✓	93	All	—	—	—	—	—	—	✓	✓	—	✓	—
	9-7275	599	✓	4	✓	4/14	—	✓	✓	93	All	—	✓	✓	✓	—	—	✓	✓	—	✓	—
	9-7276	619	✓	4	✓	4/14	✓	✓	✓	93	All	—	✓	✓	✓	—	—	✓	✓	—	✓	—
	9-7350	699	✓	4	✓	4/14	—	✓	✓	93	All	✓	—	—	—	—	—	✓	✓	—	✓	—
	9-7400	849	✓	4	✓	8/21	✓	✓	✓	93	All	✓	—	—	✓	—	✓	✓	✓	—	✓	—
	9-7145	[3]	✓	2	✓	4/14	—	✓	✓	93	All	—	—	—	—	—	—	✓	✓	—	✓	—

A guide to features (continued)

Brand and model	Price	HQ (High quality)	Heads	Sharpness control	Capability (programs/days)	On-screen programming	One-touch recording	Memory backup	Channels	Presets	Hi-fi	Linear stereo	Noise reduction	MTS decoder	Auto index	Time-left indicator	Slow motion	Frame advance	Video dub	Fine editing	Audio dub
			Picture			Programming					Tuning			Sound			Special features				
Hitachi VT-1100A	495	✓	2	—	2/14	—	✓	✓	107	All	—	—	—	—	—	—	—	—	—	—	—
VT-1110A	495	✓	2	—	2/14	—	✓	✓	107	80	—	—	—	—	—	—	—	—	—	✓	—
VT-1200A	495	✓	2	—	4/14	—	✓	✓	107	80	—	—	—	—	—	—	—	—	—	✓	—
VT-1310A	525	✓	3	—	4/14	—	✓	✓	107	80	—	—	—	—	—	—	✓	✓	—	✓	—
VT-1410A	565	✓	2	—	4/14	—	✓	✓	107	80	—	✓	✓	—	—	—	—	—	—	✓	—
VT-1450A	629	✓	3	—	4/14	—	✓	✓	119	All	—	✓	✓	✓	—	—	✓	✓	—	✓	—
VT-1370A	660	✓	3	—	4/14	—	✓	✓	107	80	—	—	—	—	—	—	✓	✓	—	✓	—
VT-1710A	749	✓	3	—	4/14	—	✓	✓	119	All	✓	—	—	✓	—	✓	✓	✓	—	✓	—
VT-1570A	799	✓	3	—	4/14	—	✓	✓	119	All	—	✓	✓	✓	—	—	✓	✓	—	✓	—
VT-1800A	999	✓	5	—	8/365	✓	✓	✓	169	All	✓	—	—	✓	✓	✓	✓	✓	—	✓	✓
JVC HR-D170U	449	✓	2	—	4/14	—	✓	—	111	All	—	—	—	—	—	—	—	—	—	✓	—
HR-D180U	499	✓	4	—	4/14	—	✓	—	111	All	—	—	—	—	—	—	—	✓	—	✓	—
HR-D370U	749	✓	2	✓	4/14	—	✓	✓	111	All	✓	—	—	✓	—	✓	—	✓	—	✓	—

HR-S100	799	—	4	—	8/14	—	—	✓	139	All	—	—	—	—	—	—	—	—	—	✓	✓
HR-D470U	849	✓	4	✓	8/14	—	✓	✓	181	All	✓	—	—	✓	—	—	—	✓	—	✓	—
HR-D566U	899	✓	4	✓	8/14	—	✓	—	181	All	✓	—	—	✓	—	✓	—	✓	—	✓	✓
HR-S200	1195	—	4	—	8/14	—	—	✓	181	All	✓	—	—	✓	—	—	—	✓	—	✓	✓
HR-D756U	1295	✓	4	✓	8/14	✓	✓	✓	181	All	✓	✓	✓	✓	—	✓	✓	✓	—	✓	✓
Kenwood KU-926HF	880	✓	2	✓	4/14	—	✓	✓	111	All	✓	—	—	✓	—	✓	—	✓	—	✓	✓
KV-917HF	1000	✓	4	✓	8/14	—	✓	✓	181	All	✓	—	—	✓	—	✓	—	✓	—	✓	✓
Kodak MVS-5380 (8mm)	1600	—	4	—	8/21	✓	✓	—	169	All	✓	[1]	—	✓	—	—	—	✓	—	—	✓
Magnavox VR9510AT	$ 315	✓	2	✓	2/14	—	✓	—	68	14	—	—	—	—	—	—	✓	✓	—	✓	—
VR9520AT	330	✓	2	✓	4/14	—	✓	✓	93	All	—	—	—	—	—	—	✓	✓	—	✓	—
VR9522AT	[3]	✓	2	✓	4/14	—	✓	✓	93	All	—	—	—	—	—	—	✓	✓	—	✓	—
VR9530AT	355	✓	2	✓	4/14	—	✓	✓	93	All	—	—	—	—	—	—	✓	✓	—	✓	—
VR9550AT	395	✓	4	✓	4/14	—	✓	✓	93	14	—	—	—	—	—	—	✓	✓	—	✓	—
VR9540AT	505	✓	2	✓	4/14	—	✓	✓	93	All	✓	—	—	[2]	—	—	✓	✓	—	✓	—
VR9547AT	[3]	✓	2	✓	4/14	—	✓	✓	93	All	—	✓	✓	✓	—	—	✓	✓	—	✓	—
VR9558AT	595	✓	4	✓	4/14	✓	✓	✓	93	All	✓	—	—	[2]	—	—	✓	✓	—	✓	—
VR9560AT	725	✓	4	✓	8/21	✓	✓	✓	93	All	✓	—	—	✓	—	✓	✓	✓	—	✓	—
VR9565AT	855	✓	4	✓	8/21	✓	✓	✓	155	All	✓	—	—	✓	✓	✓	✓	✓	—	✓	✓
Mitsubishi HS-337UR	400	✓	2	✓	8/14	—	✓	—	107	16	—	—	—	—	—	—	✓	✓	—	✓	—
HS-339UR	480	✓	4	✓	8/14	—	✓	—	107	100	—	—	—	—	—	—	✓	✓	—	—	—
HS-411UR	700	✓	4	✓	8/14	—	✓	—	107	16	✓	—	—	✓	—	—	✓	✓	—	✓	—
HS-421UR	850	✓	4	✓	8/14	✓	✓	✓	139	All	✓	—	—	✓	✓	—	✓	✓	—	✓	—
HS-430UR	1000	✓	4	✓	8/14	✓	✓	✓	139	All	✓	✓	✓	✓	✓	✓	✓	✓	✓	✓	✓

A guide to features (continued)

Picture | Programming | Tuning | Sound | Special features

Brand and model	Price	HQ (High quality)	Heads	Sharpness control	Capability (programs/days)	On-screen programming	One-touch recording	Memory backup	Channels	Presets	Hi-fi	Linear stereo	Noise reduction	MTS decoder	Auto index	Time-left indicator	Slow motion	Frame advance	Video dub	Fine editing	Audio dub
NEC N-915	449	✓	2	✓	4/21	—	✓	—	110	20	—	—	—	—	—	—	—	—	—	✓	—
N-925	549	✓	4	✓	4/21	✓	✓	—	110	20	—	—	—	—	—	—	✓	✓	—	✓	—
DX-1000[6]	699	✓	2	✓	4/21	✓	✓	✓	110	40	—	—	—	—	—	—	✓	—	—	✓	—
N-945	699	✓	2	✓	4/21	✓	✓	—	110	20	—	✓	✓	✓	—	—	—	—	—	✓	—
N-955	899	✓	2	✓	4/21	✓	✓	✓	110	20	✓	✓	✓	✓	—	—	—	—	—	✓	—
N-965	1149	✓	4	✓	8/21	✓	✓	✓	140	All	✓	✓	✓	✓	✓	✓	✓	✓	—	✓	—
DX-2000[6]	[3]	✓	2	✓	4/21	✓	✓	✓	110	40	✓	✓	—	✓	—	—	✓	—	—	✓	—
Panasonic PV-1362	475	✓	2	✓	4/14	—	✓	—	93	14	—	✓	✓	—	—	—	✓	✓	—	—	—
PV-1364	500	✓	2	✓	2/14	—	✓	—	93	All	—	—	—	—	—	—	✓	✓	—	—	—
PV-1366	550	✓	2	✓	4/14	✓	✓	—	93	All	—	—	—	—	—	—	✓	✓	—	—	—
PV-1460	625	✓	2	✓	4/14	✓	✓	—	93	All	—	✓	✓	[2]	—	—	✓	✓	—	—	—
PV-1462	750	✓	2	✓	4/14	✓	✓	—	93	All	✓	—	—	[2]	—	—	✓	✓	—	—	—
PV-1562	750	✓	4	✓	4/14	✓	✓	—	93	All	—	✓	✓	[2]	—	—	✓	✓	—	—	—

PV-1564	850	✓	4	✓	4/14	✓	✓	—	93	All	✓	—	—	✓	—	—	✓	✓	—	—	—
PV-1642	1000	✓	4	✓	8/21	—	✓	✓	93	All	✓	—	—	✓	—	—	✓	✓	—	✓	—
PV-1742	1250	✓	4	✓	8/21	✓	✓	✓	155	All	✓	—	—	✓	✓	✓	✓	✓	—	✓	✓
PV-1360	[3]	✓	2	✓	2/14	—	✓	—	68	14	—	—	—	—	—	—	✓	✓	—	—	—
PV-1361	[3]	✓	2	✓	2/14	—	✓	—	93	14	—	—	—	—	—	—	✓	✓	—	—	—
PV-1560	[3]	✓	4	✓	4/14	—	✓	—	93	14	—	—	—	—	—	—	✓	✓	—	—	—
Pioneer VH-900	999	✓	5	—	4/14	✓	✓	✓	119	All	✓	—	—	✓	✓	—	✓	✓	—	✓	—
VE-D70 (8mm)	1450	—	2	—	6/21	—	✓	—	181	All	✓	[1]	—	✓	—	✓	✓	✓	—	✓	✓
Quasar VH-5162	439	✓	2	✓	2/14	—	✓	—	68	14	—	—	—	—	—	—	✓	✓	✓	✓	—
VH-5163	450	✓	2	✓	4/14	—	✓	—	93	14	—	—	—	—	—	—	✓	✓	—	✓	—
VH-5260	560	✓	4	✓	4/14	—	✓	—	93	14	—	—	—	—	—	—	✓	✓	✓	✓	—
VH-5261	560	✓	4	✓	4/14	—	✓	—	93	14	—	—	—	—	—	—	✓	✓	✓	✓	—
VH-5168	630	✓	2	✓	4/14	—	✓	—	93	All	—	✓	✓	✓	—	—	✓	✓	✓	✓	—
VH-5268	730	✓	4	✓	4/14	—	✓	—	93	All	—	✓	✓	[2]	—	—	✓	✓	✓	✓	—
VH-5665	1050	✓	4	✓	8/21	✓	✓	—	93	All	✓	—	✓	✓	—	—	✓	✓	✓	✓	✓
VH-5865	1290	✓	4	✓	8/21	✓	✓	✓	155	All	✓	—	—	✓	✓	✓	✓	✓	—	✓	✓
Radio Shack Model 14 (Realistic)	330	—	2	—	3/14	—	—	—	105	12	—	—	—	—	—	—	—	—	—	✓	—
Model 18	350	—	2	—	4/14	—	✓	—	105	12	—	—	—	—	—	—	—	—	—	—	—
Model 31	500	✓	2	—	5/14	—	✓	—	110	All	—	✓	✓	✓	—	—	—	—	—	—	—
Model 40	700	—	2	—	3/14	—	✓	—	105	12	✓	✓	✓	✓	—	—	—	—	—	✓	—
Model 41	600	✓	2	—	5/14	—	✓	—	110	All	✓	✓	✓	✓	—	—	—	—	—	—	—

A guide to features (continued)

				Picture	Programming				Tuning		Sound				Special features						
Brand and model	Price	HQ (High quality)	Heads	Sharpness control	Capability (programs/days)	On-screen programming	One-touch recording	Memory backup	Channels	Presets	Hi-fi	Linear stereo	Noise reduction	MTS decoder	Auto index	Time-left indicator	Slow motion	Frame advance	Video dub	Fine editing	Audio dub
RCA VMT285	329	✓	2	—	2/14	—	✓	—	107	80	—	—	—	—	—	—	—	—	—	✓	—
VMT295	399	✓	2	—	2/14	—	✓	—	107	80	—	✓	✓	—	—	—	—	—	—	✓	—
VMT385	399	✓	3	—	4/14	—	✓	✓	107	80	—	—	—	—	—	—	✓	✓	—	✓	—
VMT390	449	✓	3	—	4/365	✓	✓	✓	119	All	—	—	—	—	—	—	✓	✓	—	✓	—
VMT395	499	✓	3	—	4/365	✓	✓	✓	119	All	—	✓	✓	✓	—	—	✓	✓	—	✓	—
VMT590	549	✓	5	—	4/365	✓	✓	✓	119	All	—	—	—	—	—	—	✓	✓	—	✓	—
VMT595	599	✓	5	—	4/365	✓	✓	✓	119	All	—	✓	✓	✓	—	—	✓	✓	—	✓	—
VMT630HF	699	✓	3	—	4/365	✓	✓	✓	119	All	✓	—	—	✓	—	—	✓	✓	—	✓	—
VMT670HF	899	✓	5	—	8/365	✓	✓	✓	169	All	✓	—	—	✓	✓	—	✓	✓	—	✓	✓
Sanyo VHR-2250	$ 330	✓	2	—	4/14	—	✓	—	107	All	—	—	—	—	—	—	—	—	—	—	—
VHR500	330	✓	2	—	4/14	—	—	—	107	16	—	—	—	—	—	—	—	—	—	—	—
VHR-2350	350	✓	2	—	8/14	—	✓	—	107	All	—	—	—	—	—	—	—	✓	—	—	—
VCR7250 (Beta)	380	5	2	—	8/14	—	✓	—	105	16	✓	—	—	—	—	—	—	—	—	—	✓
VHR-2550	400	✓	2	✓	8/14	—	✓	—	110	All	—	✓	✓	—	—	—	—	✓	—	—	—

VHR1600	400	✓	3	✓	8/14	—	✓	—	107	All	—	—	—	—	—	—	✓	✓	—	—	—
VHR2700	500	✓	2	—	8/365	✓	✓	—	111	All	—	✓	✓	✓	—	—	—	✓	—	—	—
VHR2900	650	✓	2	✓	8/365	✓	✓	✓	110	All	✓	—	✓	✓	—	—	—	—	—	—	—
VHR1900	750	✓	4	✓	8/14	—	✓	—	107	All	✓	—	—	✓	—	—	✓	✓	—	—	—
Sears 53283	270	—	2	—	2/14	—	✓	—	105	12	—	—	—	—	—	—	—	—	—	—	—
53292	320	✓	2	—	4/14	—	✓	✓	107	80	—	—	—	—	—	—	—	—	—	—	—
53312	340	✓	2	—	4/14	—	✓	✓	107	80	—	—	—	—	—	—	—	✓	—	—	—
53350	370	✓	2	—	4/14	—	✓	—	119	All	—	✓	✓	✓	—	—	—	✓	—	—	—
53351	440	✓	2	—	4/14	—	✓	✓	119	All	—	✓	✓	—	—	—	—	✓	—	—	—
5345	490	✓	2	—	4/365	✓	✓	✓	119	All	—	✓	✓	—	—	—	—	✓	—	—	—
5351	540	✓	4	—	4/365	✓	✓	✓	119	All	—	✓	✓	[2]	—	—	✓	✓	—	—	—
5352	640	✓	3	—	4/365	✓	✓	✓	119	All	—	✓	✓	✓	—	—	✓	✓	—	—	—
Sharp VC-682U[4]	430	—	2	—	4/14	—	—	—	110	16	—	—	—	—	—	—	—	—	—	—	—
VC-6846UB	550	✓	2	—	4/14	—	—	—	110	16	—	—	—	—	—	—	—	—	—	✓	—
VC-6847UL	570	✓	2	—	4/14	—	✓	—	110	16	—	—	—	—	—	—	—	—	—	✓	—
VC-685U	600	✓	2	—	4/14	—	✓	—	110	16	—	✓	—	✓	—	—	—	—	—	—	—
VC-T64U	600	✓	2	—	5/14	—	✓	—	110	16	—	—	—	—	✓	—	✓	✓	—	—	✓
VC-686U	620	✓	4	—	4/14	—	✓	—	110	16	—	—	—	—	—	—	—	✓	—	—	—
VC-687U	700	✓	4	—	5/14	—	✓	—	110	16	—	✓	—	✓	—	—	—	✓	—	—	—
VC-H64U	700	✓	4	—	4/14	—	✓	—	110	16	✓	—	—	—	—	—	—	✓	—	—	✓
VC-H65U	820	✓	4	—	6/14	—	✓	—	110	All	✓	—	—	✓	✓	—	—	✓	—	—	✓

A guide to features (continued)

Brand and model	Price	Picture			Programming				Tuning		Sound				Special features						
		HQ (High quality)	Heads	Sharpness control	Capability (programs/days)	On-screen programming	One-touch recording	Memory backup	Channels	Presets	Hi-fi	Linear stereo	Noise reduction	MTS decoder	Auto index	Time-left indicator	Slow motion	Frame advance	Video dub	Fine editing	Audio dub
Sony SL-250 (Beta)	400	[5]	2	—	6/7	—	—	—	148	14	—	—	—	—	—	—	—	—	—	—	—
EVA80 (8mm)	550	—	3	✓	3/7	—	—	✓	148	14	✓	[1]	—	—	—	—	✓	✓	—	✓	—
SL-700 (Beta)	550	[5]	3	✓	6/7	—	—	—	148	All	—	—	—	—	—	—	✓	✓	—	✓	—
SL-HF450 (Beta)	750	[5]	2	✓	6/7	—	—	—	148	All	✓	—	—	✓	—	—	✓	✓	—	✓	—
SL-HF650 (Beta)	900	[5]	3	✓	6/7	—	—	—	148	All	✓	—	—	✓	—	—	✓	✓	—	✓	—
SL-HFT7 (Beta)	1000	[5]	2	✓	6/7	—	—	—	148	All	✓	—	—	✓	—	✓	✓	✓	—	✓	—
SL-HF750 (Beta)	1300	[5]	4	✓	6/21	✓	✓	—	181	All	✓	—	—	✓	✓	✓	✓	✓	—	✓	—
EV-S700U (8mm)	1395	—	3	✓	6/21	—	✓	✓	181	All	✓	[1]	✓	✓	—	✓	—	✓	—	✓	✓
SL-HF900 (Beta)	1500	[5]	4	✓	8/21	—	✓	—	181	All	✓	—	—	✓	✓	✓	✓	✓	—	✓	✓
SL-HF1000 (Beta)	1700	[5]	4	✓	8/21	✓	✓	✓	181	All	✓	—	—	✓	✓	✓	✓	✓	✓	✓	✓

Toshiba M2330	430	✓	2	✓	4/14	—	✓	✓	117	100	—	—	—	—	—	—	✓	✓	—	—	—
M4330	500	✓	4	✓	4/14	—	✓	✓	117	100	—	—	—	—	—	—	✓	✓	—	—	—
M-2200	560	✓	2	—	4/7	—	✓	✓	117	16	—	—	—	—	—	—	✓	✓	—	—	—
DX3 [6]	600	✓	2	✓	4/14	—	✓	✓	124	16	—	—	—	—	—	—	✓	✓	✓	✓	—
M-4200	660	✓	4	✓	4/7	—	✓	✓	117	16	—	—	—	—	—	—	✓	✓	✓	—	—
M-2430	680	✓	2	—	4/7	—	✓	—	117	16	—	✓	✓	—	—	—	✓	✓	—	—	—
M-5530	700	✓	4	✓	4/7	—	✓	✓	117	16	—	—	—	—	—	—	✓	✓	✓	—	—
M-2700	780	✓	2	—	4/7	—	✓	✓	105	16	✓	—	✓	—	—	—	✓	✓	—	—	—
M-4500	880	✓	4	✓	4/7	—	✓	✓	117	16	—	✓	✓	✓	—	—	✓	✓	✓	—	—
DX-7 [6]	950	✓	4	✓	4/14	—	✓	✓	117	16	✓	—	✓	✓	✓	✓	✓	✓	✓	✓	✓
M-5900	1000	✓	4	✓	4/7	—	✓	✓	117	16	✓	—	—	✓	✓	—	✓	✓	✓	✓	✓
Zenith VR1810	459	✓	2	—	4/14	—	✓	—	108	All	—	—	—	—	—	—	—	—	—	—	—
VR1820	549	✓	4	—	4/14	—	✓	✓	108	All	—	—	—	—	—	—	✓	—	—	—	—
VR1870	599	✓	4	✓	4/14	✓	✓	✓	178	All	—	—	—	—	—	✓	✓	—	—	—	—
VR2220	699	✓	2	✓	4/14	—	✓	✓	108	All	✓	—	—	✓	✓	—	—	—	—	—	—
VR2300	799	✓	4	✓	4/14	✓	✓	✓	178	All	✓	—	—	✓	✓	✓	✓	—	—	—	—
VR3300	899	✓	4	✓	4/14	✓	✓	✓	178	All	✓	—	—	✓	✓	✓	✓	✓	—	—	✓
VR4100	1099	✓	4	✓	8/14	✓	✓	✓	178	All	✓	✓	✓	✓	✓	✓	✓	✓	✓	✓	✓

[1] *PCM (pulse-code modulation) sound; a version of the sound-processing method used for compact discs.*

[2] *Requires separately purchased adapter for MTS.*

[3] *No suggested retail price available from mfr.*

[4] *Has wired remote control.*

[5] *SuperBeta picture.*

[6] *Digital VCR.*

■ VCR reliability records ■

The repair index shown here is based on the experiences of more than 60,000 *Consumer Reports* readers, as reported on the Consumers Union 1985 Annual Questionnaire. Their responses to questions about VCRs purchased between 1982 and 1985 gave enough information to generate repair indexes for the fourteen brands shown. (The number of brands available has increased significantly since the survey was done.)

Overall, 11 percent of the VCRs in the survey needed repairs at one time or another. The older a VCR was and the more often it was used, the more likely it was to need repairs.

Fixing a VCR is expensive; the median cost was $80, according to the survey. A VCR's recording and playback heads have a reputation for being trouble-prone. But only 2 percent of the VCRs in the survey ever needed their heads repaired or replaced.

Repair index based on Consumers Union's 1985 annual questionnaire

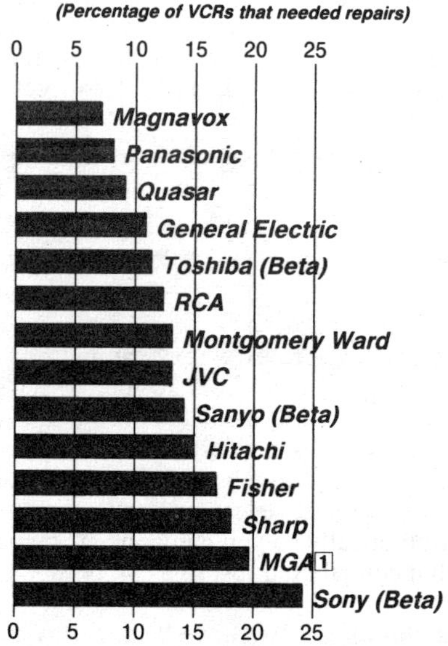

1 MGA now known as Mitsubishi.

CAMCORDERS

In one neat, compact package, the camcorder gives you everything you need to shoot a scene and play it back on your TV set.

Camcorders are a natural adjunct to videocassette recorders. And, compared with old-fashioned super-8 movies, video is inexpensive and fast; it can be screened immediately. An equivalent amount of super-8 movie film would cost more than $500 to buy and develop. That's a long way from $5 for two hours' worth of videotape.

Camcorders, like VCRs, are available in different but incompatible tape formats. For camcorders, as for videocassette recorders, the most widely used format is VHS. There are actually two varieties, VHS and VHS-C. The VHS models—the heaviest and bulkiest—use a standard tape cassette that can be played on any VHS videocassette recorder. You can shoot at least two hours of material on a single cassette. The VHS-C models are fairly compact and use a smaller cassette, which fits into a special adapter for playback on a VHS recorder. Maximum recording time is sixty minutes.

Competing with VHS is 8mm. The 8mm format has some advantages—you can record up to two hours of material, and 8mm tape can record high-fidelity sound. Some 8mm camcorders are as small and light as the VHS-C units. But if you want an 8mm camcorder, you have to buy an 8mm VCR or connect the camcorder directly to the TV set for playback.

Using a camcorder

You can't make video home movies with the point-and-shoot simplicity common to snapshot cameras. Some camcorders are bulky and loaded with buttons, knobs, and other controls. Fortunately, the machines are sensibly designed and easy to use. For example, they have a strap or a pistol grip that puts the key controls at your fingertips. Other controls are clearly labeled and logically arranged on the camera. There are some differences in design, as you might expect, but they don't seem to matter. After a little practice, you should be able to operate any camcorder quite easily. Further, camcorders generally handle focusing, exposure, and color balance automatically so you can concentrate on the action.

The features that contribute most to a camcorder's ease of use include these:

Electronic viewfinder. This is, in effect, a tiny black-and-white TV screen that displays what you're shooting. It lets you judge framing, focus, and exposure quite precisely. Indicators in the viewfinder provide important information—whether you're recording or on standby, for example, or if you're about to run out of tape. Some camcorders may use an optical viewfinder, like the one in an SLR camera. But an electronic viewfinder is so much more versatile that it's a must.

In most models, the viewfinder can also provide an instant review. A push of a button lets you glimpse the last few seconds of shooting.

Most viewfinders are convenient to use. Generally, they can slide out so you can look into the viewfinder with either eye. The viewfinder on many models also tilts; that's handy if you mount the camcorder on a tripod. Some viewfinders allow focusing to compensate for a user with moderate nearsightedness or farsightedness.

Autofocus. An automatic-focus feature is essential in a camcorder. Some models focus by *infrared triangulation,* a technique that involves bouncing a narrow beam of infrared light off the subject: A sensor slightly to one side of the light source on the camcorder measures the angle at which the beam is reflected back; electronics inside the camcorder use that angle to calculate distance and adjust the lens accordingly. The other autofocus method is *contrast;* with this technique, the camcorder keeps adjusting the focus until the contrast of the video images is at a maximum.

Both focusing methods usually work satisfactorily, although certain types of scenes and shooting styles can pose problems. Infrared focusing can be foiled by black or angled surfaces that reflect little or no light back to the camera, by subjects too tiny to reflect the entire beam, and by off-center subjects that miss the beam entirely. Camcorders with contrast focusing can be confused by subjects such as foliage, by fine repetitive patterns running in a particular direction, and by subjects partly obscured by a foreground object, such as a chain-link fence. Dim light and subjects with low contrast may also cause problems.

Further, neither autofocus method is instantaneous. That may present a problem if you pan the camera—move it from side to side to follow action. Panning subjects that are moving or that are scattered—some near and some far—can lead to an unpleasant hunting effect in which the picture goes in and out of focus. The closer the subject is to the camera, the more annoying the problem.

Fortunately, autofocus camcorders let you switch to manual focusing. If your subject is off-center, for example, you can aim the camcorder so the subject is centered, switch to manual mode to lock in the proper focus, and reaim to the original framing. The picture will remain in focus as long as the distance between lens and subject stays about the same. Most camcorders have a button that temporarily puts you back into autofocus so you can refocus to a new distance. A few models provide "zone focusing," which lets you temporarily change the size of the central area used for focusing.

Auto exposure. Accurate exposure is especially critical with videotape. A slight overexposure that would be harmless with color-print film can spell disaster in video.

Many camcorders have an automatic exposure control that adjusts the lens opening, or aperture, to admit more or less light. (Most of these models provide an aperture range of f/1.2 to f/22, wide enough to cover most normal lighting situations.) The exposure control works adequately with normal, fairly flat and uniform lighting. But with uneven lighting, the control has no way of knowing which part of the image is most important. Camcorders' exposure controls tend to average the lighting of the entire scene.

If the viewfinder image indicates that the automatic exposure setting is unsatisfactory, it's nice to be able to make manual corrections. Several models let you do that. But others give you less latitude for coping with problematic lighting—they have only a backlight compensation switch. It's most useful for scenes in which a strong light source is behind the subject; by pushing a button, you can increase the average brightness slightly to enhance detail in the foreground. But that's a partial solution; the fixed and limited change in the overall brightness provided isn't adequate for all situations. It won't provide the proper corrections at all if you're trying to capture a brightly lit subject in front of a dark background.

Zoom and macro lenses. A motorized zoom lens (found on many camcorders) is another essential. It lets you make the subject larger or smaller in the frame without actually moving the camera.

Most camcorders provide a 6:1 zoom ratio. That is, the image at one end of the zoom range is six times the size of the image at the other end of the range. The larger the ratio, the more extreme the telephoto and wide-angle effects that are available.

Most camcorders also have a macro-focusing range, which lets you get very close to the subject. As a practical matter, the shadow of the camcorder generally limits how close you can get.

Sensor. Behind the lens on most camcorders is a sensor—a sort of digital electronic retina, which converts the light into the electronic signals that make up the TV image. There are two common types—"CCD" (for charge-coupled device) and "MOS" (for metal oxide semiconductor). The type of sensor has some effect on picture quality. But a camcorder with a CCD sensor is not inherently superior to one with an MOS sensor, or vice versa.

The colors

Daylight has a slightly bluish cast, while artificial tungsten lighting tends to be reddish. The human eye automatically compensates for those differences, so a white surface actually looks white in both natural and artificial light. But the camcorder's "eye" needs some help, either through colored filters or electronic color adjustments. That's known as white balancing, and it's essential. Without it, the colors in the picture would be off.

All camcorders provide some way to set white balance. Automatic white balancing works satisfactorily when the overall color balance of the scene is fairly neutral, as is often the case. It works less well when the scene is predominately one color, particularly when the colors change from moment to moment. That's where models with manual adjustment save the day.

Some camcorders have a switch or dial with distinct settings for various types of light. That kind of control, though simple, is effective when the type of lighting is obvious. Other camcorders also provide a very accurate, semiautomatic method of white balancing that involves the use of a white card. You place the card so it's illuminated by the same type of light as the scene you want to shoot. Then you aim the camcorder at the card and press a button, locking in the white balance.

A few models offer continuous manual adjustment; you tune the white balance by monitoring the video signal on a color TV screen, adjusting until you get the best possible color. That gives the most precise white balance of all—but, of course, it ties you to the TV set.

Picture quality

The key measure of a camcorder is the quality of the pictures it produces. Several factors must be considered:

Clarity. Images should be sharp with clean edges, free of halos, spots, streaks, and other extraneous color effects. If a camcorder has two recording speeds, the slower speed tends to make picture defects more noticeable.

Color accuracy. Colors should be realistic under both daylight and tungsten lighting.

Low-light performance. This is an important consideration for people who want to use a camcorder indoors or in dark, shady spots outdoors.

The difference between a good and poor camcorder is that a poorer unit produces a barely acceptable picture. Overall, full-sized VHS units are better than the VHS-C units at their fastest speed, over a wide range of lighting conditions. The best 8mm models are on a par with full-sized VHS models, while the worst are outclassed by the VHS-C machines.

Camcorders that use an MOS sensor are particularly good at rendering colors accurately. Models that use a CCD sensor produce the best picture in low light.

Actual low-light performance may not correlate with the "lux" number manufacturers use to describe low-light capability. (The lower the lux number the better, at least in theory.)

Special effects

The following features aren't essential, but they can add a touch of professionalism to your videotapes.

"Seamless" editing. Some camcorders are designed to provide a smooth, nearly invisible transition from one scene to the next. (Some manufacturers use the term "flying erase heads" to call attention to a camcorder that can produce seamless edits.)

Fade control. This can provide a nice transition from one scene to another. You press a button and the scene gradually darkens to black. To reverse the procedure, you press the button again. (You can get the same effect on camcorders with a manual lens-aperture control.)

Wipe control. You can create a colored border around the edges of the frame and make it sweep in toward the center until it swallows up the image, or sweep back out.

High-speed shutter. If, for example, you're an avid golfer who wants to work on your swing, you may have wished for a way to tape your swing for later stop-frame analysis. The older video cameras couldn't adequately freeze the rapidly moving club, but a number of camcorders can. They have what amounts to a high-speed shutter-setting equivalent to an exposure of about 1/1000 second, giving sharply frozen images. But it won't work in dim light.

Sound quality

A camcorder has a microphone mounted somewhere on the front, usually just above the lens. It is satisfactory for recording speech, but isn't able to provide high-fidelity musical recording. The 8mm models are capable of hi-fi recording, but not with their built-in microphone.

Some manufacturers claim that their microphones are highly directional, giving preference to sounds directly in front of the camcorder. A directional microphone is desirable because it can reduce unwanted off-screen sounds, especially in noisy environments. However, the "directional" mikes on camcorders aren't likely to provide anything more than a modest reduction of unwanted sound.

Recommendations

The basic things to look for are:

- an electronic viewfinder designed to be used with either eye
- controls that automatically adjust the white balance of the picture, plus a back-up white-balancing system, such as indoor and outdoor buttons, to handle special situations
- controls that let you adjust the focus and lens opening manually; the automatic controls on many camcorders aren't ideal for every situation

The camcorder you choose should also be easy to hold, with controls placed where they are convenient for you to use. A large model, which may weigh close to seven pounds, may seem forbiddingly bulky and cumbersome at first. But it is meant to rest on a shoulder, which tends to make it easier to hold the camera steady; its controls also tend to be larger and more widely spaced. A hand-held camcorder is certainly easier to carry, but it may yield shakier pictures. It is neither easier nor harder to use than a larger model.

Finally, there's the format. The most sensible choice is a full-sized VHS camcorder. For one thing, the majority of videocassette recorders are VHS machines, too. More importantly, VHS machines provide the longest recording times. And VHS machines have the best picture quality. Machines that use the small VHS-C cassette produce the worst pictures, overall.

Ratings of camcorders

Listed in groups in order of picture quality, based on clarity, color fidelity, and low-light performance; within groups, listed alphabetically. Sound quality was not a factor in the Ratings. Prices are as published in a **November 1987** report.

◉ Better ← ◓ ○ ◒ → Worse ●

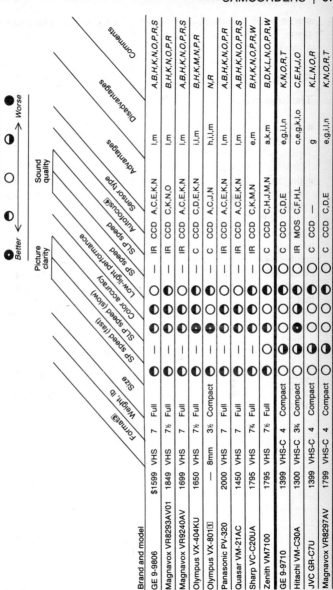

Brand and model	Price	Format[3]	Weight, lb	Size	Picture clarity: SP speed (fast)	Picture clarity: SLP speed (slow)	Color accuracy	Low-light performance	Sound quality: SP speed	Sound quality: SLP speed	Autofocus[4]	Sensor type	Advantages	Disadvantages	Comments
GE 9-9806	$1599	VHS	7	Full	◓	—	◓	◓	○	—	IR	CCD	A,C,E,K,N	l,m	*A,B,H,K,N,O,P,R,S*
Magnavox VR8293AV01	1849	VHS	7½	Full	◓	—	◓	◓	◓	—	IR	CCD	C,K,N,O	l,m	*B,H,K,N,O,P,R*
Magnavox VR9240AV	1699	VHS	7	Full	◓	—	◓	◓	○	—	IR	CCD	A,C,E,K,N	l,m	*A,B,H,K,N,O,P,R,S*
Olympus VX-404KU	1650	VHS	7½	Full	◓	—	◉	◓	◓	—	C	CCD	C,D,E,K,N	i,l,m	*B,H,K,M,N,P,R*
Olympus VX-801[1]	—	8mm	3½	Compact	◓	—	◉	○	◓	—	C	CCD	A,C,J,N	h,i,l,m	*N,R*
Panasonic PV-320	2000	VHS	7	Full	◓	—	◓	◓	○	—	IR	CCD	A,C,E,K,N	l,m	*A,B,H,K,N,O,P,R*
Quasar VM-21AC	1450	VHS	7	Full	◓	—	◓	◓	○	—	IR	CCD	A,C,E,K,N	l,m	*A,B,H,K,N,O,P,R,S*
Sharp VC-C20UA	1795	VHS	7¾	Full	◓	—	◓	◓	◓	—	IR	CCD	C,K,M,N	e,m	*B,H,K,N,O,P,R,W*
Zenith VM7100	1795	VHS	7½	Full	◓	○	◓	○	◓	○	C	CCD	C,H,J,M,N	a,k,m	*B,D,K,L,N,O,P,R,W*
GE 9-9710	1399	VHS-C	4	Compact	○	◒	○	○	◓	○	C	CCD	C,D,E	e,g,i,l,n	*K,N,O,R,T*
Hitachi VM-C30A	1300	VHS-C	3¾	Compact	○	◒	◉	◒	◓	○	IR	MOS	C,F,H,L	c,e,g,k,l,o	*C,E,H,J,O*
JVC GR-C7U	1399	VHS-C	4	Compact	○	◒	○	○	◓	○	C	CCD	—	g	*K,L,N,O,R*
Magnavox VR8297AV	1799	VHS-C	4	Compact	○	◒	○	○	◓	○	C	CCD	C,D,E	e,g,i,l,n	*K,N,O,R,T*

Ratings of camcorders (continued)

Listed in groups in order of picture quality, based on clarity, color fidelity, and low-light performance; within groups, listed alphabetically. Sound quality was not a factor in the Ratings. Prices are as published in a **November 1987** report.

⊙ ◓ ○ ◒ ●
Better ⟵⟶ *Worse*

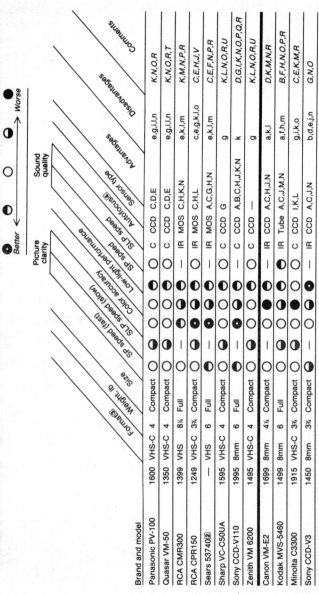

Brand and model		Format[3]	Weight, lb	Size	Picture clarity: SP speed (fast)	Picture clarity: SLP speed (slow)	Color accuracy	Low-light performance	Sound quality: SP speed	Sound quality: SLP speed	Autofocus[4]	Sensor type	Advantages	Disadvantages	Comments
Panasonic PV-100	1600	VHS-C	4	Compact	○	◒	○	○	◓	○	C	CCD	C,D,E	e,g,i,l,n	*K,N,O,R*
Quasar VM-50	1350	VHS-C	4	Compact	○	◒	○	○	◓	○	C	CCD	C,D,E	e,g,i,l,n	*K,N,O,R,T*
RCA CMR300	1399	VHS	8¼	Full	○	—	◓	◒	◓	—	IR	MOS	C,H,K,N	e,k,l,m	*K,M,N,P,R*
RCA CPR150	1249	VHS-C	3¾	Compact	○	◒	⊙	◒	◓	○	IR	MOS	C,H,L	c,e,g,k,l,o	*C,E,H,J,V*
Sears 53740[2]	—	VHS	6	Full	◓	—	⊙	◒	◓	—	IR	MOS	A,C,G,H,N	e,k,l,m	*C,E,F,N,P,R*
Sharp VC-C50UA	1595	VHS-C	4	Compact	○	◒	○	○	◓	○	C	CCD	G	g	*K,L,N,O,R,U*
Sony CCD-V110	1995	8mm	6	Full	◓	—	⊙	◒	◓	—	C	CCD	A,B,C,H,J,K,N	k	*D,G,I,K,N,O,P,Q,R*
Zenith VM 6200	1495	VHS-C	4	Compact	○	◒	○	○	◓	○	C	CCD	—	g	*K,L,N,O,R,U*
Canon VM-E2	1699	8mm	4¼	Compact	○	—	○	●	◓	—	IR	CCD	A,C,H,J,N	a,k,l	*D,K,M,N,R*
Kodak MVS-5460	1499	8mm	6	Full	○	◒	○	◒	◓	◓	IR	Tube	A,C,J,M,N	a,f,h,m	*B,F,H,N,O,P,R*
Minolta C3300	1915	VHS-C	3¾	Compact	○	◒	○	●	○	○	C	CCD	I,K,L	g,i,k,o	*C,E,K,M,R*
Sony CCD-V3	1450	8mm	3¾	Compact	◒	—	○	◒	⊙	—	IR	CCD	A,C,J,N	b,d,e,j,n	*G,N,O*

[1] *A later version is* **VX-802,** *$1750.*

[2] *A later version is* **53741,** *$1300.*

[3] *VHS-C accepts smaller cassette than VHS, which can be played back on a VHS VCR.*

Specifications and Features

All: • Can play back directly to a TV or a VCR. • Have viewfinder eyepiece whose focus can be adjusted for moderate nearsightedness or farsightedness. • Have automatic and manual focus. • Have automatic shutoff in VCR mode. • Signal low battery in viewfinder.

Except as noted, all have: • Power and manual zoom. • Viewfinder that can be tilted or rotated. • Accommodate left-eyed and right-eyed users. • A "quick review" feature that lets you see the last few seconds of recorded tape. • "Freeze" frames with little or no visible "noise" bars. • 6:1 lens zoom ratio. • Backlight-compensation feature. • Viewfinder indicator showing white-balance setting. • Viewfinder indicator showing when tape is about to run out. • Headphone jack. • Macro focus.

Key to Advantages

A–Can perform "seamless" editing.
B–Can perform picture "wipe."
C–Can temporarily engage autofocus while in manual mode.
D–Size of region in center of frame that is sensed for autofocusing is adjustable.
E–Has high-speed shutter for "freezing" action.
F–Has self-timer and wind-noise suppressor.
G–Comes with carrying case.
H–Has manual aperture control.
I–Autofocus works even in macro; has two zoom speeds; has exposure lock.
J–Can record expanded-bandwidth audio through line input.
K–Internal clock can record time and date.
L–Lens assembly is well-protected.
M–Can record in stereo.
N–Has mounting shoe for accessory movie light.
O–Has 8:1 zoom ratio.

Key to Disadvantages

a–Viewfinder inconvenient to use with left eye.
b–Has only manual zoom.
c–Unbalanced on tripod with ac power supply mounted on camcorder.
d–Lacks tape counter.
e–Lacks push-button fade control.
f–Unbalanced in use; controls are inconvenient.
g–Freeze frame very noisy at SP speed.
h–Lacks "quick review."
i–Viewfinder cannot be tilted or rotated.
j–Has 2.5:1 zoom ratio.
k–Lacks backlight compensation.
l–Lacks viewfinder indicator for white balance.
m–Lacks "tape out" viewfinder indicator.
n–Lacks headphone jack.
o–Lacks macro focus.

Key to Comments

A–Has video dubbing; lets you record new images without disturbing original sound track.
B–Has audio dubbing; lets you record new sound without disturbing the images on tape.
C–Has only power zoom.
D–Has switch to increase sensitivity for dim light.
E–Has power focus control for manual focusing.
F–User must set date; model lacks clock.
G–Although camcorder can't record at LP speed, it can play back tapes recorded at that speed.
H–Has standby mode to conserve power.
I–Has time-lapse picture feature.
J–Closing lens cover turns power off.
K–Has jack for remote Pause/Record control.
L–Has focusing-aid indicator in viewfinder for manual focus.
M–Can accept optional add-on lenses.
N–Front of lens is threaded for accessories.
O–Has "tally light," which lets subject know when camcorder is recording.
P–Designed to rest on shoulder during recording.
Q–Can perform titling.
R–Has memory or reset for tape position.
S–Essentially similar to **Panasonic PV-320.**
T–Essentially similar to **Panasonic PV-100.**
U–Essentially similar to **JVC GR-C7U.**
V–Essentially similar to **Hitachi VC-C30A.**
W–Comes with stereo microphone

VIDEOTAPES

The type of VCR you own determines what tape format you buy. Most brands come in VHS and Beta, and several come in the newer 8mm format. The length of tape you buy may be decided by where you buy it. Many stores sell only the most popular lengths—T-120 for VHS and L750 for Beta.

The confusion sets in when you try to choose from the various *grades* of videotape. Nearly all brands offer at least two grades, and the biggest companies offer at least five. Typical designations are standard, high grade, super high grade, hi-fi, and professional, but companies can call their products whatever they wish.

The grade designations promise differences in quality, particularly since the list prices step up grade by grade, from "standard" to "camera" and "pro." The premium grades presumably carry the highest markup and bring in the highest profits. Manufacturers had once hoped that the premium grades would capture as much as half of all tape sales. Advertising pitches appealed to the home-video archivist with pro and camera grades. They appealed to the sound-conscious with special hi-fi grades. The standard grade, with its low profit margin, was to be a backup to snare diehard bargain-hunters.

But most consumers have been choosing the standard grade. And that is an intelligent response to advertising: The grade designations are quite useless to anyone trying to buy the best tape. One company's "super high grade" is often no better than another company's—or even its own—high grade. The performance of each brand's various grades is often indistinguishable, as shown in the example on page 94.

The overall quality of brand-name videotapes is quite high, and Consumers Union's tests didn't turn up a single instance of a lower-grade tape being significantly worse than a more expensive brandmate.

The person in search of a tape that will make the perfect video recording is on a vain quest anyway. In the recording process, both the VCR and the tape limit perfection in their own ways. The differences between tapes are fairly small. Even the best of them can't fix picture degradation that comes from the limitations of your VCR.

Tape details

In concept, a videotape is a fairly simple product: a polyester ribbon with a surface coating made up of millions of tiny magnetic particles bonded in place. The tape must be flexible enough to come in full contact with the VCR heads, which record and read the information. The tape must also be tough enough to survive constant tension and flexing.

Corresponding VHS and Beta tapes generally use the same coating. The thickness of the polyester may differ from one format to the other, but that shouldn't be a significant performance factor.

When you record a program, the tiny particles on the tape are magnetized in a particular pattern. When you record another program over the same length of tape, you simply change the pattern of magnetization. The tinier the particles and the more densely they're packed, the better the video and audio quality of the tape is.

The particles are primarily oxides of iron and chromium. But there's enough leeway in the formulation for tape manufacturers to announce periodically miraculous new substances with impressively technical names. The art of tape formulation is already very advanced, and further significant improvements will be hard to come by. So you should generally be skeptical of claims for miracles.

Gaps, scratches, or a poorly applied coating cause the intermittent flecks and streaks called dropouts, which show up as intermittent horizontal streaks on the TV screen. Dropouts are rare even with the worst tapes. Some are so short that the human eye can't see them. Additionally, all VCRs have special electronic circuitry that replaces missing bits of picture lines with corresponding bits from the line above—a trick that ordinarily goes undetected by the viewer because of the general similarity between any two successive lines.

Video color noise—the type known as AM chrominance noise—is noticeable even on the best tapes. It can make large, vividly colored parts of the picture seem to ripple and wriggle. In red areas, where the effect is most noticeable, modest differences can be detected between the best and worst tapes. But even the worst tape won't be very bad.

All along, VCRs have used a method known as linear-track recording to record sound. A narrow strip is reserved along the edge of the tape for the sound track. That method leaves much to be desired. The low speed of the tape makes good audio quality difficult to obtain.

A few years ago, high-fidelity sound was introduced in VCRs. It uses an FM recording technique that lays the sound information across the entire tape with the video information, producing sound nearly as good as that of compact discs.

In response, tape manufacturers have introduced new grades of tape whose names might make you think they're engineered especially for hi-fi VCRs. The implication seems to be that you need such special tapes to make full use of your VCR's hi-fi potential. That's not so. Just about

Ratings of videotapes

Listed in order of estimated quality. Models with equal scores are listed alphabetically.

Overall score. Based primarily on video performance. The difference between the best and the worst performer was noticeable but not great. Differences of less than 10 points were judged not very significant. Prices are as published in a **November 1986** report.

◉ ◓ ○ ◒ ●
Better ⟵ ⟶ *Worse*

Brand and model	Price range: List	Price range: Lowest paid	Overall score	Video: Dropouts	Video: Color noise	Video: Dynamic range	Audio: Bandwidth
Fuji Beridox Super XG	$18.99	$ 9.99	98	◉	◉	◉	◉
Kodak Hi-Fi Stereo	13.99	9.00	96	◉	◉	◉	◉
Maxell RX Pro	19.99	9.49	96	◉	◉	◉	◉
TDK E-HG Super Avilyn	11.80	6.89	96	◉	◉	◉	◉
TDK Extra High Grade Hi-Fi Super Avilyn	14.70	6.49	95	◉	◉	◉	◉
TDK HD Pro Super Avilyn	17.40	8.29	95	◉	◓	◉	◉
Maxell HGX Gold Hi-Fi	16.99	7.99	94	◉	◉	◉	◓
Scotch EXG	9.99	6.99	94	◉	◓	◉	◉
BASF Chrome Super HG Hi-Fi	12.95	7.99	92	◉	◓	◉	◉
Scotch EXG Hi-Fi	10.99	7.89	92	◉	◓	◉	◉
Scotch EXG Pro	12.99	9.99	92	◉	◓	◉	◓
RCA Super High Grade Stereo Hi-Fi	11.95	6.25	91	◉	◓	◉	◓

Scotch EG	6.99	4.99	91	◉	◒	◉	◉
Scotch EXG Camera	11.99	12.00	91	◉	◒	◉	◉
Memorex Pro Cam	11.99	8.50	90	◉	◒	◒	◉
Panasonic Premium Standard	9.00	4.39	90	◉	◒	◒	◒
Polaroid Supercolor	8.99	3.75	90	◉	◒	◒	◉
Fuji Beridox	9.99	4.89	89	◉	◒	◒	◉
Scotch EG+ High Grade	8.99	5.99	89	◉	◒	◉	◉
TDK HS Super Avilyn	9.40	6.29	89	◉	◒	◒	◉
Maxell EX	9.99	5.44	88	◉	◒	◉	◒
JVC Dynarec	9.50	5.49	87	◉	◒	◒	◒
Maxell HGX High Grade	12.99	5.99	87	◉	◒	◉	◒
Quasar Super GT	6.99	3.99	87	◉	◒	◒	◒
Radio Shack Supertape Gold High Grade	9.99	9.99	86	◉	◒	◉	◒
RCA Premium Quality Stereo Hi-Fi	8.95	4.99	86	◉	◒	◉	◒
Kodak	8.99	4.49	85	◉	◒	◒	◒
Konica Super SR High Performance	13.95	4.12	85	◉	◒	◒	◉
Memorex HS	5.99	4.59	85	◒	◒	◒	◒
Radio Shack Supertape	7.99	7.99	84	◉	◒	◉	◒
SKC Standard Grade Stereo Hi-Fi	4.99	3.50	84	◉	◒	◒	◒
BASF Chrome Extra Quality	8.95	4.79	83	◉	○	◒	◉
Certron High Grade	8.99	4.57	80	◒	◒	◒	◉

any tape should give excellent audio results with any hi-fi VCR. When using an older VCR with linear-track recording, however, you may find slight but discernible audio differences from one tape to another.

Dynamic range measures the range in volume between the loudest sound that can be reproduced without serious distortion and the level of noise present when there is no recorded sound. The difference between the best and the worst tape might be 4 decibels (dB), a small but perceptible increment if the subject being recorded covers a wide dynamic range.

Recommendations

You won't go far wrong with any brand-name tape. Differences in performance are discernible, but only under certain conditions and with a high-quality video signal. Any of the tapes would be quite acceptable for casual time-shifting—taping a TV program so you can watch it later. If the signal you're recording shows even a little noise or your VCR isn't in good condition, the differences between tapes may not be noticeable at all.

Try to avoid unlicensed tapes (see page 95). Since licensed tapes are often available for as little as $4 on sale, unlicensed tapes offer little saving and a lot of potential trouble.

■ High grade ■

Sometimes, the only thing higher about a higher-grade tape is its price. Among the tapes Consumers Union tested, that shows up most clearly in how the various grades of *Scotch* videotape scored.

Brand and model	List price	Low price paid	Overall score	Dropouts	Color noise	Dynamic range	Bandwidth
Scotch EXG	$ 9.99	$ 6.99	94	◉	◓	◉	◉
EXG Hi-Fi	10.99	7.89	92	◉	◓	◉	◉
EXG Pro	12.99	9.99	92	◉	◓	◉	◓
EG	6.99	4.99	91	◉	◓	◉	◉
EXG Camera	11.99	12.00	91	◉	◓	◉	◉
EG+ High Grade	8.99	5.99	89	◉	◓	◉	◉

The six *Scotch* grades tested are listed in Ratings order. Two pairs—the *EXG Hi-Fi* and the *EXG Pro,* the *EG* and the *EXG Camera*—scored identically. Only five points separate the scores of the highest-rated and lowest-rated *Scotch* products—a range so small you wouldn't be able to tell one grade from another. While quality and price in most other brands showed more of a correlation, none of the differences were large enough to be very significant.

■ Tapes to avoid ■

When you shop for tapes, you may see some with unfamiliar names. Some have no obvious brand name. What may make such tapes appealing is their price: $3 to $4. But beware. Such tapes' quality may be poor and they might even damage your VCR.

These products may be unlicensed videotapes. The VHS format was originated by a company called JVC, which licenses other manufacturers to produce VHS tapes. Similarly, Sony originated the Beta format, and it allows various manufacturers to make Beta tapes under license. Both JVC and Sony have quality-control procedures that must be followed by licensees using the VHS or Beta logo.

If the brand isn't one of the familiar ones, or if it isn't apparent who made the tape, look at the packaging carefully. Some unlicensed tapes avoid using the official VHS logo. Others skirt the legal edge by using the logo in a sentence rather than having it stand alone, as it does at least once on licensed tapes. Another tipoff is the absence of an address for the manufacturer or distributor.

In testing samples of these tapes, Consumers Union found that they all fared much worse than the licensed tapes in video performance tests. All had higher—some, markedly higher—dropout counts. Only one of the eight unlicensed models had as little noise as the very poorest licensed tape.

But worst of all, one or more of the tapes appear to have damaged the recording heads on the VCR used in the tests. That's an expensive drawback indeed.

CABLE TELEVISION

The first full year that Congress allowed deregulation of cable TV rates was 1987; now cable companies, not municipalities, determine how much to charge for basic service. And in 1987 the government increased the significance of cable TV in the formula used to calculate the Consumer Price Index.

Cable TV has taken forty years to entrench itself firmly in America's living rooms and consciousness. In the beginning, cable TV wasn't even called that. It was community-antenna television, or CATV, and its sole purpose was to bring broadcast channels to places that otherwise couldn't get them.

The Federal Communications Commission, at the behest of movie-house owners and television networks, clamped severe regulatory curbs on the movies and sports that cable TV could carry and forbade cable from carrying advertising. Movie-house owners and television networks had recognized the new technology as potential competition when some CATV systems started "importing" distant stations for their subscribers. As part of the campaign against cable, those interests lobbied the FCC for protection, raising the specter of eventual "pay TV" for all Americans.

The growing cable industry battled back in court and before the FCC. By 1980, the regulatory curbs were overturned.

Meanwhile, technology had caught cable TV in another snare. In the early days of cable, the way to relay distant signals was with microwaves transmitted and received with ground-based antennas. That sort of transmission worked well but wasn't practical for long distances. *Home Box Office,* developed in the early 1970s, first transmitted from the top of a skyscraper in New York City and reached only into Pennsylvania with its signal—clearly limiting any visions of large-scale national cable networks.

When RCA launched its Satcom I communications satellite, in 1975, the picture brightened considerably. A cable channel like *HBO* could beam its microwave transmissions to Satcom and relay them back to virtually anywhere on the ground. Dish antennas at local cable systems could pluck the signals from the air, amplify them, and shunt them over coaxial cables directly to subscribers' television sets. Nationwide distribution of cable channels became a reality.

HBO signed on to Satcom in 1975. Today, satellites orbiting the earth beam more than fifty cable channels to the 7,800 cable systems that dot the United States. The programs range from TV dramas that first aired more than thirty years ago to the most contemporary of rock videos. They include the flatly commercial, such as nonstop home-shopping. And they include a good deal of programming that has difficulty finding a home on commercial broadcast channels.

If the channels and their programs are cable TV's product, the local cable systems are the pipeline. In its 1986 Annual Questionnaire, Consumers Union asked *Consumer Reports* readers to rate both. Nearly 150,000 subscribers, 90 percent of them cable customers, responded. They detailed their life with cable TV—the channels, the service, the rates, the changes that the medium wrought at home. Much of the information in this chapter is based on that survey.

The channels

Apart from cities slow to approve cable systems—New York, Philadelphia, and Washington, D.C., are three—most of the areas that will ever be wired for cable have already been wired. Now, the industry can grow mainly by selling cable to people who have so far chosen not to subscribe. To lure them—and to hold onto its 43 million current customers—cable must offer more, better, and different programming than what you find on "regular" television.

So the industry has begun to produce more and more of its own programming. Pay channels, which charge a monthly fee but carry no advertising, can use their cut of that fee to help finance their programming. *HBO,* for instance, takes a bit more than half of what the local cable company charges you, the subscriber.

Each "basic" cable channel—channels included as part of the standard cable service—earns revenue from commercials and from monthly fees paid it by the cable-system operators. Such fees range from 5 to 25 cents per household. The nickels and quarters add up. In 1986, basic channels spent more than $1 billion on their programming, much of it on original material.

Here's a rundown of the kinds of channels you can expect to find as you flip the dial:

Pay movie. These channels' staple is fairly recent films, unedited and uninterrupted by commercials. *HBO* is the biggest by far, with almost three times as many subscribers as its rival *Showtime.* Both air scores of movies every month (with some overlap), and both offer occasional concerts and comedy specials. *HBO* also covers sports, like the Wimbledon tennis tournament and World Championship Boxing.

Cinemax is a sister to *HBO* (both are owned by Time Inc. and share films). *Festival,* a new *HBO* sibling, will show movies that are rated G and PG, not R. *The Movie Channel,* a sister to *Showtime* (both are owned by Viacom, an entertainment conglomerate), shows only movies, old ones included. *American Movie Classics* features just that.

Cultural. The movies these channels show are apt to be foreign films or other repertory-house fare. *Bravo,* a pay channel, airs theater, opera, music, and dance as well. *Arts & Entertainment,* a basic channel, imports similar programming, including series from the British Broadcasting Corporation.

Sports. Ted Turner has promoted the Atlanta Braves as "America's team" on his station *WTBS. The Nashville Network* shows tractor pulls. Some of the channels devoted to sports are pay channels. *ESPN,* a basic channel, covers everything from cheerleading championships to yacht racing. Its coverage of the America's Cup live from Australia last winter helped turn that elite activity into a spectator sport for the first time. *ESPN* truly ascended to the big time when it signed a deal to cover NFL Sunday football in the fall of 1987.

Children. *The Disney Channel,* a pay channel, shows a mix of original films, new features and series, and old-time favorites from the Disney Studios. *Nickelodeon,* a basic channel, offers reruns along with new programming—talk, cartoons, and educational game shows. Both channels try to present shows that the entire family can watch together.

Music. *MTV: Music Television* pioneered the music video and offers rock videos, concerts, and related programs twenty-four hours a day. Other formats offered include country and western from *The Nashville Network* and soft rock from *VH-1 (Video Hits One).*

News and weather. Ted Turner's *Cable News Network (CNN)* bills itself as "the world's most important network." The channel offers news and features twenty-four hours a day, often filling gaps in coverage left by the networks (live, gavel-to-gavel coverage of the Iran/Contra hearings during the summer of 1987, for instance). A spinoff service, *CNN Headline News,* gives a fast-paced news update every half-hour. The *Financial News Network (FNN)* covers business and the stock-market beat. *The Weather Channel* prognosticates.

Special audience. *Lifetime* features shows on health, fitness, fashion, and family, aimed at a female audience. *The Playboy Channel* aims at men, with soft-core "adult entertainment." (That channel is offered on relatively few systems, which generally will provide subscribers with a block-out device to limit who watches.) Other special-audience channels include *The Discovery Channel* (it features programs on science and nature), *Black Entertainment Television,* and *National Jewish Television.*

General audience. These cable networks appeal to the broadest audiences with diverse programming—movies, action series, sitcoms, sports. Cable networks include *USA Network* and *CBN Cable Network,* owned by the Christian Broadcasting Network. (Three-fourths of *CBN's* programming is nonreligious fare, especially reruns of old television series.)

Superstations. The "super" in superstations refers to their wide reach, not to anything special in their programming. They are ordinary independent broadcast stations, piped in from out of town. They include Atlanta's *WTBS,* Chicago's *WGN,* and New Jersey's *WWOR.* If you come originally from their broadcast area, the local news and announcers' accents may make you homesick. Look for lots of sports, syndicated shows, and old movies.

Public affairs. *C-SPAN* broadcasts congressional hearings, press conferences, panel discussions, and call-in shows with public officials and newsmakers. About one-fifth of the country's cable systems also offer "public access" channels that televise council meetings and other area events such as Little League games, festivals, even homework hotlines.

Home shopping. More than a half-dozen channels devote themselves to selling—essentially, nonstop direct merchandising—and others (such as *The Travel Channel)* use home shopping as part of their appeal. On some, personable hosts demonstrate the goods—cubic zirconium rings and Italian glass figurines are hot—and chat with callers on the air. Others resemble endless low-budget commercials, with plenty of "buy now—this offer won't be repeated" incentives. Cable operators have good reason to offer shopping channels: They typically pocket a 5 percent sales commission on the orders.

Pay-per-view. Cable systems with PPV show recent films when they first appear in videocassette, typically six to nine months before they appear on the pay-movie channels. For about $5, the company sends a movie to your TV at a specified time. Such channels also offer sports and other specials. PPV requires "addressable" circuitry so operators can control channels to individual homes.

CATV service

The most sophisticated cable operation doesn't amount to much if you're frequently blacked out or if you can't get the company on the phone.

Three out of five readers in the cable-TV survey reported that they had experienced one or more service disruptions in a single year; one in five, in the month before the survey. An unlucky one in ten said they had endured more than a half-dozen such disruptions.

Service outages tend to be fairly short. But during the outages all channels on the system are usually affected.

Picture quality is reported as generally good, with some variability from channel to channel.

Billing can be a problem. The biggest peeves are the high cost of adding or dropping pay channels, confusing or inaccurate bills, and the hard-to-understand package deals that cable companies promote.

Ratings of cable-TV channels

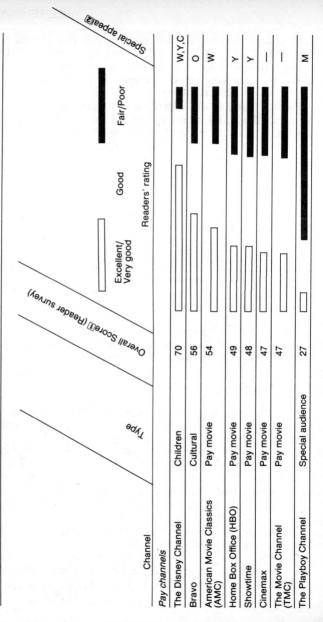

Channel	Type	Overall Score[1] (Reader survey)	Readers' rating (Excellent/Very good · Good · Fair/Poor)	Special appeal[2]
Pay channels				
The Disney Channel	Children	70		W,Y,C
Bravo	Cultural	56		O
American Movie Classics (AMC)	Pay movie	54		W
Home Box Office (HBO)	Pay movie	49		Y
Showtime	Pay movie	48		Y
Cinemax	Pay movie	47		—
The Movie Channel (TMC)	Pay movie	47		—
The Playboy Channel	Special audience	27		M

Basic channels			
Cable News Network (CNN)	News and weather	64	—
CNN Headline News	News and weather	62	—
Arts & Entertainment (A&E)	Cultural	56	W,O
ESPN	Sports	55	O
WTBS	Superstation	55	—
WGN	Superstation	54	—
The Weather Channel	News and weather	54	—
Nickelodeon	Children	52	W,Y,C
USA Network	General audience	48	—
Lifetime	Special audience	46	W
Financial News Network (FNN)	News and weather	43	O
C-SPAN	Public affairs	42	O
CBN Cable Network	General audience	42	W
The Nashville Network	Music	36	O
MTV	Music	35	W,Y

1 *Scale: Excellent = 100; Poor = 0*

2 *M—Men rated the channel's quality significantly higher than women did. W—Women rated the channel's quality significantly higher than men did. O—Readers aged 45 and over rated the channel significantly higher than younger adults did. Y—Younger adults rated the channel significantly higher than people aged 45 and over did. C—At least half of the respondents with children at home called the channel a favorite with their youngsters.*

■ The CATV habit ■

Being hooked up to cable TV makes you watch more television, according to half of the *Consumer Reports* readers surveyed.

To make room in a busy life for all those new channels you must curtail other leisure-time activities, like going to the movies and reading. Concerts, hobbies, and other pastimes also suffer, but to a lesser extent.

Most of those who took part in the survey were seasoned observers, having been hooked up longer than two years. Most also received at least one pay channel, like *HBO* or *Showtime,* which costs perhaps $10 or $12 a month over the basic rate.

Each month the average reader wrote out a $23 check to the cable company, a fee that may be going up with rate deregulation. Add up the checks for a year, and you could buy a basic videocassette recorder or rent quite a few feature films if you already own a VCR.

So why sign up for cable?

The main reason: more movies. A second reason: extra programming that cable brings, including receiving the so-called superstations—ordinary TV from out of town. All told, cable customers in the survey usually receive at least ten channels more than they can get with an antenna.

Reasons for subscribing to cable

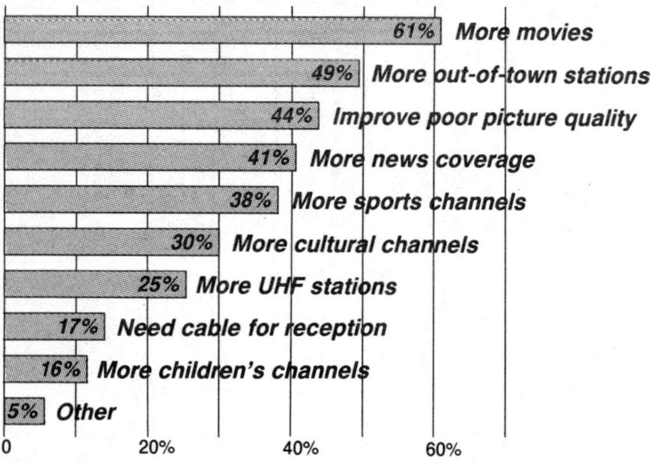

Figures add to more than 100% because respondents could check several choices.

Leisure activities curtailed since receiving cable TV

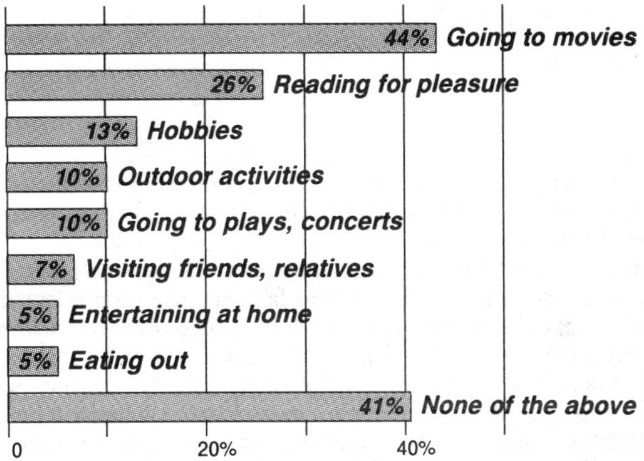

Figures add to more than 100% because respondents could check several choices.

Home computers

Computers are still finding their way into homes. Most personal computers (PCs) are bought for use in small businesses, including businesses conducted from home.

Home uses of PCs are many, ranging from inventory to entertainment, education to word processing. Some families use computers much as businesses do: to help draw up a budget; to keep track of spending, savings, and investments; to balance the checkbook; to make out tax returns. Many find them useful for word processing; computers make it easier to compose letters and school papers that are perfectly typed and free of spelling and grammatical errors. Enthusiasts claim that computers can make reading, writing, and arithmetic fun for children. Older children can use them to help learn skills, such as programming musical compositions and typing. Home use of computers to prepare cross-indexed lists—of names, addresses, phone numbers, and birthdays; of books, records, tapes, and recipes; of the contents of your closets and your safe-deposit box; and so on—has not been as popular as originally predicted.

Telephone lines can link your computer with other computers. You can send and receive mail electronically, both personal letters and business transactions. The telephone-computer link gives the family access, for a price, to central data banks containing the kind of information you now receive from newspapers, periodicals, encyclopedias, and even libraries.

Computers can also entertain—and not merely with the well-publicized video games. There are hundreds of ready-made programs good for such standbys as poker, backgammon, and chess at varying skill levels.

THE COMPUTER SYSTEM

The core element of a computer is the central processing unit, or CPU, and its microprocessor, located on a circuit board (the "motherboard"). The CPU is something like a brain capable of certain tricks but unable to remember what those tricks are. The CPU goes into action only when you load a program into the computer's memory.

Once "educated" by a program (formerly a cartridge but nowadays a flexible, or "floppy" disk), a video game machine can remember what tricks it's supposed to perform. But, like other computers, it usually doesn't do anything useful until you give it directions through an "input device." The game machine's input device is typically a joystick or perhaps a joystick plus a keypad with numbers on it, and maybe a pointing device called a "mouse." Because a game machine does simple tasks, a simple input device is sufficient for all the commands the game machine needs.

The input device of a personal computer, however, is a keyboard much like that of a typewriter but with additional symbols. The keyboard lets you "input" the more detailed information required by complex programs as well as write your own.

To show you what it's doing or to ask for additional directions and information, the CPU must be connected to an "output device." For a game machine, the output device may be an ordinary TV set. Although many personal computers can also use a TV set, computers that are working tools typically use a video display terminal (VDT), or monitor, as the output device.

The central processing unit and its memory, the program and data storage source, the keyboard (or input device), and the display (or output device) make up a computer system. There's additional equipment—called peripherals—that you're likely to want for a home computer. The most common is a printer (see section on printers, page 109).

CPU and memory

There are two kinds of computer memory: Read Only Memory, or ROM, and Random Access Memory, or RAM.

ROM. Read Only Memory contains the programs built into the computer by the maker. If you think of a computer as a brain, ROM is innate intelligence. Ordinarily, you can't change it, add to it, or subtract from it. In some machines, ROM provides the computer with just enough intelligence to do a few simple tasks, such as suggest that you load a program. In others, ROM provides enough intelligence to "understand" a programming language so that you can write your own programs or copy a program from a book or magazine.

RAM. Random Access Memory is far more important to the user. It's the amount of RAM that usually defines the computer's maximum capacity to receive information and process it efficiently.

Memory is measured in kilobytes, or thousands of bytes. One byte is what it takes to store a single character in memory. So a kilobyte (1K) could contain a document of about 1,000 characters (1,024, to be precise).

The importance of RAM becomes clear when you consider a complex computer application such as word processing—the application considered most important by many owners of a home computer. Let's say a college student uses a word-processing program to write a long paper using a personal computer with 16K of RAM, which is very low by today's standards. Once loaded into the CPU, the program itself might take up 10K of RAM, leaving 6K for the student to use. The remaining 6K of RAM translates into about 6,000 characters and spaces, or roughly four double-spaced pages. The student who wanted to continue after four pages would then have to take the first part of the paper out of memory (by saving it on a floppy disk) and thus free the 6K of RAM the words had occupied. Because the first part of the paper was no longer in RAM, the student couldn't read or edit it until the continuation of the paper had been transferred to disk and the first part of the paper loaded back into RAM. While this isn't an impossible way to work, it is certainly inefficient.

The same 10K program in a 48K computer would leave 38K of memory still available for work—enough to let the student write, edit, and rearrange more than twenty pages at a sitting. But 48K of memory became too little as the sophistication of programs increased. Nowadays, more than ten times as much memory is commonplace.

The nominal RAM is usually higher than the RAM actually available to the user. The amount of RAM available to the user differs among machines because manufacturers make different decisions about how much of the necessary "housekeeping" programs should be permanently supplied in Read Only Memory and how much should be loaded along with a programming package.

Programming and storage

To use a computer, you would normally buy ready-made applications programs. They are usually stored on a floppy disk.

Disks. A disk can store a ready-made program or your own information. Almost all personal computers can use disks as the program-loading and data-storage device. Disks can load and store programs and data very quickly. The student writing a paper longer than four pages can save it on a disk or load it from the disk into RAM in a matter of seconds.

Just as important is the speed with which you can search a disk for a specific file. (A file is any body of data—a letter, a budget, the moves of a chess game in progress—that you've indexed in storage.) It takes a computer only a fraction of a second to search a disk.

Disk drives. There are two kinds: floppy and hard disk drives. A "hard" disk drive is so named because unlike the "floppy" drive, disks are not removable for file or substitution. But, in lieu of that feature, you gain the hard disk's capacity to hold a great deal of data. You must have a floppy disk to supplement a hard disk for transferring programs or data.

On a floppy-only system it is best to have two drives. One can hold your program and the other can hold the data you generate. PC programs are becoming increasingly elaborate. A word processor these days is likely to come with a spelling checker disk and a thesaurus disk. Using a program like this with two drives means that when you want to perform a spelling check or use the thesaurus you need to substitute that disk for the program disk. This is necessary because the standard PC floppy disk can only hold 360K (each K holds roughly 1,000 characters) of information.

A hard disk can hold at least 20,000K (more commonly referred to as 20 megabytes, or MB) of data or programs.

A 20MB drive is a fairly standard size. Hard disks generally come in multiples of 10. The 10MB size is the smallest; 30MB is the largest in common use. At about $300, a hard disk is worth considering at the outset. Such a disk can hold all of your programs and data, saving you the trouble of locating and inserting floppy disks, and the hard disk is also considerably faster than a floppy in saving and loading data.

The keyboard

Most keyboards look and feel like the keyboard of a typewriter, with some exceptions.

On some inexpensive computers, the keyboard is laid out like a typewriter's but looks and feels more like a calculator's keypad. Such keyboards are much easier to use than a membrane keyboard, though less desirable than a standard typewriter style.

Numeric keypad. Computers differ not only in the type of keyboard but in the number and usefulness of keys. The keyboards of costly computers, for example, usually have a separate numeric keypad in addition to the usual numerals in the top row of any keyboard. The numeric keypad is helpful in business applications, where numbers are entered more often than in typical home use.

Special function keys. A keyboard may also have several programmable special function keys. As the name suggests, these are keys that you can use for a purpose of your own choosing. For example, if a command for a word-processing program requires you to hold down the computer's control key and then punch two letter keys, you might designate a special key for that command, replacing three keystrokes with one. Or if you were learning a programming language such as BASIC, you might designate a key for a frequently typed phrase, such as GOTO, replacing four keystrokes with one. Often, applications programs will themselves designate various keys (even those that aren't designated as special function keys) for commands that would otherwise require two or more keystrokes.

Display

The way a computer displays its information has an important bearing on how you can use it. Some computers—the portables, in particular—come with a built-in monitor. Others operate with a freestanding optional monitor, usually permitting a larger display. (See the section on monitors, page 132.)

Inexpensive computers are sold without a monitor: You attach the computer to your own TV set with the "RF modulator" usually included in the basic price. Though this holds down the initial investment in a computer system, a TV set is not an adequate display for some of the things you may want to do with a home computer.

Drawbacks of TV. Although a TV set is sharp enough for the "moving pictures" of television, an image displayed on a TV screen is inherently less sharp than the same image displayed on a computer monitor. The difference is important because you don't "look at" a monitor the same way you watch a TV screen. Rather, you read a monitor as you read a book or magazine. Fuzziness in the characters displayed on a TV screen quickly leads to eyestrain.

A second drawback is the limited number of columns—that is, characters per line—that a TV screen can display clearly. The maximum practical column display for a TV screen is about forty characters. A forty-column display is barely adequate for word processing: It's like working on a typewriter with a maximum margin setting of forty characters. Some computers intended for use with a TV set are designed for even fewer characters per line, displaying large characters to compensate for the TV set's relative fuzziness.

Optional monitor. The advantages of the optional monitor are added sharpness of the characters, the additional convenience of a screen permanently devoted to computer uses, and the ability to reproduce more than forty columns when used with a computer that can generate a larger number of columns. You might buy such a computer in a package that includes a monitor and a plug-in circuit board that increases the display to eighty columns.

Color. The main advantage of a TV screen as a monitor is its ability to reproduce color, which, of course, adds to the fun of games. It is essential for color-graphics work.

You can buy a color monitor as an option, but they are expensive. A high-resolution color monitor is called an "RGB monitor." Inexpensive color monitors are more like TV sets without a tuner than true computer monitors.

Number of rows. A typical monitor has a screen that measures 11 or 12 inches diagonally. An exception would be the small built-in monitor that comes with some portable computer systems.

Displays, whether on a TV screen or a monitor, typically show twenty-three to twenty-five rows. A model that displays only sixteen rows is inconvenient in word processing or when writing programs because the shallow screen forces you to scroll a lot. (You scroll a computer screen to display information that has been entered but is no longer visible on the screen.)

Upper- and lower-case. Most computers display both upper- and lower-case characters—desirable for word processing—but some display capital letters only.

Printer

Although a printer is a peripheral (an extra piece of equipment), almost all computer applications other than games and educational programs will require a printer sooner or later. You can't mail a letter you have written with a word-processing program until you print it out on a piece of paper. You can't usually set the computer down next to the stove after you've located a recipe on a disk. And you can't send the IRS a computer disk in lieu of a tax form. For more information, see the section on printers.

Recommendations

When shopping for the elements of a computer system, you should sort out what's needed for which purposes. It's important to distinguish between a computer system and a piece of hardware that might be sold as a computer.

Of the uses a computer is likely to be put to at home, word processing is probably the one that would demand the most RAM. At least one disk drive is recommended for machines that don't already include disk drives as part of the basic hardware.

■ How to shop for a computer ■

If you're clear about why you want a home computer and the uses to which it will be put in your household, you will decrease the possibility of making a costly error when you shop for a computer.

Learning computers. With a relatively inexpensive computer system, you can play games—perhaps more enjoyably than with a game machine, because a number of games sold for computers are more interesting than those sold for the game machines. In addition, such a system can help you and your family become computer-literate. You could use the keyboard to run educational programs or delve into the operation of a computer language such as BASIC. Programs available for inexpensive computers can let you investigate the uses of a computer for home accounting and, to some degree, for word processing.

Given the limitations of the relatively inexpensive models among the home computers and the programs available for them, however, doing the task may prove more valuable than the result of the work. The experience will help prepare you to make an informed choice in the future, should you decide to move up to a more versatile computer.

Convenience

When narrowing your choice of computer to a final few, you should remember that a computer can dazzle in one role and disappoint in another. To avoid disappointment, it's best to ascertain not only the things you want a computer to do, but also how much you're willing to put into the task of mastering the computer.

If games and educational applications are your main interests, a computer is not hard to learn. Sophisticated tasks, however, are another matter.

The ability to make a computer do what you want it to do is a learned skill, much like driving a car or playing winning bridge. Motivation is important, and the more complicated systems and software will require time to learn—that's all the more reason not to buy an expensive computer and complex programs unless you have clearly defined needs. Your high-priced investment will gather dust if you abandon it because it's hard to learn.

Assembly. Except for some portable units, you will need to assemble your computer before you can operate it. Even with a portable unit, if you want to use an optional monitor, a connection must be made to the display. A more complicated system will need all the parts to be connected.

Documentation. The instruction manuals that accompany computers and computer software are called "documentation."

Documentation should be written clearly and should be easy to understand. Yet, some is so impenetrable that a whole industry has sprung up to provide manuals that are better written than the originals. On the shelves of computer stores you'll find books, audio tapes, and the tutorial floppy disks devoted to the uses of individual computer models and to the functioning of particular word-processing and spreadsheet programs.

Using programs. Computers in the Apple and IBM price class come with a manual (often more than one) that is hundreds of pages long. Even so, the manual only teaches you how to use the computer and its disk-operating system (DOS), not how to perform tasks with an applications program. The most motivated of writers or office workers will need a month or more to absorb the material in the instruction manual for WordStar, one popular word-processing package. (However, a complex program may have more "powers," or functions, than you will use. You don't need to learn all of the program if you're only going to use part of it.)

Many programs, moreover, lack the quality called "friendliness." To make them work you have to decipher arcane instructions on the screen and key in seemingly illogical commands. For example, using WordStar, you might want to move a block of text from one place to another. You have to mark the beginning of the text block by holding down the computer's control key and typing KB, then mark the end of the block by depressing the control key and typing KK. And that just marks the block—you still haven't moved it.

It is sometimes possible to configure the computer's programmable keys to reduce the frequent commands to a single keystroke—"mark block," for instance. But many computers have only a few such programmable keys and, given the state of home computers, there is an advantage to a program that utilizes arbitrary letter combinations plucked from the full range of the keyboard: The number of possible combinations creates a powerful and very flexible program. Convenience of use all too often involves a trade-off in power and versatility.

Modern software is usually "menu driven." You need only pick what you want to do from a list, or menu, of alternatives. The trade-off is that you may be forced to wade through cumbersome lists to get to the exact job you want performed. Some programs let you bypass the menu to "command-mode" if you are an experienced user.

Using BASIC. BASIC is the most widely used language for personal computers. Learning a little programming using BASIC is helpful in understanding how the computer works and in figuring out what you may have done wrong. What's more, the study of programming itself can be very enjoyable, especially for those who like to solve problems requiring the rigorous application of logic.

Hardware and software compatibility

Making decisions about the programs you want can limit you to a very narrow range of computers capable of running the programs, and vice versa.

Should you buy the computer hardware you like best, or can best afford, and assume that suitable programs are available for it? Or should you first choose the specific applications programs you want to use, then buy the hardware that will run those programs?

When shopping for a computer, you have to consider hardware and software together. The program made for one computer generally will not run on another. (But see the section on IBM-compatible computers, page 115.)

It is because of this basic incompatibility that any decision about buying a computer must rest on prior decisions about what you want to use the computer for. After you decide which one or two applications are most important to you, you must then zero in on applications software: which word-processing program, which financial-management program, which educational programs, which games. In this area, you will have to make compromises. A system good for word processing may be poor for games. A system good for games may be poor for word processing, and a system good for both may cost more than you can afford.

How a DOS works. The greatest number and variety of programs are on disks. To run a disk, a computer needs not only a disk drive but a group of programs called a "disk operating system," or DOS.

The disk operating system is sometimes stored in a computer's permanent memory. Sometimes, it's written on the same disk as the applications program itself and loaded into temporary memory along with the applications program. Sometimes it's on a separate disk.

The typical first step in working with a computer, then, is to load the DOS programs (this is called "booting up"). That done, the operating system functions like a traffic cop, regulating the transfer of information from the disk to the central processing unit and back again, and from the CPU to the monitor and the printer.

Differences in DOS. Almost every computer for home use was originally designed for its own proprietary DOS. A program written for one DOS will not run on another. And sometimes, because of hardware incompatibilities, a program written for the DOS of one brand of computer will not run on another brand with the same DOS.

Disk operating systems differ from one another in speed of operation and in their degree of friendliness (for example, the amount of on-screen help they provide an inexperienced user). But the most important consideration for the buyer is the number and quality of applications programs available for the DOS. Over time, the serious computer user will invest hundreds of dollars in software. For that reason, the DOS can become a stumbling block.

First, the DOS ties you to the software available for it. Later, if you start thinking about upgrading your hardware, the software already collected may restrict your hardware choices to the one brand, or small number of brands, that can run your programs.

It's becoming increasingly likely, however, that marketplace forces already at work will considerably ease the problem of incompatibility. Although incompatibility does complicate the buying decision, the marketplace itself is finding imperfect cures. There are three main paths to easing incompatibility.

A good program written for one DOS tends to inspire other publishers to write similar (or improved) programs for other DOS. Thus the *VisiCalc* spreadsheet for Apple DOS inspired *SuperCalc. SuperCalc* was in some respects an improvement over the original *VisiCalc,* no doubt inspiring the creation of still more spreadsheets, such as *Multiplan,* and *Lotus 1-2-3* with additional enhancements. In the same vein, popular games—such as *Breakout,* or the *Pac Man* maze, or *Adventure* (the first deduction game)—inspire other games of the same genre for every DOS in wide enough use to represent a good source of software sales.

Because computer owners are always looking for something else to do with their hardware, news of a worthwhile program, whether a game or a business application, tends to spread fast. This creates a demand for the program and an incentive for the publisher to write it for more than one DOS.

Computer manufacturers recognize that a single proprietary DOS can be a liability as well as a benefit, discouraging a sophisticated computer buyer who seeks maximum choice in software. As a result, they provide additional DOS choices.

How and where to buy

Computer stores. A computer store is likely to deal in only a small number of brands, and the goal is to sell one of them. Also, don't expect every salesperson to be intimately familiar with every computer and accessory carried in the store, and certainly not with the details of every applications program the store sells.

The best approach in trying to locate a helpful computer store is to ask friends to recommend stores especially good at helping novices. Then call one of the stores and ask for an appointment with the staff person most familiar with your particular need. If you simply walk in off the street, you may wind up with a salesperson who can't answer all your questions yet is reluctant to turn you over to another person for fear of losing a commission.

You will need plenty of time, both with the salesperson and alone with the computer, to try out various combinations of equipment and software, to get used to the feel of a keyboard and the readability of the monitor, and to evaluate the usefulness of the disk operating system and the documentation. It's a good idea to buy from a store that can set up the exact system you ultimately buy, with several choices of software.

Other stores. These days, computers of all kinds are sold in department stores, camera stores and consumer electronics/appliance outlets. In these stores low price takes precedence over service. You are apt to encounter less knowledgeable help than in a computer store, in exchange, possibly, for a bargain price.

Checking with the school. If you want to buy a computer as a teaching tool for your child, check with the school first. Does it use computers? Which models does it use and what programs does it run on them? Because hardware and software incompatibility is virtually complete among inexpensive computers, you may want to match your equipment to the school's.

Buying hardware by mail. The price advantage of ordering hardware by mail is undercut by several disadvantages.

The relationship between buyer and seller is reduced essentially to a telephone call or an order form. While some mail-order firms will discuss your purchase by phone with you at great length, it's not the same as having a salesperson at your elbow to point out a mistake or a feature.

Ordering by mail runs the risk of damage in transit.

Problems are harder to iron out. If you buy a printer locally and it doesn't seem to work with your computer, you can cart the whole system back to the store for help. Repairs don't come by mail.

Buying software. At the computer store, check the range of software carried. Are several vendors represented for diversity? For applications such as word processing, you will want to try more than one software package.

Although you may be in a hurry to make your computer "do something," it's wise to delay the purchase of applications programs unless you know exactly what you want to do. Get used to the computer first. Work with the manual and the system's software until you are comfortable with them. If you want to do something in a hurry, delve into BASIC. Meanwhile, read computer magazines—especially publications devoted to your own make of computer—keeping an eye out for reviews of programs that seem worthwhile.

IBM-compatible computers

IBM did not introduce its personal computer until 1981, but when it did, it revolutionized the business end of the personal-computer market. A host of me-too competitors quickly capitalized on the IBM entry, and the expression "IBM compatible" entered everyday use. IBM compatibles often designated "XT," from IBM's XT model, are compatible, more or less, with the large library of IBM programs, and with IBM accessory equipment.

IBM compatibles, or "clones," have changed quite rapidly over just a few years, with brands appearing and disappearing regularly. Nevertheless, IBM clones are an important marketplace factor because their prices are much lower than IBM's, placing a computer within the reach of large numbers of consumers who would not otherwise consider investing in an expensive machine for home use. What's more, an IBM clone may improve on the IBM product in certain design particulars.

Software. There are hundreds of business programs available for the IBM-PC and compatibles. The main attraction of this mountain of software is that it's there. The owner of an IBM-compatible enjoys a wide choice among the basic productivity programs and first crack at interesting new ideas such as outlining programs or utility programs that help keep track of complex projects.

To take advantage of the large software library, an IBM-compatible must be able to run all or most of the available programs without a lot of difficulty.

Bundled programs. A number of IBM-compatibles bundle programs for word processing, spreadsheets, filing, and other applications into the price of the computer. If the programs are close to what you want, the package can be very attractive. However, if your goal is to choose the best from the big IBM library, the value of the bundled programs—not usually of the first rank in quality—diminishes.

Slots. Virtually all clones have a number of slots into which you can plug expansion cards—or circuit boards—for increasing the computer's capabilities. It is important to have *some* slots, but the need for *many* slots is not as great as it used to be. Previously, the typical use for expansion slots was to increase the memory size of the machine. These days, clones are likely to come with 512K or 640K of built-in memory. The practical limit of memory for IBM-type machines is 640K. Units that come with 512K probably can be expanded to 640K without using any of the machine's expansion slots. There are cards to expand the user memory beyond the normal limit of 640K, on up to the megabyte region, but most home-computer users don't need such powerful options.

You might want to add a modem (a device used to communicate with other computers over telephone lines), or a special graphics card.

Compatibility. All the features in the world will do no good if the clone you buy does not run the programs you want. To test for compatibility of software, Consumers Union engineers ran a gamut of programs (some with a reputation for troubling clones) through a number of IBM clones. They included *Lotus 1-2-3, Supercalc 3, Microsoft Word, Webster's New World Writer, WordStar, Microsoft Windows, Gem Desktop, Flight Simulator, NFL: Challenge, Sargon III* and *Advanced Flight Simulator.* The programs ran with virtually no problems.

Speed. The engineers ran the IBM clones through a series of tests using the BASIC language, a spreadsheet and a word processor. They found no significant differences in speed among them in the normal IBM emulation mode. However, a few of the units had a "Turbo" mode which increased the systems' clock frequency from the normal 4.77 MHZ (Megahertz) to a higher frequency, about 7 or 8 MHZ. This Turbo mode did, in fact, increase the processing speed of the machines that had it: A spreadsheet that took 12 seconds to calculate in the normal mode took 7 seconds in the Turbo mode. A word processor that took 36 seconds to change the margins on a 10-page document took 22 seconds using the Turbo mode. Turbo may upset the timing of some software. Therefore, each of the machines that has it includes a keyboard option or front panel control to switch back to the normal mode. Turbo is a useful feature to have, especially if you deal with large spreadsheets or word processing documents.

Service. When you buy an IBM-compatible computer at a discount, the store may not offer carry-in service, or they may offer it for a limited time. Typically, when you buy a clone you will first operate under a store guarantee. This means that if the unit breaks down within a short period, say seven to ten days, you can return the entire unit to the store and get a total replacement. After that period you may have to deal with the manufacturer's service provisions. That could mean shipping the unit somewhere for repair—and waiting patiently. Before buying an IBM clone—or any computer—check at the point of sale about your service options.

Recommendations

The IBM clones tested by Consumers Union (Beltron, Blue Chip, Commodore, PC Source, Vendex) did a good job of emulating their IBM progenitor in compatibility and performance. Since the testers generally opted for rock-bottom prices, they wound up in some cases with relatively unknown brands. Nevertheless, except for some initial start-up difficulties, all worked well. If you want more "peace of mind" you should buy a more established brand, such as Epson or Leading Edge.

Some stores or mail order ads will list a "come-on" price for the computer while others offer a more complete package price. As an example of a "come-on," you might see an ad for an IBM PC compatible for, say, $499. At that price you get a system unit, a keyboard and one disk drive. You will still need a video display card of one kind or another, a monitor, a second disk drive and a copy of MS-DOS, the operating system. These necessary additions will run the price up to about $800 or so. For the most part, though, ads list the price for a complete or almost complete package. The one item that tends to be missing is the MS-DOS operating system, which can add $60–$90 to the cost.

If you want a Turbo mode, or beginner's software, you'll have to pay more. Given limited funds, it would be best to opt for a hard disk–equipped clone over these features. If you are a rank beginner, try the software package *Microsoft Works*. It is a good beginners' package and has a lot of useful information on how to get started with MS-DOS.

Interpreting advertising terminology

Computer terminology can be all but undecipherable to neophytes. Here are some "claims" from a real advertisement, and some explanations.

"8088-1 CPU, 4.77/10MHZ." The *8088-1* CPU refers to the microprocessor used in the PC (developed by the Intel Corp.). Sometimes the ad will specify an NEC (Nippon Electric Corp.) V-20 CPU. This is a version of the *8088*, claimed to be slightly faster than the *8088* for some operations. "4.77/10MHZ" tells you the clock speed of the system; 4.77MHZ is standard; the 10MHZ option is a higher clock speed. This higher clock speed sets off the advertised model as one of the many Turbo models of the PC.

"640K RAM on board, 8 slots." The original *IBM PC* had 64K of RAM (Random Access Memory) on the motherboard. Any more RAM that was needed required using one of the slots in the PC for this memory expansion. As programs for the *PC* proliferated, they tended to require more and more RAM. There are many programs that will run on 128K, 256K and 384K, but memory in recent years has become cheaper and cheaper and the use of "accessory" programs, such as *Sidekick* and *Prokey,* which reside in RAM at the same time as regular programs, has put pressure on the RAM bankbook. Most clones offer 640K or 512K as standard. If the clone you are interested in is offered in a 512K version, you might just as well have the dealer put in the extra 128K, at the store. The $30 to $50 extra cost is worth it.

"On board" means that all of this memory is on the motherboard and therefore does not require usage of the machine's slots. The "8 slots" refers to the number of slots in the machine (but not necessarily the number of slots that are available for expansion) into which you can add circuit boards. On the original *IBM PC* you needed to use one of these slots to add memory. The *PC* also needed a display card of one sort or another to run a monitor (some clones have this display circuitry on the motherboard; some use a slot). And the *PC* did not come with a parallel or serial port (for plugging in peripheral equipment) as standard equipment. Therefore, these add-ons required slots. As an example, the *Leading Edge Model D* is advertised as having four slots. When you open the case you, indeed, find four *empty* slots. The memory, display circuitry, parallel and serial ports and floppy disk controller are all on the motherboard. Based on Consumers Union's experience with clones such as the one in this ad's claims, of the eight slots advertised, five of them are empty. For most home and small business users this is more than enough for future expansion needs.

"Parallel/serial (2nd optional), game." A parallel port is necessary in order to connect a printer; nowadays, virtually all clones come with this port, and most printers are of the parallel variety. You may come across some printers with a serial input. Thus many, but not all, clones have a serial input (port) built in. A serial port is needed for an external modem or, in some instances, to add a mouse. "Game" refers to a port for connecting a joystick for game use.

"150-watt power supply." This is *not* the AC power that the computer requires (as in a light bulb rating) but rather the internal DC power that the machine is capable of producing to supply expansion needs. Each time you add a board or add, say, a hard disk to the system, the internal DC power supply is required to furnish power to the device. The original *IBM PC* had an internal DC power supply rating of only 63.5 watts. This caused some problems with system expansion. Clones typically have a rating of over 100 watts, which should pose no difficulties for home or small business uses.

"AT style 84 key keyboard" refers to the physical layout of the keyboard. *AT* is different from the *IBM PC* keyboard. If you are used to the layout of the *AT*, this can be an important consideration. The overall feel of the keyboard can be just as important as the layout.

"Monographic system." The phrase "monographic system" has come to mean, in almost all cases, a high resolution Hercules graphics card. This option is the one Consumers Union engineers recommend. It gives you high resolution text and the ability, depending on the software, to view monochrome graphics.

"RGB color system." This option provides medium resolution text, coarse in comparison to the previous option. What you pick up here is the ability to run games in color. The RGB color system can also display monochrome graphics, and is sometimes confused with the "monographic system." The difference is that the RGB can display monochrome graphics in medium resolution *only*.

"EGA color system." This option, though expensive, can display text in high resolution as well as being able to display color. However, the resolution may still not be satisfactory for demanding users.

APPLE MACINTOSH COMPUTER

Once you learn how to navigate the keyboard, operate the disk system, what each application program does—and which keys get it to do those things—it's easy to run a computer.

Computer manufacturers and program designers have tried a number of devices to make the learning simpler.

The easiest programs to learn use a menu command system. The program presents you with a menu—a multiple-choice list of things you can do, written in plain English. By typing an item from the menu, or just the first letter of the item, or sometimes by moving the cursor next to the item and hitting a key, you choose the function you want. In complex programs, you may be led step by step through a series of submenus or prompted by yes/no choices.

Computer manufacturers help with expanded keyboards that may include keys programmed by software to enter a complex command or series of commands with a single stroke.

But even with all the help—all the "user friendliness," as the jargon has it—the newcomer to computing can look forward to a long period of study and of trial and error. This is more than a little daunting to many potential computer users. Executives accustomed to delegating the details to others want to use a computer, not learn a new discipline. People working at home, without the office whiz kid to turn to, may be leery of spending $2,000 or more on a learning experience when they could be making a living with a typewriter or a calculator.

It is to such people that Apple directed its *Macintosh.*

User friendliness

The *Macintosh* is Apple's idea of the ultimate user-friendly computer. The designers have in many places substituted symbols, called "icons," for menus and commands that must be explicitly typed in on other computers. They boldly simplified the keyboard, replacing most function and special-purpose keys with a "mouse." And they designed a unique operating system that invites all programmers for the *Macintosh* to integrate the mouse and the icons into the command structure of their programs.

The mouse is a hand-held unit about the size of a cigarette pack. Rolling it about on the tabletop next to the computer moves a cursor around on the screen. You move the mouse to position the cursor on an icon; by clicking a button on the mouse, you execute the chosen activity.

Bundled into the price of the *Macintosh* is a word-processing program called *Macwrite* and a drawing program called *Macpaint.* Those programs exemplify the ease of use designed into the *Macintosh.*

In the *Macintosh* system, a trash-can icon represents a delete file command, a manila-folder icon represents data files, and a sheet-of-paper icon represents a text document. To delete a file, you "grab" its icon with the mouse, move it over to the trash can, and let it drop. Such tasks as taking letters you have written and grouping them into a single file folder is a simple matter of moving icon symbols around on a screen that is itself an icon for a desktop. With other personal computers, moving or copying documents often requires you to wade through menus or enter special codes while trying not to inadvertently delete the document you are moving.

Macintosh programs also use "windows." These show up as rectangular boxes that afford a glance at one or more projects at a time. For instance, you can display the contents of two files side by side, and cut and paste between them. Using the mouse, you can make the windows smaller or larger, or make them disappear. You can also use a window to display a working calculator or a clock somewhere on the screen while you work on another project.

Finally, *Macintosh* programs display a horizontal menu bar across the top of the screen. The bar contains headings naming the various menus. When you "click" one of these headings with the mouse, the menu "pulls down" like a window shade to reveal a series of choices. In word processing, there are menus for File, Edit, Fontsize, and Style, among others.

Choose Edit with the mouse, and a menu pulls down with such choices as Cut, Copy, and Paste.

Instead of searching through an instruction manual, you can easily experiment with different commands to see their effects on the program. If a command is not appropriate at a particular stage in the program, its menu listing is dimmed. This helps to remind novices what they might, and might not, do next. A handy "undo" feature allows you to rescind an order you made rashly. For instance, you can use it to restore the last thing you deleted. You can also retrieve things from the "trash can," provided you have not "emptied" it.

Anyone accustomed to operating a computer by mastering an extensive keyboard may at first be doubtful about a computer that makes you leave the keyboard for a mouse. In word processing especially, you might want your fingers on the keys so you can make corrections during the regular flow of typing.

But you will soon learn to abandon that way of working. When you type first and make corrections later, you will undoubtedly find it fast, natural, and pleasant to whip around the screen with the mouse.

Memory

There are drawbacks to the *Macintosh* system. All the little frills that make the *Macintosh* so inviting are extremely memory-intensive, leaving less memory available to the user than may be necessary or desirable.

A simplified keyboard

The *Macintosh* has a standard fifty-eight-key typewriter-style keyboard with very few extra keys for computing. The keyboard feel is excellent. There are no cursor keys, since cursor functions are shouldered by the mouse. All the appropriate keys repeat when held down.

A function key with a cloverleaf design activates some other keys to perform double duty. For instance, hitting the clover key plus *E* will eject the disk. Most of these are short-cut keystrokes for common commands that can be a little wearisome to do with the mouse.

There is no separate numeric keypad—something that will be missed by those who work a lot with numbers—but you can buy one as an add-on.

Microfloppy

The *Macintosh* uses a microfloppy disk sealed in a rigid, 3½-inch-wide plastic envelope. These disks have certain advantages over 5¼-inch disks. They are only about half the size, they fit into a shirt pocket, and they are rather easy to load and unload from the disk drive.

But most important, the hard-shell covering protects them from bending, fingerprints, and to some extent from dust and dirt—the bugbears of 5¼-inch disks. Nearly everyone who uses the standard disks experiences disk failure now and then—either some dirt gets onto the surface or the disk gets bent, resulting in a damaged disk and lost data. The greater measure of protection afforded by the hard-shell microfloppy may avert some of those painful catastrophes.

Painting words

The *Macintosh* comes with a built-in high-resolution black-and-white monitor. It measures 9 inches diagonally, compared with 11 inches for the typical computer monitor. The relatively small screen is no drawback, since both letters and pictures are unusually sharp. Most computers address text to the screen using a mode known as block-mapping. But the *Macintosh* doesn't use a block-mapped text mode. Instead, all text and graphics are bit-mapped. In essence, each letter and line is drawn on the screen with lots of fine dots.

Block-mapped screens can display only those characters that are present in permanent memory, whereas bit-mapped screens can display a much wider variety of type styles. Thus, with the *Macintosh,* you can choose and see displayed something approaching the type selection available at a small print shop.

If you want to italicize a word for emphasis, for example, the italics appear on the screen as italics. You can also create distinctive posters, if you wish.

Another unusual feature of the *Macintosh* display is that it appears black-on-white rather than green (or amber) on black. Consequently, writing or drawing with the *Macintosh* is much more like working with a sheet of white paper than on the "blackboard" familiar to most computer-terminal users.

Special printer

Because both text and graphics are bit-mapped, you can't use just any printer with the *Macintosh.* You need a graphics printer specially adapted to the computer. With *Macpaint* drawings, the *Imagewriter* printer produced pictures that were near replicas of what was displayed on the screen.

Print generated by the *Macwrite* and *Macpaint* programs does not try to imitate the standard pica typewriter print. The system's closest approximation of typewriter type is a bit larger. The print is proportional both on the screen and on paper, and it looks quite good.

Accessories are available that allow you to hook up the *Macintosh* to a formed-character daisy-wheel printer. However, with a daisy wheel you won't be able to use the print enhancements that make *Macwrite* so unusual.

Recommendations

The *Apple Macintosh* is a computer that's relatively easy to learn and use. The combination of mouse, pull-down menus, windows, and icons is a logically thought-out system that deserves the careful consideration of anyone about to buy a computer to work on at home or in a small business, away from formal training programs and office gurus.

■ Software for Apple Macintosh ■

When the *Macintosh* was introduced, there was very little software available for it, except for the *Macwrite* and *Macpaint* programs supplied with the machine, but a great deal of software has emerged.

While the library is growing, it's still smaller than that available for the *Apple II* series or the IBM-compatible family. For one thing, software companies have to adapt their wares to the *Macintosh*'s unusual display and to its mouse. And vendors of some of the large business programs apparently waited for the 512K version of the *Macintosh* to catch on before committing themselves to adapting those programs for the *Macintosh.*

Writing with Macwrite

Macwrite is a word-processing program available for the 128K *Macintosh.* It's a reasonably complete program in the same general league with such heavyweights as *WordStar* and *Multimate* for other computers.

Macwrite is far easier to learn and use than any other program of similar complexity. It takes full advantage of the *Macintosh*'s built-in graphics abilities and mouse-oriented, menu-driven functions. Such features as boldface, block moves, and partial deletions are much simpler than on most word processors. Pull-down menus allow for such special effects as varying type faces and character size.

The typesetting features can be fun to play with. However, except for those who want to turn out posters or handbills, the ability to salt a document with Old English script probably won't play a major part in the buying decision.

One nice aspect of *Macwrite* is that setting margins and moving tabs around is very simple. Just move a tab icon with the mouse. As you move a tab, any columns of figures you have on the screen move with them. The system also supports decimal tabs, so numbers with decimals line up with the decimal points. As a result, it's quite easy to work with tables and columns of figures.

The find-and-replace function, which appears on a pull-down menu, is very simple to use. It allows you to find and replace words or groups of words. If, for example, you wanted to change the spelling of a word throughout a long document, the computer will find the word and change it according to your instructions.

The program doesn't show the cursor position in terms of line and page numbers, as most word processors do. Instead, *Macwrite* shows you approximately where you are by means of a vertical "scroll bar" at the edge of the screen. You can see only that you are roughly halfway down the page, or close to the end of the document. To find what page you're on, you have to scroll to the nearest footer or header and take a look.

Macpaint

Macpaint is a drawing program that is a tour de force of *Macintosh* technology. Using the mouse, you can create computer-aided freehand drawings on the screen. With associated on-screen "tools," you can fill them in, add shading and shadows and numerous canned patterns, such as cross-hatching and checkerboards. You can also enlarge, shrink, flip, and rotate a drawing with a mere click of the mouse. The computer enhancements make even simple doodles look interesting.

Apple's advertisements and the *Macpaint* instruction guide are illustrated with very accomplished drawings made on the *Macintosh* with *Macpaint*. But if you're not already an artist you won't be able to create anything close to those alluring designs.

Drawing on the screen by rolling a mouse around on the tabletop requires a good deal of hand-eye coordination. Also, the simulated pencil and paintbrush don't act quite like their real-life counterparts. Consequently, it takes a lot of practice to draw as naturally on the screen as with pencil and paper.

Apple released a later version of *Macpaint* with the 512K "Fat Mac." This version, called *Macpaint 1.4*, works considerably faster in some operations than the older version. It also allows you to fit more drawings onto a disk. For that reason, it might be worthwhile to acquire the 1.4 version even for a 128K machine.

Doodling on a computer may not be your cup of tea. But the range of creative tools offered by the program, and the ease with which they can be used, can make *Macpaint* great fun.

COMMODORE AMIGA AND ATARI 520 ST

The home-computer boom has declined. Computer games have lost their appeal, and many of the uses to which home computers were put have suffered from loss of interest.

Commodore and Atari have drawn on the *Apple Macintosh* for inspiration. Like the *Macintosh,* these computers depend on a "user interface" (computerese for the connection between you and the machine) consisting of a hand-held pointing device called a mouse; icons, or graphic symbols; and pull-down windows. You manipulate the mouse to point an arrow or cursor at the symbol of what you want to do (a brush for painting, for example, or a file folder for filing); or you press a button to pull down a menu of function choices, which may also be described in symbols, and then point to the desired function. It's all much easier to do than to describe—provided that software programs take advantage of the mouse and its attendant iconography. Not all do.

To that basic *Macintosh* approach, Commodore and Atari have added large palettes of color, assuming no doubt that games in color might be of interest to home users and that many users might also want to explore the computer as a substitute for paint and canvas.

Display

The basic claim to fame of both these computers is that they're like the *Apple Macintosh* but with color. As color computers, both excel. They can display a wide variety of colors and shades simultaneously, permitting pictures of great range and subtlety when viewed on the analog RGB display monitors for which these computers are intended. And both can also display a variety of type styles, which you can select. However, text reproduced on the color monitors has the relatively coarse appearance characteristic of an IBM-type computer in its graphics mode rather than the crisp, easy-on-the-eye look of the IBM in monochrome mode or of the *Apple Macintosh.* The resolution is acceptable if you don't spend hours daily doing word processing or processing numbers. Here's how the two computers compare.

COMMODORE AMIGA. The Commodore analog RGB monitor *1080* can display eighty columns of text in its medium-resolution mode. The text, though quite legible, is hard on the eyes over long periods. A high-resolution mode is available. But text displayed in high resolution flickers so much as to be greatly disturbing. The display is far more impressive with games and paint programs, which show off a vast assortment of colors and shades.

ATARI 520 ST. the *Atari SC1224* analog RGB color monitor is a cut above the Commodore monitor. Text display, although also of only medium resolution, is more readable, and the colors themselves appear a bit more vivid. The choice of colors, however, is not as wide as the Amiga's.

A monochrome Atari monitor, the *SM 124,* makes use of the *520 ST*'s high-resolution mode. The image is smaller than the monitor screen, roughly 10½ inches diagonally. The display is sharp enough for extended use, though not as sharp as an IBM monochrome display. With the monochrome monitor, the Atari is worth considering as a work machine.

Disk capacity

Programs and the work created using programs are stored on floppy disks. In the *Commodore Amiga* and the *Atari 520 ST,* as well as in the *Apple Macintosh,* they're 3½-inch microfloppy disks.

COMMODORE AMIGA. The *Amiga* comes with the double-sided disk drive built into the main system unit. It can record up to 880K of information on a disk—the equivalent of about 300 typed pages. A second, external disk drive can be connected through a port in the rear.

ATARI 520 ST. The *Atari* disk drive is a separate, stand-alone unit. It is a single-sided drive, capable of writing 360K of information. You can substitute or add a double-sided *Atari* disk drive with a capacity of 720K. The *Atari 1040 ST*'s built-in drive is of the double-sided type.

Keyboard

Even when pull-down windows, icons, and a mouse replace typed commands, an adequate keyboard is essential for serious use with word processing, spreadsheet, and filing programs. Commonly used keys, such as cursor keys and special-function keys, should be placed in a logical, convenient location. Number-heavy programs all but demand a separate numeric keypad.

COMMODORE AMIGA. The *Amiga* has a full-sized full-travel keyboard with a standard typewriter layout, ten function keys, and a separate numeric keypad. Cursor keys are in a sensible diamond array over on the right. Keyboard feel is very good.
ATARI 520 ST. The *Atari* keyboard is of a piece with the system unit, a low, wedge-shaped affair with a stylish, swept-back look. The keyboard has a standard typewriter layout with sculptured, full-travel keys and a separate numeric keypad. The cursor keys are in a modified diamond array, with the south-pointing key placed between the east and west keys. The keyboard feels less crisp than the Commodore's.

Sound

The ability to deal with sound is more important for computers that entertain or teach than it is for computers that merely work. Both of these computers have extensive built-in sound capability, useful for music and some education programs.
COMMODORE AMIGA. The *Amiga* has four electronic "voices" that it can reproduce over two channels. That gives the computer a potential stereo effect. (To get stereo, you need to add a stereo amplifier and loudspeaker.) The *Amiga* also has a speech synthesizer that speaks typed words in either a male or female synthetic voice.
ATARI 520 ST. The *Atari* has three computer "voices" that it pipes over one channel. But it has no separate sound output. Instead, the sound comes out through the monitor speaker.

Ports

Peripheral equipment must all plug into the computer somewhere, through what are called ports. The number and type of ports determine how and in what ways you can expand a computer system. These machines offer a bundle of choices.
COMMODORE AMIGA. The *Amiga* comes with one Centronics parallel and one RS232 serial port. Those ports let the *Amiga* run any standard brand of printer off the shelf, although you'll need to connect the printer with special Commodore cables. The unit also has an expansion connector on its right side that permits you to add extra hardware such as a hard-disk drive and memory-expansion devices. The *Amiga* has two joystick connections (for game controllers), two audio output jacks, and a standard video output connection.
ATARI 520 ST. The *520 ST* also has one Centronics parallel and one RS232 serial port, a hard-disk interface, two joystick connections, and a

TV output jack. In addition, there are MIDI (Musical Instrument Device Interface) ports to connect a music synthesizer.

Documentation

The instruction manuals that come with computer gear should be clearly written and understandable even to beginners. Although the design of these two computers should make them easier than most to learn and use, especially with programs that take full advantage of the mouse-and-icon approach, no computer's operation is exactly transparent to newcomers.

COMMODORE AMIGA. The user guide is easy to read and understand; the guide to BASIC, less so.

ATARI 520 ST. The Atari BASIC manual is quite complete and provides many helpful examples. The system user guide is clear, as are the manuals accompanying the bundled applications programs.

Recommendations

Commodore and Atari set out to market computers that would be as easy to learn and use as the *Apple Macintosh* and that would also excel at reproducing color graphics and sound. From a technical point of view, both machines succeed: They're easy to use and graphically dazzling.

There's a more serious side to graphics and sound, of course: These machines may well prove attractive to artists and musicians who want to explore the electronic possibilities of their art.

For the typical buyer the decision will finally hinge on how well these computers perform the workaday tasks of word processing, number handling, and filing. For those jobs, the color capability that makes these computers so seemingly attractive is an impediment, since it also results in a display that's less than ideal for text characters.

From a technical standpoint, the *Commodore* outpoints the *Atari* simply because it has more of what makes both computers special—more colors in its color display, better animation capabilities, more voices (and in stereo, if you want to add the loudspeakers to hear it), and more speed. Color-graphics buffs will especially like the *Commodore.*

The *Atari* offers fewer colors but plenty of them, sound that's more than adequate for those who have less than a professional or hobbyist interest in synthesized music, and sufficient speed for any home use. Its color display is better for text work than the *Commodore.*

The *Atari* is also an interesting choice for those who merely want an inexpensive computer for work tasks. Then, you might equip it with a monochrome monitor instead of a color monitor. Add a printer, and you'd have a complete system, including software, at less than half the price of an *Apple Macintosh,* but one just as easy to use.

MAGNAVOX VIDEOWRITER

The *Videowriter* (about $800) is a computer set up for word processing and no other function. You can use it without learning an operating system—a program computers use to perform many housekeeping chores—or bothering with any other aspect of computing.

The system comes in two pieces—a keyboard and a unit containing a monitor, disk drive, printer, and computer electronics. A word-processing program is permanently etched on memory chips inside the machine. The *Videowriter* weighs 21 pounds and occupies about as much space on a desktop as an office typewriter. To use the machine, you just unpack it and plug it in.

The screen

The monitor screen measures only 8½ inches diagonally. Still, it displayed 80 columns and 18 rows of crisp amber characters. The only size-related problem is that commas and periods are hard to distinguish.

A handy "help" window occupies a narrow strip down the right-hand side of the screen. The help window, which stays on whenever the machine is running, offers prompts for starting or ending a document and suggestions that walk you through various procedures, such as preparing a new disk or printing out a document.

Word processing

The *Videowriter*'s word processor is simple to learn, practically self-evident. Since you choose what you want to do from the on-screen menu of possibilities, there's nothing to memorize. The program itself is moderately full-featured. It can perform all the tasks that make word processing so much more convenient than typing: It can move or copy text from one place to another; change text by typing over what's written or by inserting new material; erase text a letter or a block at a time; search a document for words or phrases and replace them on command with other words or phrases; assign page numbers; and change margins with a few keystrokes. It can't handle certain specialized word-processing functions, however. For example, it doesn't generate an index or insert footnotes automatically. And it can't triple space when printing.

Information storage

The *Videowriter's* single disk drive uses 3½-inch microfloppy disks to record and store the documents created on the machine. Those are the kind of disks used by the *Apple Macintosh,* among other computers. They are sturdy pocket-sized packets with a hard plastic shell. Blank disks hold up to 300 kilobytes of data, or approximately 150 pages of documents.

One peculiarity that can affect the number of documents you can store electronically is the *Videowriter's* way of checking spelling. A special disk with a spelling program on it comes with the system. If you want the computer to check for spelling and typing errors, you must create your document on the spelling disk itself. The spelling disk has room for about 100 pages. When you have filled it up, you can either delete old documents to make room for new ones or you can buy another spelling disk from the manufacturer. (They cost about $8 each, compared to $3 for a blank disk.) But with the unit tested by Consumers Union, you couldn't copy a document from the spelling disk to another disk.

Nor could you make backup copies of disks, whether it's the spelling disk or a document disk. The best you could do was copy a document to another place on the same disk. If the disk itself were to become damaged, you could lose everything on it permanently.

The spell-check program itself is not as useful as it could be. It requires *you* to ask for a check of each word you suspect might be misspelled. So if you don't know that "wierd" is weird, the program won't help you. Most computer spell-checkers point out suspected errors to you rather than the other way around, operating as proofreaders as well as dictionaries.

The *Videowriter* saves the document you're working on—that is, records it on the disk—in any of several situations. It does so when you reach the end of a page, or if you do nothing for two minutes, or if you hit keyboard keys that start a new operation.

While the frequent saves to disk are a nice safety feature, there's a slight penalty for having the document mostly on disk at all times. With documents longer than a page or so, making changes that affect the whole document can take time. For instance, it takes about 1½ minutes to change the margins of a 10-page document. Personal computers make that sort of change almost instantly when the document is in memory rather than on disk.

The keyboard

The *Videowriter* keyboard is compact. The space bar is a little stiffer than the other keys, but the keyboard feel is very good overall. The key layout is very straightforward.

Sixteen clearly labeled special function keys are arrayed across the top and down the sides of the keyboard. The keys call up specific labeled activities, such as "Menu," "Print," "Do," and "Undo." (The Undo key lets you rescind an order you may have made in error.)

Two screens in one

Hitting a key labeled "Split" breaks the screen horizontally into two windows, so you can display parts of two separate documents simultaneously. Though the upper and lower windows show only eight lines of text, you can scroll through either document and easily compare, move, or copy text from one to the other.

Paper and printing

The *Videowriter* has a built-in thermal-transfer printer that works by melting ink from a plastic ribbon onto the paper. The printer has both a fast "draft" and slower "near letter quality" mode, but neither is as fast as a typical dot-matrix printer. Printing out a 170-word business letter took sixty-six seconds in draft mode and ninety-one seconds in "near letter quality" (NLQ) mode. Comparable times for a typical nine-pin dot-matrix printer would be twenty and sixty seconds, respectively.

The NLQ print quality was comparable to that of a good nine-pin dot-matrix printer in NLQ mode (see page 139). That's good for a computer printer but not the quality you'd get from an electric or electronic typewriter.

The *Videowriter*'s printer can use any smooth paper, but is geared to 8½-x-11-inch sheets. The print mechanism types out a maximum of 80 characters across the page and 55 lines down.

Paper handling is limited to one sheet (or envelope) at a time, as with a typewriter. A tractor feed for continuous paper is not available. The thermal transfer printing uses up a lot of ink cartridges. Each $6 cartridge lasts for about 50 pages, which means that each page could cost 12 cents just to print. There's no way to hook up the *Videowriter* to a different printer, should you want to make a substitution.

Recommendations

The *Magnavox Videowriter* is exceptionally easy to learn and use, and its built-in word-processing program works fine for all but the most specialized of uses—a great advance over creating documents with an ordinary typewriter.

The system is about two-thirds the price of an *IBM-PC* clone fully equipped with a printer and a word-processing program, but substantially more expensive than an electronic typewriter that can perform some word-processing functions.

Drawbacks such as single-sheet feeding, relatively slow printing, and rather costly ribbon cartridges do tilt the *Magnavox* toward those who write short documents, like letters or school reports, rather than lengthy manuscripts.

Some electronic typewriters have a memory that stores anywhere from a paragraph to a few pages before typing out. So if all you really need in a word machine is the ability to correct spelling or typing errors without retyping whole pages, an electronic typewriter can produce documents that look better than the *Magnavox*'s computer-printed documents.

One limitation with most electronic typewriters is the small display (typically less than a line long) of what you're typing into memory. The display can't give you much idea how the document will look in its final form and doesn't lend itself to the sort of heavy editing involved in moving paragraphs around.

At the other end of the word-machine scale is the general-purpose personal computer. For approximately $1,200, you can put together an *IBM-PC* clone that consists of a computer, two disk drives, monitor, software, and a decent printer. With such a system you can choose any of a number of word-processing programs that offer advanced features, if that's what you need, and, if you wish, explore the many other uses of a computer.

COMPUTER MONITORS

A computer used at home as a work tool needs a monitor screen that can display eighty characters on a line, just as an office computer does. (An ordinary TV set can only display forty characters on one line.)

For a computer devoted to work, a monochrome monitor is all you are likely to need. Monochrome monitors are far less expensive than color monitors. And typically, though not invariably, monochrome monitors display eighty-column text more clearly than color monitors do.

There are two common types of monochrome monitor. One type, introduced by IBM for its own personal computers, is called a "monochrome display" monitor—mainly to distinguish it from the other type, called "monochrome composite."

Games and educational programs don't absolutely require color graphics, but they are certainly enhanced by color. And there are certain specialized business and professional computer applications that require color graphics.

A computer owner who needs a color monitor has three basic choices: First is a composite color monitor, which is similar to a color TV set in its limited resolution, but unable to receive a TV signal. Composite monitors are relatively inexpensive as color monitors go, and their display can offer some improvement over the display of the same signal routed through a TV tuner. So a composite monitor may be of interest to those who are content to use the TV set for games if only it didn't interfere with using the TV set for TV.

A second possibility is a device called a "monitor/receiver." As the name implies, it can both receive TV signals and produce computer-generated color graphics and, frequently, eighty columns of text. It can be the monitor of choice if you would rather watch TV where the computer is than move and use the computer where the TV set is.

The third possibility is an RGB monitor. This is a costly, high-resolution color monitor that can reproduce eighty characters of text clearly and provide the sharply defined color graphics necessary for a good display of tables and charts for business purposes or for using the computer as a design tool.

The monochrome monitor

Any monochrome monitor should be able to display at least eighty characters of text across the screen. The main difference is between the types of monitor rather than between monitors within types. The IBM monochrome display type crams more scan lines into the screen and uses more dots to form each character. As a result, such a monitor achieves greater resolution than a composite monitor, producing a text display that looks more like type than like an array of dots.

Text quality is the most important criterion for distinguishing one monochrome monitor's performance from another's.

The brighter the characters on the screen, the more visible they will be in bright ambient lighting. But text appearance is also intimately related to contrast. Other things being equal, the greater the contrast, the better the text appearance. Adjusting the brightness and contrast controls helps to achieve the best image.

The color monitor

Pictures of text or other images depend on a video signal sent from the computer to the monitor. The most common type is called a "composite signal," because it is an amalgam of brightness, synchronization, and sometimes color information. Color composite signals are used in color TV sets, composite monitors, and monitor/receivers.

For technical reasons, the color composite signal can't produce eighty columns of clear text. Nor can it produce very sharp graphics. Instead, color monitors that use composite signals can display about forty characters of text and the "moving pictures" of TV programming and video games, which don't demand that the eye focus sharply on a fine shape, such as a letter or a line in a chart.

In judging a color monitor, one must use a different judgment scale for judging composite graphics than for RGB graphics because RGB graphics are almost always sharper and more vivid. Likewise, judgments for eighty-column text cannot be directly compared to those for forty-column text.

Color composite monitors. A color composite is similar to a color TV set but without the tuner found in TVs. Freed of the need to process broadcast signals, it can sometimes offer a small but significant improvement in picture clarity, but it can't reproduce an RGB signal.

Monitor/receivers. When you attach a computer to a TV set, the signal passes to the screen indirectly, through the tuner, thereby suffering some degradation. A computer attached to a monitor delivers its signal directly; thus, the reproduction on the screen can be of a higher quality. A monitor/receiver is, in effect, a composite monitor with a TV tuner. It differs from ordinary TV sets in that signals from a computer need not pass through the tuner.

RGB monitors. The RGB monitor takes its name from the red, green, and blue color signals used in its display. It is the best form of color monitor available.

The RGB video signal can have a broader bandwidth than a television video signal since the color information doesn't have to be multiplexed—that is, squeezed in with other video information such as luminance and synchronization. The color information arrives at the set unhindered by various filtering processes necessitated by broadcast TV.

Ratings of computer monitors

Listed by types; within types, listed in order of estimated quality, based primarily on text and graphics appearance. Prices are as published in a **July 1985** report.

◉ ◓ ○ ◒ ●
Better ⟵ ⟶ *Worse*

Brand and model	Price	Weight (lb.)	Dimensions, H × W × D (in.) [1]	Measured screen size (in.)	Display color	80-column text	40-column text	Composite color graphics	RGB color graphics	Screen brightness	Audio	Advantages	Disadvantages	Comments
IBM monochrome display-type monitors														
Amdek Video 310A	$230	19	12 × 15 × 14	11¼	Amber	◉	—	—	—	○	—	C,D	—	*C,H,I*
Quadram Amberchrome	250	19	12 × 15 × 14	11¼	Amber	◉	—	—	—	◓	—	C	—	*C,H,I*
PGS MAX-12	249	25	13 × 15 × 15	11½	Amber	◉	—	—	—	◓	—	—	—	*C*
IBM 5151	275	19	11 × 15 × 14	11¼	Green	◓	—	—	—	◓	—	—	b	C
Monochrome composites														
Amdek Video 300	179	19	12 × 15 × 14	11¼	Green	◓	—	—	—	○	—	C,D	—	*H*
Panasonic TR120M1PA	219	18	11 × 14 × 14	11½	Green	◓	—	—	—	◉	✓	—	—	*B,H*
Zenith ZVM123A	140	15	12 × 13 × 13	11½	Green	○	—	—	—	◓	—	C,E	—	*A*
Electrohome EDM926	195	13	9 × 9 × 12	8½	Green	○	—	—	—	◉	—	—	—	*B,H*
NEC JB1201M(A)	179	16	12 × 15 × 13	11½	Green	◒	—	—	—	◓	✓	—	—	*B,H*

[1] *Rounded up to nearest in.* **Panasonic DTH103** *includes removable base. Allow about 3 in. in rear for cable connections for all monitors.*

[2] *The* **GE** *and the* **Amdek Color 300** *yielded slightly improved graphics quality when their Luma/Chroma inputs were used.*

[3] *Judgment of the* **Sears** *is for its green-on-black mode.*

[4] *Differences in score between white-on-black and green-on-black modes were, in general, insignificant. Judgment for the Panasonic is for its reverse video mode (black-on-white), which was preferred.*

Ratings of computer monitors (continued)

Listed by types; within types, listed in order of estimated quality, based primarily on text and graphics appearance. Prices are as published in a **July 1985** report.

◉ ◓ ○ ◒ ●
Better ⟵⟶ *Worse*

Brand and model	Price	Weight (lb.)	Dimensions, H × W × D (in.)[1]	Measured screen size (in.)	Display color	80-column text	40-column text	Composite color graphics	RGB color graphics	Screen brightness	Audio	Advantages	Disadvantages	Comments
Color composites														
Amdek Color 300	349	24	13 × 15 × 16	13¼	—	—	◓[2]	◓	—	◓	✓	C,F	c	*J*
Panasonic DTS101	339	16	11 × 12 × 14	10¼	—	—	◓	◓	—	◓	✓	F,G	c	*B,H,K,L*
BMC BMAU9191U	370	26	14 × 15 × 17	13¼	—	—	○	◓	—	◓	✓	F	—	—
NEC JC1215MA	399	20	13 × 15 × 15	11¾	—	—	◒	◓	—	◉	✓	—	—	*B,H,K*
Zenith ZVM131	379	35	15 × 16 × 17	13¼	—	●	◒	○	●	◓	✓	—	—	*B,D,E,G,K*
Monitor/receivers														
Sears Cat. No. 4084	340	27	14 × 15 × 15	13¼	—	○	◓[3]	◓	◒	◓	✓	E	—	*D,F,G,I,P*
Panasonic CTF1465R	650	35	14 × 16 × 17	14	—	○	○	◓	◒	○	✓	F	—	*B,F,G,H, K,N,O*
Mitsubishi AM1301	520	31	15 × 15 × 16	13	—	○	○	◓	○	○	✓	C	a	*G,J,K,M, N,P*
GE 13BC5509X	490	29	14 × 19 × 17	13	—	—	◒[2]	◓	—	○	✓	H	—	*B,K,Q*

RGB monitors

Panasonic DTH103	749	21	13 × 12 × 14	10½	—	◓ [4]	—	—	◓	◓	—	A,B,C, E,F,G	—	*E,F,G*
Quadram Quadchrome II	599	30	12 × 16 × 17	12½	—	◒	—	—	◓	○	—	C	—	*H*
IBM 5153	680	28	12 × 16 × 17	12¾	—	○	—	—	○	◓	—	C,E	—	—
Zenith ZVM136	799	32	14 × 16 × 17	13	—	○ [4]	—	—	◒	◓	—	C	b	*D,G*
PGS HX12	695	28	11 × 16 × 17	11½	—	◒	—	—	○	◉	—	—	—	*G,H*
Amdek Color 700	649	28	13 × 15 × 16	13	—	◒ [4]	—	—	○	○	—	C	—	*D,E,H*

Key to Advantages
A–Can be switched to reverse video.
B–Has swivel and tilt base for more convenient viewing.
C–Has high-contrast screen.
D–Has matte-finish antiglare mesh over screen, which reduces glare significantly.
E–Has user control for adjusting height of display.
F–Has user control for centering display.
G–Has user adjustment for vertical centering.
H–Display width can be reduced by 10% to compensate for overscan.

Key to Disadvantages
a–Cannot produce highlighted text, a significant disadvantage; can only reproduce 8 colors in IBM RGB signal; has annoying buzz in RGB mode even with volume at minimum.
b–Some users may find long-persistence screen annoying during scrolling.
c–Color and tint are inconveniently located.

Key to Comments
A–Has 40/80 column switch.
B–Has a loop-through video input.
C–Designed to work only with **IBM-PC** and **PC** compatibles.
D–Can be switched to green display.
E–Manufacturer says can directly reproduce color from RGB-equipped **Apple II** series computers with appropriate cable.
F–The "brown" in the 16-color IBM set displays as yellow.
G–RGB cable is optional accessory.
H–Display height adjustable with screwdriver (technician's adjustment).
I–Display centered with screwdriver (technician's adjustment).
J–Has headphone jack.
K–Has sharpness control.
L–Character mode judged of no advantage.
M–Has two sets of video/audio inputs.
N–Has a composite output derived from antenna (RF) signal.
O–Has a quartz tuner; full-function remote control; on-screen display of channel, time and program source. Has a sleep timer, last-channel recall, and channel bypass; can receive 69 cable channels.
P–Has a varactor tuner.
Q–Has 2-knob mechanical tuner.

The picture tube of an RGB monitor is designed to take advantage of the broader bandwidth. The practical result is a much sharper and more vivid color display than is possible with a composite monitor receiving composite signals.

COMPUTER PRINTERS

Almost every computer system eventually needs a printer. There are several types of printers, but one—the dot-matrix printer—has become almost the universal choice for home use. Dot-matrix printers are relatively inexpensive and very fast. The best of them can produce something quite close to the print quality of a good electric typewriter.

When operating at their fastest speed, dot-matrix printers typically produce jagged-looking "computer print." That's good enough for first drafts, and probably good enough for school papers. But you want something better for correspondence or formal reports. Most dot-matrix printers can be set to a slower speed, at which they seek to emulate the high-quality print of an electric typewriter. This "best" print mode is often called "letter quality" (LQ) or "near letter quality" (NLQ).

In the past, few dot-matrix printers did a really convincing job of imitating typewriter print. Recent technical improvements, however, have made it possible for the best dot-matrix printers to produce print quality that is virtually indistinguishable from an electric typewriter's.

Other types of printers can turn out quality print. One promising alternative is an *ink-jet* printer, which squirts little droplets of ink onto the page. An ink jet can go as fast as a dot-matrix. And, since there is no impact of print head against ribbon and paper, ink jets are fairly quiet. (See page 146.)

Office workers are likely to use a *daisy-wheel* printer. Daisy-wheel print looks just like the print from an electric typewriter because the design of the print head is the same as that used by many electric typewriters—a wheel of formed characters. But daisy-wheel printers are slow and noisy, and they can't produce high-quality charts and graphs. These days, *laser* printers are rapidly replacing daisy wheels in offices. Laser printers produce a whole page at a time, much as photocopy machines do. They can reproduce graphics or whatever kind of type the computer can create—including type fonts and sizes previously reserved to professional print shops. They are fast and almost silent. But they're expensive, priced at $2,000 to $3,500 and up.

Print quality

A good dot-matrix model produces print near in quality to that of a daisy-wheel. Only on close examination can you distinguish between the best of pages produced by the two methods.

This high quality is due in part to improvements in the technology of dot-matrix printing. The print head of a dot-matrix printer has at least one vertical column of tiny pins. As the print head sweeps across the page, these pins pop out, drawing letters by hitting a ribbon and the paper underneath. To create darker, better-quality type, the print head may make more than one pass across the line, firing a different matrix of pins. Or it may hit the paper half a pin-width to the side of the first impression. Or the paper feed may move the paper up a tiny bit between passes. Those maneuvers have the effect of creating a tight cluster of print impressions.

Most dot-matrix printers used to have nine pins. Over the last few years, print heads with eighteen and twenty-four pins have become more common. An eighteen-pin or twenty-four-pin printer can make much finer characters than a nine-pin version. With more pins, the print head doesn't need to make multiple passes over a line, so these printers tend to be faster than nine-pin models.

Print speed

The time it takes a printer to do its work is more significant in business use, where time is money, than in home use. Still, the less time it takes to print out a letter, a school paper, a budget, or a phone list, the better.

Printer manufacturers rate the speed of their machines in characters per second, or cps. But those ratings are not always a good guide to the comparative speed of competing printers. First, the cps figure given often applies only to a printer's draft mode, ignoring the important, and slower, best-quality mode. Secondly, the manufacturer's cps ratings may clock the speed for printing one horizontal line but fail to consider how fast the printer advances the paper or returns the carriage. There may be differences of as much as 20 percent between printers with identical cps ratings.

To create a more useful measure of printing speed, Consumers Union testers had each printer type out a sample 170-word business letter in both draft and near-letter-quality (NLQ) mode. That puts the emphasis on how long it takes each printer to finish a job.

The best fast printers can rattle off a 170-word letter in twenty-eight seconds or less at their slowest speed. And they are among the fastest when delivering their best quality. The next-fastest printers take about fifty seconds. That's still a good deal faster than most daisy-wheel printers.

All the printers are fairly quick in their draft mode. The fastest cranks out a 170-word letter in ten seconds, and even the slowest takes only twenty seconds. For a home user, those differences don't have much practical significance.

How quiet?

Any printer that depends on the hammering of metal against ribbon and paper is going to make some noise. To gauge how much, Consumers Union testers set up printers in a simulated living room and measured their sound output with a sound-level meter.

An ordinary printer is noisy enough that it would be hard to carry on a phone conversation if you were sitting next to it. A printer can be quieted by enclosing it in an acoustically tight case. If the manufacturer hasn't provided an enclosure, one may be available at a computer supply store.

Some technicalities

Not all printers can be used with all computers. One potential source of incompatibility concerns the type of connection—or interface—between computer and printer. The two types of interface are called "serial" and "parallel."

A serial interface is generally required if the printer is farther than about fifteen feet from the computer. It is standard equipment with some brands, such as the popular Apple computer line. *IBM PC*s and most other personal computers come with a parallel interface as standard.

The parallel interface (sometimes called the *Centronics* interface) has the virtue of being the more "standardized," so there's almost no likelihood of error in matching a printer with a given computer.

If the printer you want has the wrong interface for your computer, you can usually buy an interfacing device for $50 to $80. But you'll probably need some expert help to make the interface work properly.

Another source of incompatibility between computer and printer has to do with the software you use for word processing or other functions. Beyond printing the roman alphabet, punctuation, and numerals, printers are often theoretically capable of such enhancements as boldface, underlining, and foreign characters. But not all software programs have the right codes to activate these features. Your word-processing program, for example, may have a print menu that lists proportional spacing. But if your printer doesn't allow it, or if its software doesn't recognize the word processor's command, you won't get proportional spacing.

Most word processors and other applications programs list the printer brands they support. The software has an installation command that supposedly tells the computer which printer is hooked up.

Major print enhancements include such nice features as italics, superscripts, and subscripts in the NLQ mode. Most up-to-date software can handle those features.

Buffer size. When a document is sent to the printer, it runs through a buffer (an electronic memory in the printer), where all or part of it is held while the printer does its work. When the buffer has accepted the last of the document, it frees the computer for other tasks. A buffer of 5,000 bytes is equivalent to about two full pages of text. One smaller than 1,000 bytes is too small to be of much use.

Download characters. Musical notes, checkmarks, and other special characters may not appear in the printer's built-in character set. If you need to use such symbols, you can get computer software that helps you design the special character and sends it to a memory location in the printer (that's known as downloading), and reconfigure the keyboard.

IBM characters. A printer may have built-in software to create the IBM special character set, which has, among other things, many graphic symbols.

Paper handling

A typical printer can handle single sheets of paper 8½ inches wide, the standard size of typewriter paper. That's essential if you're using letterhead paper. However, only a few are wide enough to handle a sheet of typing paper the long way. On most printers, you roll the paper in sheet by sheet. Some printers offer a "cut-sheet feeder," an accessory that feeds single sheets automatically.

To handle continuous-feed paper, these machines are designed to use either a tractor-feed (T) or pin-feed mechanism (P). The more versatile sort is a tractor feed—a pair of sprockets on a drive shaft independent of the printer's platen. Tractor-feed mechanisms are adjustable to a variety of paper widths, from narrow mailing labels all the way up to the width of the platen.

The handiest continuous paper is fanfold paper, which has sprocket holes along the sides. This paper is commonly 9½ inches wide, with perforations that allow the sprocket holes to be stripped off, and pages separated, to yield 8½-x-11-inch pages.

A spin-feed mechanism is less desirable. It has sprockets at the ends of the platen. They can be adjusted inward or outward only a fraction of

Ratings of computer printers

Listed by types. Within types, listed in order of estimated quality.
Prices are as published in a **May 1987** report.

Brand and model	List price	Weight, lb.	Print quality: NLQ	Print quality: Draft	Speed: Mfr. claim, cps	Noise	Special print modes: Draft italics	NLQ italics	Draft sub/superscript	NLQ sub/superscript
Dot-matrix printers[4]										
✓ NEC P660 (P)	$699	20	◉	◉	◓	216	○	✓	✓	✓	✓
✓ Epson LQ-800 (P,S)	799	13¾	◉	◉	◓	180	◒	✓	✓	✓	✓
✓ Okidata 292 (P)	749	13	◓	◉	◉	240	○	✓	✓	✓	✓
C. Itoh Prowriter C310 (P)	699	21	◓	◓	◉	300	◓	✓	✓	✓	—
Canon A-60 (P)	679	14	◓	◓	◓	200	○	✓	✓	✓	✓
Toshiba P321 (P,S)	699	22	◓	◉	◓	216	○	✓	✓	—	—
Panasonic KX-P1092 (P)	499	20	◓	○	○	180	◓	✓	✓	✓	[1]
Imagewriter II (S)	595	20¾	○	◓	◉	250	○	—	—	✓	—
Manesman Tally MT85 (P)	499	18	○	◓	◓	180	◉	✓	✓	✓	✓
IBM Proprinter (P)	549	17¾	○	◓	◓	200	◒	—	—	✓	—
Brother M-1409 (P)	479	13¼	○	○	○	180	◓	✓	✓	✓	✓
Juki 5510 (P)	479	20½	◓	●	○	180	◓	✓	—	✓	—
Citizen MSP-20 (P)	499	11	◒	◓	◓	200	○	✓	✓	✓	✓

[1] *Does not print subscripts or superscripts at all in NLQ mode.*

[2] *Weight with batteries; 3 lb. without batteries.*

[3] *No NLQ mode available. Best mode called "normal print."*

[4] *The letter in parentheses indicates the type of interface post—parallel (P) or serial (S)—the printer uses to connect with a computer.*

● ◒ ○ ◓ ⬤
Better ⟷ Worse

Buffer size, bytes	Draft	NLQ	IBM characters	Single-sheet width, in.	Continuous width, in.	Paper feed	Life, pages	Price	Cost/100 pages	Advantages	Disadvantages	Comments
	Download characters			Paper handling								
2000	128	128	✓	10	5–10	T	253	$16	$6.32	A	a	*A*
6640	96	96	—	10	4–10	T	221	15	6.79	G	a	*B,C*
8160	160	96	✓	9½	9½–10¼	P	303	11	3.63	E	—	*B*
3200	95	14	✓	11	4½–11	T	200	8	4.00	C,E,G	—	
8400	213	—	—	10	9½–10	P	205	12	5.85	B,G,H	a	—
1840	—	—	—	11	4–10⅝	T	308	12	3.90	F,G	a	*B,D*
7120	256	161	✓	10	4–9	T	453	9	1.99	B	a	*E*
2000	175	—	—	10	3½–10	T	200	8	4.00	C	—	—
3360	256	—	✓	10	4–10	T	258	8.50	3.29	E,G,H	a	—
560	94	94	✓	11	3–10	T	316	10	3.16	C,D	b,e	—
3200	256	—	✓	13⅞	4½–13	T	239	12.50	5.23	—	c	—
3040	256	—	✓	11	3⅞–10¼	T	271	7.50	2.77	—	—	—
7520	242	—	✓	10	3–10	T	283	10	3.53	—	f	—

Ratings of computer printers (continued)

Listed by types. Within types, listed in order of estimated quality.
Prices are as published in a **May 1987** report.

Brand and model	List price	Weight, lb.	Print quality	Speed: NLQ	Speed: Draft	Speed: Mfr. claim, cps	Noise	Special print modes: Draft italics	Special print modes: NLQ italics	Special print modes: Draft sub/superscript	Special print modes: NLQ sub/superscript
Ink-jet printers[4]											
Canon BJ-80 (P)	679	13¾	●	●	○	220	●	—	—	✓	✓
HP QuietJet (P,S)	595	11¼	○	◓	○	160	●	—	—	—	—
Diconix 150 (P)	479	3¾[2]	○	◓	○	150	●	✓	✓	✓	—
HP Thinkjet (P)	495	7½	◒[3]	—	○	150	●	—	—	—	—

Key to Advantages

A–Ten-position front-panel switch controls print mode and pitch.
B–Front-panel switch controls pitch.
C–Can print on single sheets with fanfold paper still in machine.
D–Has front-panel opening for single sheets.
E–Printer parameters set from front panel via printed menu.
F–Above-average print quality in draft mode.
G–Can accept font cartridges.
H–Interface is removable circuit card.
I–DIP switches, used to set machine's various print modes, accessible from front panel.

Key to Disadvantages

a–Cable connector intrudes in paper path.
b–Loading fanfold paper more difficult than most.
c–Attaching tractor feed more difficult than most.
d–Best print quality requires special paper.
e–Lacks proportional spacing.
f–DIP switches located inside printer housing.

Key to Comments

A–Tractor-feed mechanism, $55 extra.
B–Tractor-feed mechanism, $60 extra for **Epson,** $99 for **Toshiba,** $45 for **Okidata.**
C–IBM special characters, $80 extra.
D–IBM special characters, $49 extra.
E–Model discontinued at the time the article was originally published in Consumer Reports. The information has been retained here, however, for its use as a guide to buying.

⊙ ◓ ○ ◒ ●

Better ←→ *Worse*

Buffer size, bytes	Draft	NLQ	IBM characters	Single-sheet width, In.	Continuous width, In.	Paper feed	Life, pages	Price	Cost/100 pages	Advantages	Disadvantages	Comments
	Download characters			Paper handling								
1920	—	—	✓	8½	9½	P	989	14	1.42	—	a,d,e	—
800	9	4	✓	8½	4–9¾	T	459	10	2.17	B,I	a,d,e	—
2700	—	—	✓	8⅞	9½	P	405	10	2.47	—	d	—
880	—	—	—	8¾	9½	P	465	10	2.15	—	a,d,e	—

an inch, so they don't accept a roll of narrow paper or mailing labels. You may be able to buy a tractor feed for a machine that lacks it.

Recommendations

Dot-matrix printers are versatile performers. They print acceptably fast and usually come with a lot of handy features.

One of the significant virtues of dot-matrix printers is that they can switch automatically from regular letters to, say, compressed letters or italics; daisy-wheel printers can't. All should be able to provide underlining, bold print, compressed print, and expanded print. Most also offer proportional spacing.

Most modern word-processing programs have the codes needed to access special print modes such as italics, superscripts, and subscripts in NLQ and draft modes. If you're using older software, be sure it will work with the printer you want.

■ Printing with ink jet ■

Ink-jet printers offer an alternative to the conventional dot-matrix or daisy-wheel machines. Ink-jet printers run fast, tend to be compact, and are relatively quiet.

Strictly speaking, an ink jet *is* a dot-matrix printer in that it prints characters on a page in the form of tiny dots clustered together. But the print head consists of a matrix of tiny holes through which ink squirts onto the paper. When the ink runs out, you simply snap out the entire print-head unit and snap in a new one. As there is no impact of metal against paper, the only sounds are the carriage returning and the paper advancing.

Ink-jet printers use special smooth paper, so that the liquefied ink doesn't blur. The special paper costs more than twice as much as plain paper. Further, the paper has a grayish tint that may not be to your liking. And the paper may not tear as neatly as ordinary fanfold paper.

Printing speed in NLQ and draft is comparable to that of the dot-matrix printers. Early ink-jet print heads tended to clog at unpredictable intervals. Evidently, manufacturers have solved that problem.

High-Fidelity Equipment

An inexpensive compact stereo sound system includes, in a single package, an AM/FM radio, a record player, a cassette deck, and speakers. It might sell at discount for about $200 to $400 (not including a compact disc player, one component you will probably want). But it's unlikely that a compact system will match the sound reproduction of a more expensive and top-quality component system. Nor will such a system capture the sound offered by compact disc technology.

For many people, however, a low-priced component system will fill the bill. Such a system has enough power to fill a room of moderate size with loud sound that is relatively free of distortion.

Systems at the higher end of the price range can fill a large room with loud music. They can reach deep into the bass, reproducing at reasonably loud levels the bass tones of, say, a pipe organ. They can deliver more power (perhaps more than you need) and offer more up-to-the-minute features, such as connections for two tape decks.

It's easy to spend more than $2,000 for a top-quality system—one that includes a pair of loudspeakers, an AM/FM stereo receiver, a compact disc player, a turntable, a phono cartridge, and a cassette tape deck. If you do, fine performance is virtually assured. So is a multitude of features, controls, and conveniences. You could also spend even more and improve performance or versatility by adding an FM roof antenna, an audio equalizer, stereo headphones, a pair of microphones, and, if you edit tape, an open-reel tape deck.

LOUDSPEAKERS

It won't matter how good the other components in your music system are if the loudspeakers don't reproduce music accurately, or if they can't reproduce the bass reasonably fully.

You don't have to spend a lot of money for high-fidelity loudspeakers. Decent sound is available for less than $300 or so a pair—sometimes for much less. You do, however, give up some niceties with modestly priced speakers. In a very large room, they generally won't deliver very loud sound without distortion or delve deeply into the bass.

Those shortcomings may not matter if you don't listen to music at window-rattling volume or in a huge room. Similarly, there's little point in buying costlier speakers that can let you hear the very deepest notes of an organ or an electronic synthesizer if you listen mainly to violins or vocals.

Some people, however, want more volume and more bass than an inexpensive speaker can provide. Medium-priced speakers with list prices of about $400 to $600 a pair should fill the bill. No one but a devout audio buff could ask for more fidelity, bass, or loudness than most of these speakers provide.

Accuracy

Accuracy is the touchstone of all quality loudspeakers, regardless of price. The sounds you hear should correspond as closely as possible to the message carried by the electrical signals that come from the receiver or amplifier. Accurate speakers respond uniformly and smoothly to every part of the music spectrum.

A loudspeaker with perfect accuracy would reproduce all the frequencies at the same relative volume at which they had been recorded or broadcast. It would neither emphasize nor deemphasize any part of the sound spectrum. It would respond smoothly and uniformly to every sound in the music spectrum, from the deep bass of the organ to the shimmer of the cymbals. That spectrum ranges from 30 Hertz (Hz) in the bass to about 14,000 Hz in the treble, the range over which most musical instruments put out their fundamental tones and combinations of overtones.

Consumers Union engineers used a computer to collect data on the acoustic power output of loudspeakers within that frequency range. They used the computer to modify the measurements to simulate the effect of a speaker's position in a typical listening room; the computer also converted the data into measurements of loudness as the ear hears it.

The engineers analyzed the data and scored the results as a percentage of "perfection." The scores, along with graphs, show where in the music spectrum each model deviated from a perfectly uniform response.

To check the test results, the engineers listened to a variety of music played through each pair of speakers in a room furnished to match the acoustics of a typical listening room. Listening-test results tallied nicely with the laboratory findings.

Tests have shown that even an experienced listener can't consistently tell which of two speakers is the more accurate when their scores differ by eight points or less. Differences in accuracy are insignificant among a number of models, giving you a wide choice.

Speaker location

Because a room's walls, floor, and ceiling reflect sound, bass response depends heavily on a speaker's location. Putting a speaker on the floor in a corner maximizes the bass; putting the speakers on a high table in the middle of the room provides the least boost for the bass. The position of listeners in the room also affects the way the bass sounds.

The speakers' location and the listener's position, as well as the size and shape of the room, its construction and furnishings, all affect the bass much more than they affect the mid or high range of sound.

Most manufacturers recommend putting lower-priced speakers off the floor and away from corners. Many manufacturers also recommend installing the speakers so that the tweeter (the small speaker element that reproduces high notes) will be at the listener's ear level. Following these recommendations usually gives good results. Because many low-priced speakers are not very large, installation at ear level is relatively simple to arrange if you have adjustable shelves.

When listening to a pair of conventional speakers, it is reasonably easy to pinpoint the position of specific instruments and voices. The image is usually stable—that is, instruments stay put. A trumpet on the left will not seem to move to the center or right.

In a typical room, speakers provide the best stereo effect when they're placed about ten feet from each other.

Other characteristics

Factors other than accuracy can affect your satisfaction with a set of loudspeakers.

Bass harmonic distortion. This kind of distortion can make music muddy or even break up bass sounds. The problem is often caused by the loudspeaker's woofer—the large element that reproduces low notes. A loudspeaker with very good bass reproduction has an advantage if you like to play organ or synthesizer music at high volume. Otherwise, it offers no special edge.

Power. The louder the sound a speaker must produce, the more power it needs. To produce a given level of loudness, an efficient speaker needs less power from an amplifier than a low-efficiency speaker. Low efficiency doesn't mean low quality.

If you usually play compact discs and "audiophile" recordings or listen at loud levels, your amplifier will need to deliver twice as much (or more) power than with ordinary discs.

Be especially wary of setting high volume levels on the receiver when playing compact discs. Since CDs have virtually no background noise, the tendency is to turn the volume up when you start to play the disc. A sudden blast of sound can damage a loudspeaker.

Impedance. Without getting into technicalities, impedance is a speaker's resistance, in ohms, to electric current from an amplifier. Impedance is unrelated to quality; low-impedance speakers simply draw more current from an amplifier or receiver than high-impedance speakers do.

Impedance is important in determining whether a given amplifier or receiver can drive a loudspeaker adequately. If the speaker's impedance is much lower than the amplifier's minimum impedance rating, the amp may not deliver enough power to drive the speakers without distortion.

Directivity. This characteristic refers to the way a speaker radiates sound in different directions at different frequencies. Directivity can affect a loudspeaker's interaction with a room and the listeners in it.

Stereo image. This is one way to describe the listener's sense of an instrument's position in the orchestra; it is also used to characterize the spaciousness of the musical sound. Differences in stereo imaging are usually subtle.

With a pair of conventional speakers, the stereo image is most balanced when you're as far from each speaker as the speakers are from each other. The closer you get to one speaker, the more the location of the sound moves to that speaker, until eventually the sound appears almost monophonic, although the other speaker still adds to the fullness of the sound.

Cabinet buzz. When you play organ or electronic music loudly, a speaker may rattle or buzz if its cabinet, back connector, or speaker moldings are improperly glued, or if the grille is loose. Try playing a CD that features sustained, deep bass notes. If you hear a buzz or rattle that goes away when you press your hand firmly on the speaker's cabinet panels, the cabinet is poorly assembled.

Quality control. Check the two speakers you buy to be sure that they look the same and that each unit has the same identifying model designation. Once you know you have a matched pair, hook them up and listen to them monophonically to be sure they sound the same.

Design

Most good loudspeakers have two or three elements: a woofer for the bass, a tweeter for the treble, and possibly a third element for the mid-range. However, the number of elements alone is not a reliable clue to performance.

Some models have controls to adjust the relative loudness of the tweeter, or the tweeter and mid-range components, with respect to the woofer. Those models usually attain peak accuracy with the controls at their maximum position.

A few models have tweeter-protection circuitry that cuts off the tweeter signal momentarily if volume is excessive. Others have a light that goes on at high volume levels and serves as a warning to turn the volume down.

Hooking up. For connecting a pair of speakers, ordinary lamp cord does nicely for runs of up to thirty feet. When you hook up a pair of speakers, a good way to get the two speakers in phase (the cones moving in and out together) is to use the raised bead (or different colored wires) often found on one lead of lamp cord to make sure that the red and black (or plus and minus) terminals on the speakers are connected to corresponding terminals on the amplifier.

You can also check phasing with a listening test. Place the speakers side by side and either play a record monophonically or tune to an FM station with your tuner switched to mono. Listen carefully, especially to the bass (a male voice is a good source). Now reverse the leads of one speaker, and listen again. If you hear significantly more bass, the speakers are now in phase. If you hear much less bass, the speakers were in phase originally.

Recommendations

Your major considerations are accuracy, size, and price. A pair of large speakers standing on the floor in your listening room will function as furniture as well as hi-fi components. With large speakers, especially, you may want to consider styling.

And then there's price. Depending on the competitive situation in your area, you might find discounts as low as 5 percent or as high as 50 percent. A good discount might well tip the scales. Then again, you might be asked to pay full list price at a component "salon." You may even be asked to pay *more* than list price for a pair of speakers. Don't do it.

Even if you can afford expensive speakers, don't buy them without a realistic appraisal of your listening needs. The extra money may buy nothing of value to you unless you are an audiophile who wants a superb deep bass response (in general, more money does tend to buy improved bass), or your listening room is so large as to rule out a low-priced model, or you like loud music for dancing.

Chances are you'll want to hear some speakers in a showroom before you buy. If so, here are some suggestions:

Look for a dealer with good listening facilities—a room that's shielded from outside distractions and equipment that permits switching rapidly from one speaker to another. This helps you to make valid comparisons.

Compare only two different pairs of speakers at a time. Then stay with the preferred speakers and compare that pair with another.

Try to listen to the models you're comparing in the same general location so that the room's effect on them is as similar as possible.

Set the volume control of the sound source so that the two pairs of speakers sound equally loud. (The ear tends to favor the sound of the louder speakers.) Set any tweeter controls on the speakers at maximum to start with, and adjust them lower if needed.

Once you select speakers, check to make sure that each speaker of the pair sounds the same. Audible differences could mean damage or defects.

Make sure the dealer will permit you to return or exchange the speakers if they don't sound right in your listening room at home.

Ratings of loudspeakers

Listed in order of accuracy score; models with identical scores are listed alphabetically. Differences in score of 8 points or less were not judged very significant. Power figures are for fairly loud sound in a 2500-cubic-foot room having average acoustics. Impedance rounded to nearest ohm, dimensions (H × W × D) to the higher ¼ in. Except as noted, all have removable cloth grille, imitation or natural wood-grain finish, and spring-loaded, push-on connectors for wiring to amplifier. Prices are as published in an October 1986 report.

ADVENT LEGACY. $399. Accuracy: 92. 28¼ × 16¼ × 9¾ in. 41 lb. Impedance: 6 ohms. Power required: 8 watts. Freedom from bass distortion: much better than average. Top and bottom are solid wood; rest has wood-trim finish.

ALLISON CD7. $500. Accuracy: 92. 27 × 9½ × 10¼ in. 24 lb. Impedance: 4 ohms. Power required: 13½ watts. Freedom from bass distortion: average. Has plastic-mesh grille.

INFINITY RS4000. $458. Accuracy: 92. 22½ × 12½ × 9½ in. 25 lb. Impedance: 4 ohms. Power required: 16 watts. Freedom from bass distortion: average. Has rotary level control.

AR AR38BXi. $470. Accuracy: 90. 21¾ × 13 × 7½ in. 29 lb. Impedance: 5 ohms. Power required: 13 watts. Freedom from bass distortion: much worse than average. Has "banana-plug" screw-type connectors.

CERWIN-VEGA D3. $530. Accuracy: 90. 27¾ × 14 × 11½ in. 36 lb. Impedance: 5 ohms. Power required: 2½ watts. Freedom from bass distortion: much better than average. Has self-resetting tweeter protection. Has rotary level control.

EPI T/E 320. $500. Accuracy: 89. 29 × 17 × 10½ in. 45 lb. Impedance: 4 ohms. Power required: 11 watts. Freedom from bass distortion: better than average. Score was 83 when speaker was placed according to manufacturer's directions. (According to the company, this model had been discontinued at the time the article was originally published in *Consumer Reports.* The information has been retained here, however, for its use as a guide to buying.)

ADS L570. $460. Accuracy: 88. 18¾ × 11½ × 11 in. 21 lb. Impedance: 4 ohms. Power required: 12½ watts. Freedom from bass distortion: better than average. Has metal-mesh grille and "banana-plug" screw-type connectors. Tweeter has protective fuse.

BOSTON ACOUSTICS A150 II. $500. Accuracy: 88. 32½ × 16 × 8¼ in. 42 lb. Impedance: 4 ohms. Power required: 11 watts. Freedom from bass distortion: much better than average.

JBL L60T. $590. Accuracy: 88. 30½ × 12 × 11¾ in. 37 lb. Impedance: 5 ohms. Power required: 11 watts. Freedom from bass distortion: better than average. Has "banana-plug" screw-type connectors.

POLK AUDIO MONITOR 7C. $500. Accuracy: 88. 24 × 14 × 9½ in. 30 lb. Impedance: 4 ohms. Power required: 9 watts. Freedom from bass distortion: much better than average. Has "banana-plug" screw-type connectors.

SCOTT 311DC. $500. Accuracy: 88. 23 × 13 × 9¼ in. 22 lb. Impedance: 6 ohms. Power required: 3½ watts. Freedom from bass distortion: much better than average. Has rotary level control. According to the company, this model had been discontinued at the time the article was originally published in *Consumer Reports.* The information has been retained here, however, for its use as a guide to buying.

GENESIS 33. $498. Accuracy: 87. 31 × 16 × 8¾ in. 33 lb. Impedance: 4 ohms. Power required: 16 watts. Freedom from bass distortion: better than average. Comes with stand assembly.

ELECTRO VOICE INTERFACE 2 SERIES II. $546. Accuracy: 86. 24¼ × 13¾ × 10¾ in. 30 lb. Impedance: 5 ohms. Power required: 6 watts. Freedom from bass distortion: much better than average. Has rotary level control and screw-type connectors.

KEF C40. $490. Accuracy: 86. 25½ × 9¾ × 10¼ in. 20 lb. Impedance: 8 ohms. Power required: 4¼ watts. Freedom from bass distortion: much worse than average. Has "banana-plug" screw-type connectors.

TECHNICS SBX700A. $550. Accuracy: 86. 27¼ × 15¼ × 13 in. 35 lb. Impedance: 5 ohms. Power required: 11 watts. Freedom from bass distortion: much better than average. Has self-resetting speaker-overload protector, 2 rotary level controls, and screw-type connectors.

KLIPSCH KG4. $600. Accuracy: 84. 27½ × 15¾ × 11 in. 39 lb. Impedance: 4 ohms. Power required: 5 watts. Freedom from bass distortion: average. Has "banana-plug" screw-type connectors.
CELESTION DL8. $550. Accuracy: 80. 19¾ × 10¾ × 10¾ in. 22 lb. Impedance: 7 ohms. Power required: 8 watts. Freedom from bass distortion: much worse than average. Has "banana-plug" screw-type connectors.

■ How to be sure of enough amplifier power ■

It's important that your loudspeakers are able to produce music loud enough for the room's acoustics and loud enough to please the ears of most listeners. The accompanying chart will help you determine how powerful an amplifier you need.

To use the chart, first determine the volume of your listening room by multiplying the length of the room by the width and the height (in feet). Consider adjoining rooms or hallways as part of your listening room if they are connected to it by large openings.

The amplifier power you need

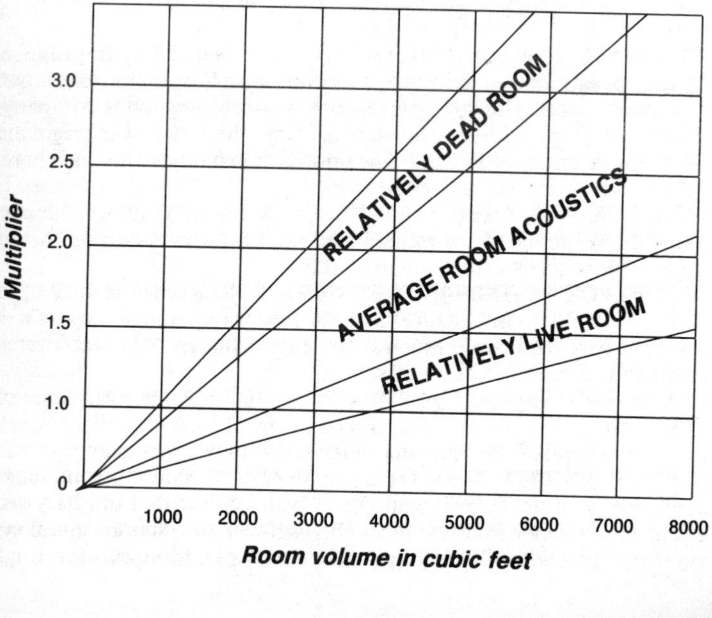

Next, figure out the room's approximate acoustic demands. A room with hard floors, several scatter rugs, plain wood furniture, and few or no draperies will be relatively "live." A room with thick wall-to-wall carpeting, heavy draperies, and upholstered furniture will be relatively "dead." (Consider any open doorway of ordinary size as an area equivalent to one covered by heavy draperies.) Acoustically, most rooms fall somewhere between these two extremes.

On the scale at the bottom of the chart, locate the room's volume. Move up that line to the region of the chart that matches your room's acoustics. Then, move across to the numbered scale on the left to get a "multiplier" figure. Next, take the power requirement of a speaker in the Ratings, and multiply it by the multiplier from the vertical scale on the chart. The result is the approximate power in watts the speaker will need to get from the amplifier.

Don't be afraid to use a more powerful amplifier than your calculations indicate. You may need more than twice the usual power to do full justice to compact discs or to audiophile records—digitally mastered and direct-to-disc, for example.

STEREO RECEIVERS

You don't have to spend a lot of money to get state-of-the-art design in a stereo receiver. Innovations in design and improvements in performance have found their way to the lower-priced models in a manufacturer's line. You can get up-to-the-minute features that were once found only on higher-priced equipment. The solid-state circuitry used in today's receivers virtually guarantees good performance from equipment at any price level.

More money does buy more versatility in the form of higher power, more accommodations for other components, and more controls. More power lets you play music at very high volume or through more than one set of speakers. You also need extra power in an unusually large room in order to keep music free of distortion when you turn the volume up.

If your system includes a compact disc player you may also want to upgrade the receiver. To bring out their best qualities, some compact discs seem to need two or three times the power that ordinary records require. More accommodations let you build an elaborate audio system that might include, for example, two tape decks. More controls might let you tune in weak radio stations swiftly and surely or listen to the radio while copying a tape. But even a "bare bones" receiver, list priced at $150 to $200, can serve as the cornerstone of a decent, although modest, hi-fi system.

Here are some of the characteristics to look for when shopping for a good stereo receiver.

Tuner

This circuitry lets you tune in stereo-FM radio stations as well as AM stations. The newest receivers will tune in stereo-AM as well, a broadcast mode that is becoming more and more available. (Check to be sure the receiver is able to reproduce the type of AM stereo signal being broadcast in your area.) An FM tuner should excel in sensitivity, selectivity, and frequency response. Important, too, are strong-signal capability, freedom from audio distortion, and freedom from tuning error. Other areas are less crucial to a tuner.

Sensitivity. A good tuner should be sensitive enough to bring in weak or distant stations without a rushing background noise. This is most critical with stereo reception, which needs a much stronger signal than monophonic listening. Of course, a good roof antenna (see section on antennas, page 200) lets you get the most out of a receiver's sensitivity, and it can improve reception with less sensitive models. If a station is weak, however, and stereo reception is still too noisy, you'll want to change from stereo to mono, using the receiver's built-in switch.

Selectivity. Good selectivity, which is as important as good sensitivity (and even more important in metropolitan areas), allows a tuner to separate a weak station you want to hear from strong stations close to it on the dial.

Frequency response. This is the ability to render the range of music frequencies faithfully and evenly. The frequency range of FM music is at least 50 Hertz (Hz) to 15,000 Hz. If a tuner renders all the frequencies of the music spectrum smoothly and uniformly, its frequency response is said to be "flat," since an engineer's graph of that response would be free from peaks and dips.

Strong-signal capability. A tuner should be sensitive yet able to handle strong signals. In large cities, it may be almost impossible to avoid distortion or interference from strong stations. Sometimes a nearby FM broadcasting station produces a signal stronger than a tuner can handle. This may increase noise or even cause the station to come in at several different places on the dial. If there are no FM transmitters near you, the ability of a tuner to resist strong FM signals from a nearby transmitter is not very important.

FM audio distortion. This can impart a rough or gritty quality to the sound. Tuners tend to vary in their ability to deliver a loud level free of FM distortion. None of them is apt to be poor, however. Indeed, some low-priced receivers are very good at keeping FM audio distortion low.

Capture ratio. The Federal Communications Commission tries to make sure that two stations assigned the same frequency are far enough apart geographically not to intrude on each other's listening area. Sometimes, though, reception conditions in a particular location can result in interference caused by a weak signal from a distant station broadcasting on the same frequency as the station you want.

Capture ratio is a measure of the tuner's ability to capture the desired station and reject the weaker one. Good capture ratio also helps a tuner to resist multipath distortion, which can occur when the signal you tune in bounces off a building or mountain, say, and reaches the receiver slightly later than the signal following a direct path.

AM rejection. Good AM rejection helps an FM receiver ignore all sorts of electrical interference—lightning or sparking from small-appliance motors or car ignitions, for example.

Airport radio rejection. If you live under a flight path or near an airport, you may pick up communication signals from powerful flight radio transmissions (when you don't want to hear them) through an FM tuner that doesn't properly reject them. This is often termed "image interference." The higher-frequency FM transmissions between the aircraft and the control tower can, if the tuner doesn't reject them, produce images of transmissions in the FM broadcast band.

Stereo separation. Sound shouldn't leak from one stereo channel to the other. Such leakage would create an undesirable "narrowing" of the stereo effect. Even low-priced receivers have excellent FM stereo separation.

Phono preamp

The phono preamp amplifies the tiny signal produced by a phono cartridge. Low-noise circuitry is used, with the specific tonal correction that is required for records.

Consumers Union's tests of phono preamps on medium-priced and low-priced receivers have shown them to be excellent or very good.

Frequency response. When a record is cut, the RIAA (Recording Industry Association of America) curve is incorporated—a special frequency response correction. The phono preamp restores the original flat response.

Signal-to-noise ratio. The electronics of a stereo system should not add annoying background noise to record playing. This factor becomes especially important if you use a so-called high-output moving-coil cartridge, which actually has a relatively low output.

Harmonic distortion. This can make music sound harsh, but it's not a problem even with lower-priced models, which deliver a clean undistorted signal at loud levels.

Phono compatibility. The sensitivity, input impedance, and overload point of any receiver's phono preamp are almost certain to be fully compatible with any phono cartridge designed for use with a magnetic phono input.

Phono input capacitance. Some cartridge models suggest a specific input capacitance, an electrical characteristic measured in picofarads (pf). The measurement is unimportant except for those who want to match the cartridge manufacturer's "recommended load capacitance." The benefits of such matching are meager at best, but if you are concerned about this, you can add a capacitor across the phono input, using a "Y" adapter. You can find the value of capacitor to use by subtracting from the cartridge's recommended load capacitance both the receiver's input capacitance and the turntable's cable capacitance (the latter is usually around 150 picofarads). If the result is 50 picofarads or less, don't even bother.

Amplifier

The audio source, be it the FM tuner, phono preamp, or an external tape deck or TV set, must have its available signal power increased to a level that will cause the cone of a loudspeaker to vibrate adequately, producing sound. The amplifier accomplishes this, and also allows you to alter the volume, tone, and left/right balance of the sound. Included in the amplifier are the necessary switches to select the program source, allow tape-recording, and turn loudspeakers on or off.

Frequency response. As with the FM tuner and phono preamp, this characteristic is unlikely to pose problems with the amplifier, even with most low-priced models.

Harmonic distortion. You can induce distortion in any amplifier by turning it up above its maximum power. A drastic increase in distortion gives a hard, gritty quality to the sound and, in time, might damage the speakers. An amplifier with adequate power, however, is unlikely to harm the speakers in the normal course of playing music and should not produce distortion that is audible.

Signal-to-noise ratio. This is the proportion of desirable sound to undesirable noise. Consumers Union's tests showed low-cost receivers to be excellent in this respect.

Loudness compensation. On most amplifiers, there's almost always a switch to provide some kind of bass boost, which works in conjunction with the volume control. The bass boost can help music sound more natural when it's played at a lower loudness level than that of the original performance. Unfortunately, there may also be an unneeded boost to the treble. There should always be a way to switch loudness compensation off. Otherwise, it's not possible to achieve real high fidelity when you want it.

Power. In assembling a hi-fi system, you need to pay heed to power output—the number of watts the amplifier circuitry in the receiver can deliver to power the speakers. The higher the output capacity, the louder you can play music without audible distortion.

A knowledge of power output can also help you match receiver to speakers. The impedance of a loudspeaker is a complex electrical characteristic measured in ohms, and can be roughly defined as the speaker's resistance to the driving current coming from the amp. Consumers Union engineers measure power output at impedances of 4 and 8 ohms, and often at 6 ohms as well. These impedances are typical of today's speakers.

On average, a low-priced receiver has about half the output power of the medium-priced models. This doesn't mean you can play a low-priced unit only about half as loud. The audible difference is about the same as turning down the volume a little. But you won't be able to fill an oversized room with loud, undistorted music—whether from audiophile recordings, compact discs, or even some regular discs. Nor will you be able to do full justice to recordings with unusually strong bass played at window-rattling levels. Nevertheless, a low-priced model should deliver adequate amounts of power for more modest musical demands.

If you try to push sound levels too high, you will overdrive the amplifier and produce a kind of distortion called "clipping," which gives music a harsh, gritty sound. Clipping, which can occur well below the maximum volume setting, may damage the speakers. In fact, any receiver that is rated to deliver about 35 watts or less may produce some degree of clipping under the severe conditions we've described. If you have reason to think your hi-fi system's power needs exceed a given receiver's output, play it safe by opting for extra power.

Amplifiers are often designed to provide extra power for driving low-impedance speakers. And some speakers rated at 8 ohms may actually present only 4 ohms for some portions of the music spectrum, since impedance and frequency are interrelated. So it's desirable for an amplifier to have at least as much power available at 4 ohms as at 8 ohms. You should be particularly concerned about sufficient amplifier power if you have (or intend to have) more than one pair of speakers driven by a single receiver.

If you want two pairs of speakers connected to your hi-fi system, the receiver's internal connection between the two speakers on each stereo channel is important. The connections are in parallel or in series, depending on the receiver. The parallel connection is preferable because, in a series connection, the signal reaching one speaker is degraded by the impedance of the other speaker in that channel when both pairs of speakers are on. (In a parallel connection, however, don't use two pairs of speakers that are *both* rated much below 8 ohms; the combination may overstress the amplifier.)

Some models will blow an internal fuse if you accidentally short-circuit the speaker wires. You'd have to take such a model to the shop for repair. It's better to have a receiver that copes with a speaker-wire short in ways that don't require service replacement parts.

Features

Since today's receivers generally perform so well, your choice of one model over another may well depend on features that suit your special needs. Here are the major features you may wish to consider.

Station selector. Although some receivers still have a conventional tuning knob and a scale for the stations on the radio band, most now come equipped with quartz tuning: Tuning errors are eliminated through the use of a microprocessor. By pushing a button, you make the tuner step up or down the radio frequencies produced by a quartz-crystal oscillator inside the tuner. Mistuning is virtually impossible. The frequencies, corresponding to the channels of the FM band, usually show as numbers on a digital display.

Thanks to the microprocessor, quartz-tuned models can be set to remember the frequency of from ten to sixteen favorite stations, bringing in the one you want at the touch of a button. With most models, a given button can preset an FM *and* an AM station.

On a quartz-tuned receiver, there's no fiddling with a tuning knob. Rather, you scan the FM band with the aid of push buttons.

Auto-scan tuning. This feature finds the next listenable station up or down the FM or AM band at the press of a button. (The scan on a few models works only on the FM band.) You can hold the scan button down until the frequency shown on a digital display is close to a desired frequency; then you release the button to catch the station. If you miss, jumping up or down a station or two is easy.

Because auto-scan is unlikely to catch a weak station, quartz-tuned receivers let you tune by hand. Each time you press a button, the receiver steps the tuning a discrete interval up or down the band. It's best if the receiver steps exactly one FM "channel" (200 kilohertz) with each press of the button. Some models step only one-half or one-quarter channel to suit the FM channel-spacing in other countries; with them, manual scanning can be tedious.

Some receivers without quartz-synthesis have a station-lock feature—a light to show that the receiver is fine-tuning a station you have come close to. These units have a switch to turn the feature off so strong signals from a nearby station can't pull the receiver away from a weak-signal station you might want to hear. With some models, the station-lock feature automatically cuts out when you touch the tuning knob, which is a plus when tuning a weak-signal station.

Signal-strength indicator. To show the strength of the signal coming in, most models have a string of little lights that go on or off as strength increases or decreases. Although the lights may help with the FM tuning, they won't help you to position an antenna for optimum reception. The indication they give isn't precise enough.

Reception aids. Given the sometimes-marginal reception conditions caused by weak signals, multipath noise, and interference, some manufacturers are providing the means to clean up received stations at the expense of reducing fidelity a bit. Alternatively, you can use the FM mono switch to reduce noise on a stereo station—at the sacrifice of stereophonic sound, of course. The same switch reduces the noise and distortion when you play old records. A *high-blend mode* is a kind of semi-mono switch; it reduces, but doesn't eliminate, stereo separation in the treble frequencies and thus also reduces noise. This feature can make all the difference when you listen to weak stereo stations.

Tape-deck accommodation. Look for at least one set of tape-recording jacks so you can include a tape deck in your sound system. On many models, you can hook up two tape decks, which then allows you to copy a tape. An unused tape facility can be used to connect an equalizer or a noise-reduction unit, or as an extra input for the audio from a TV or videocassette recorder.

Tone controls. In addition to the usual bass and treble controls, there are several other means of adjusting tonality to suit your taste or the program material. There may be a *mid-range tone control,* which allows you to alter the sound in the critical "presence" range of music and voice fundamentals. More versatile is a *five-band tone control,* which lets you adjust the tonality in five areas of the spectrum. To quiet a hissy tape or crackly record, a *high-frequency filter* may be used, though it will reduce any high-frequency sounds in the recording. If warped records cause your speaker cones to flutter in and out (reducing reserve amplifier power in the process), a *subsonic filter* will allay the problem with little

Ratings of low-priced stereo receivers

Listed in order of estimated quality. Prices are as published in a **September 1985** report.

◉ ◓ ○ ◒ ●

Better ⟵⟶ *Worse*

			FM							AM	
Brand and model	List price	Convenience and features	Sensitivity	Selectivity	Strong-signal capability	Capture ratio	AM rejection	Frequency response	Freedom from distortion	Sensitivity	Frequency response
Ultrx R35	$190	◉	◓	◉	○	◉	◉	◉	◉	◓	○
Realistic STA115	199	◓	◓	◓	○	◉	◓	◉	◉	◉	◓
Sony STRAV260	180	◓	◓	◓	○	◉	◓	◉	◉	◉	◒
Pioneer SXV200	210	◓	◓	◓	○	◉	◉	◉	◉	◓	◒
Parasound DR25	200	◓	◓	◉	○	◉	◉	◓	○	◉	◓
Sherwood S2620	180	◓	◓	◓	○	◉	◓	○	◉	○	◒
Onkyo TX17	200	◓	◓	◉	○	◉	◓	◓	◉	○	○

Marantz SR240	140	○	◓	◉	○	◉	◉	◓	◉	○	◒
Rotel RX830	199	◓	○	◉	○	◉	◉	◓	◒	◉	◓
Nikko NR320	199	◓	◓	◓	○	◉	◉	◓	◉	◓	○
Ultrx R25	140	○	○	◉	○	◉	◓	◉	◉	○	○
JVC RX110	175	◓	○	◓	○	◉	◓	◉	◉	◓	○
Sherwood S2610	150	◓	◓	◓	○	◉	◉	○	◉	◉	◒
Technics SA150	170	◓	○	◓	○	◉	◓	◓	◓	◓	●
Akai AAA1	170	○	○	◓	○	◉	◉	◉	◉	○	◓
Hitachi HTA25F	200	◓	◓	◓	○	◉	◓	◉	◓	○	◒
Technics SA120	150	○	○	◓	○	◉	◉	◓	○	◉	◒
Pioneer SX212	150	◒	○	◓	○	◉	◉	◓	◉	◉	○
Hitachi HTA2	150	○	○	◓	○	◉	◉	◓	◓	◓	◒
Vector Research VR2200	150	◓	○	◓	○	◓	◉	◓	◒	◉	○

Performance Notes

All were judged: • Excellent in FM stereo separation, amplifier signal-to-noise ratio, phono frequency response, phono freedom from distortion, and phono signal-to-noise ratio. • Excellent or very good in amplifier freedom from distortion.
Except as noted, all were judged: Excellent in amplifier frequency response

Specifications and Features

All have: • FM and AM tuners. • 1 phono input, volume and balance controls, and bass and treble tone controls. • Headphone jack. • Connection for 75-ohm FM antenna. • AM loop antenna.

or no audible effect. A *bass boost* switch can assist with small speakers or a recording that's shy in bass.

FM muting. To subdue the rushing noise that may be audible between FM stations, almost all models have a muting feature. To hear a station whose signal is very weak, you switch the feature off. Some receivers combine the muting feature with the stereo/mono switch, which eliminates the option of listening to weak-signal (and thus noisy) stations in stereo.

Filter. Many models have some kind of switchable filter. A subsonic filter is useful because it can remove inaudible low-frequency noise, which may come from a warped record, without affecting the audio range. This helps the amplifier to maintain full reserve power and makes distortion less likely when tape-recording.

Filters to reduce the treble may improve some scratchy records and hissy tapes. On some receivers, however, turning down the treble tone control produces nearly the same benefit. Filters that reduce audible bass frequencies are less useful.

AC outlets. By plugging an audio component into an AC outlet on the receiver, you can power the component through the receiver, rather than from a wall outlet. Almost all receivers have this useful feature.

Antenna connections. A simple T-shaped wire (called a "dipole") comes with nearly every receiver for use as an antenna. Most models also have a connection for 75-ohm coaxial cable, thus allowing you to hook up a cable-equipped roof antenna or a cable-TV system with FM capabilities.

Source lights. Some models have tiny lights to show the source of the sound you hear—FM, AM, phono, tape, etc. The lights are especially welcome in a receiver with push-button switches. On some of these models, you can't easily tell if a switch is on or off unless you look at the light.

Power-output indicators. A few models have a series of lights that follow changes in volume. (You can follow those changes with your ears, too, of course.)

RECOMMENDATIONS

Begin your shopping for a receiver by checking lower-priced models (providing they meet the power requirements of your sound system).

There are, of course, listening situations that might require more power than is usually available in a low-priced receiver. People who like to turn up the volume on music with unusually strong, deep bass may need extra power to keep the bass free of distortion. You may require extra power because you want to fill an unusually large room with undistorted sound, or are planning to use more than one pair of loudspeakers, or if you want to make sure music played in the living room can be heard in the dining room. Compact discs or other audiophile records played loudly may demand extra power. So may low-efficiency loudspeakers.

Choosing a receiver, then, is basically a matter of finding a model that has sufficient power, easy-to-use controls and convenience features, and an attractive price.

Be sure to shop around before you buy: In some areas, discounts run quite high.

COMPACT DISC PLAYERS

Compact discs deliver flawless, noise-free sound with a near-concert-hall dynamic range, superb frequency response, negligible distortion, and freedom from such impediments to listening pleasure as flutter and drift, as well as absence of the pops and clicks that are so commonplace with long-playing vinyl records.

Though a CD measures less than 5 inches in diameter, it's capable of providing about 75 minutes of playing time. (Its grooves, or tracks, are so fine that more than 300 of them could fit side by side in a single groove of a standard vinyl record.) Millions of tiny pits in the disc hold musical "information" encoded digitally. To convert those signals into electrical energy and thence into music, a CD player "reads" the pits, using a low-powered beam of laser light. Since the light alone tracks the grooves, there's no wear, as with a conventional phono cartridge and stylus. If handled with reasonable care, CDs can last virtually forever.

CDs have staked out a significant proportion of the audio market and pushed vinyl records on a slow but inevitable path to obsolescence. RCA Records has shut down its last vinyl-record pressing plant, although the company promised to continue marketing records "as long as the consumer requests it," even though RCA won't do the actual pressing.

While there's little doubt that CDs are replacing 33⅓ rpm LP records in the same way that LPs replaced 78s, one may wonder how long discs will remain the medium of choice for music in the home. There is a still newer medium, digital audio tape (DAT). Digitally recorded tape, packaged in a new small cassette, is supposed to match the playback performance of a CD—and can be recorded on as well. Although it's likely to be years before DAT players and their tapes become affordable, the CD industry is worrying that consumers will tape borrowed CDs rather than buy them.

Sound quality

CD players respond uniformly to frequencies from the lowest bass to the highest treble. They have virtually no audible "noise"—just the sounds actually recorded.

Overall dynamic range, the difference between inherent background noise and the loudest peaks recorded, is more than 90 decibels—wider than with most any high-fidelity gear for home use. CDs don't have significant distortion or measurable flutter, a variation in pitch that can give music a watery sound.

Under optimum conditions, any CD player sounds just fine. But there may be performance differences when a CD player is used under less-than-ideal conditions.

Bump immunity. A CD player should be resistant to shock and vibration, or the player's laser beam may mistrack and produce an audible skip in the music. The machine should stand up much better to controlled jolting than a well-designed conventional turntable.

Disc-defect immunity. CDs routinely contain tiny defects that may generate the wrong digits in playback; such defects can cause a player to mistrack or make a clicking noise. The error-correction circuitry of the players usually compensates for those shortcomings.

Disc-damage immunity. Though inherently resistant to minor scratches, CDs become unplayable if exposed to severe scratching. Most players cope with across-the-groove scratches as much as $\frac{6}{100}$ of an inch wide; others get stuck on narrower scratches. Scratching that follows the direction of the grooves is harder for a player to deal with.

Data Mud is a polish intended to rejuvenate scratched CDs. Nothing works on severe scratches. But both the *Mud* and *Rally* cream auto wax dramatically reduce the number of errors caused by light scratches.

If you do manage to scratch a CD, *Rally* wax is cheap and works well.

Fingerprints. A few light ones don't matter, but a nasty smudge can cause a blip in the music. Some players will blip at a smudge. (This isn't really a problem, since even a bad smudge can be wiped clean.)

Warped discs. It's not easy to warp a CD, because its polycarbonate plastic doesn't soften below 220° F.

Track-locate speed. The faster a CD can jump to the beginning of a desired track, the less "dead air" at the start of play and between selections.

Features

Even an inexpensive model has all the features most buyers would want. However, some people might want a more full-featured model.

Time-remaining display. This shows, in minutes and seconds, how much music remains to play. Another display gives elapsed playing-time.

Programming ability. This determines the number of tracks the model can be programmed to play. You program either by using Up/Down buttons or by punching in the track numbers on a keypad. You can program the tracks in any order.

Direct entry. Some CDs let you directly enter the track numbers from 1 to 99 with a keypad, punching in the numbers of desired tracks so that you can go directly to a selection you want to hear. On a few models, the process is more complicated.

Repeat a track. With this convenience the player can be programmed to play a given track over and over.

Repeat a segment. You can program some players to repeat a musical passage of any duration within any track. Others can be set to repeat a single track or track segment continuously.

Headphone jack. This allows you to connect stereo headphones to the player so you can listen without disturbing others. A headphone volume control permits adjusting volume at the headphones.

Index selection. Pressing a button during play takes the player from one section of a track to the next.

Clear program. Pressing a single button during play cancels the program, restoring the disc's original playing sequence.

On/Off pause. This button stops the music instantly so you can answer the phone or unlock the door—and then lets you start it again with a second push of the button. This is easier to use than the Pause-then-Play button arrangement on some models.

Audible scan. A player should have audible scan, a feature that lets you find a particular musical passage by "stepping" forward or back through a track without changing the music's pitch. The sound is choppy, but it's not the chipmunk chatter of an audio tape in fast-forward.

Some models have multiple-speed scan for speedy homing in on distant musical passages; they start slowly and speed up when you keep your finger on Scan.

Remote control. Most higher-priced models come with a full-function remote control; from almost anywhere in the room, you can make the player pause, stop, skip, select a track, and scan. Most remote controls let you program track sequence, too. Of those players with a remote, almost all let you control all the player's operations except On/Off from your easy chair.

Digital filter. Many models have a digital filter, which is supposed to be better than an analog filter at eliminating distortion resulting from

Ratings of CD players

Listed by types; within types, listed in order of overall performance score. Prices are as published in a **May 1987** report.

Brand and model	Price (list/paid)	Bump immunity	Disc-defect immunity	Disc-damage immunity	Fingerprint immunity	Track-locate speed	Time-remaining display	Programming ability	Direct entry, 1 to 99
Regular table models									
Sharp DX-611	$260	◓	●	●	●	◒	—	9	—
Sony CDP-203	500	◓	●	●	●	●	✓	20	—
Sanyo CP700	180	●	●	●	●	○	—	16	—
Pioneer PD-7030	470	●	●	●	○	○	✓	24	✓
Denon DCD-1300	550	○	●	●	◓	○	✓	20	—
Pioneer PD-5030	300	◓	●	●	○	○	—	24	—
GE 11-4800	199	◓	●	●	○	○	—	15	—
Magnavox CDB460	250	◓	●	●	○	◓	—	20	—
JVC XL-V440	360	●	●	●	○	◓	—	15	✓
Panasonic SL-P3620	300	●	●	●	○	●	✓	20	—
JVC XL-V220	275	○	●	●	○	◓	—	15	—
Magnavox CDB650	430	○	●	●	○	○	✓	20	✓
Yamaha CD-700	599	◓	●	●	○	◒	✓	12	✓
Akai CD-A30B	200	●	●	●	○	●	—	36	—
Realistic CD-2200	200	◓	○	●	○	◓	—	15	—
Yamaha CD-400	319	○	●	●	○	◒	—	9	—
DBX DX-3	499	◓	●	◓	○	○	—	9	—
Sony CDP-25	260	○	●	●	○	◓	✓	16	—
Technics SL-P310	350	○	●	●	○	●	✓	20	✓
ADC 16/1	200	○	●	●	○	○	✓	16	—
Emerson CD160	300	●	●	●	○	◓	—	15	—
Akai CD-A70B	349	○	●	●	○	◓	—	28	✓
Technics SL-P110	320	◓	●	●	○	●	✓	20	—
Teac PD-100	299	◓	●	○	○	●	—	29	—
Hitachi DA6000	400	●	●	◒	○	◓	—	15	—
Teac PD-400	349	◓	●	○	◓	●	—	29	✓

⦿ ◒ ○ ◓ ●
Better ⟷ Worse

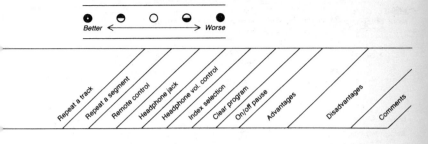

Repeat a track	Repeat a segment	Remote control	Headphone jack	Headphone vol. control	Index selection	Clear program	On/off pause	Advantages	Disadvantages	Comments
—	—	—	—	—	—	—	—	A,G	b,c,g,h,j	*C,J*
✓	✓	✓	✓	✓	✓	—	✓	A,F,I,K,O	j	*C,G*
—	—	—	—	—	—	—	✓	G,H	j	*C*
—	✓	✓	✓	✓	✓	—	✓	A,C,G,H,K,Q	g	*C,G,J*
—	✓	✓	✓	✓	✓	✓	—	E,H,K	g	*C*
—	—	—	✓	—	—	—	✓	A,C,G	g	*C,G,J*
—	✓	—	—	—	—	—	✓	A,C,G,H	d	*K*
—	—	—	—	—	✓	—	✓	G	d,i	*A*
—	—	✓	✓	✓	—	—	—	C,G,K	g,i,j	*B,CJ*
—	✓	—	—	—	✓	—	—	A,C,G,K,O	g	*A,C,J*
—	—	—	✓	✓	—	—	✓	C,G,K	g,i	*B,C,J*
✓	✓	✓	✓	✓	✓	—	✓	B,C,G,K,O,Q,R	i	*A,D,E*
—	✓	✓	✓	✓	✓	—	—	C,G,H	d,f,g,i,j	*C,J*
—	✓	—	—	—	—	✓	✓	A,B,C	k	*C,J*
—	✓	—	—	—	—	—	✓	A,C,G,H	d	*C,K*
—	—	—	✓	—	—	—	—	C,G,H	d,g,i,j	*C,J*
—	—	—	—	—	—	—	—	G,H,J	f,g,i,j	*C*
✓	—	—	✓	—	—	—	✓	A,H	d,j	*C,J*
—	—	✓	✓	—	—	—	✓	A,B,G,M	c,g,j	*C*
✓	—	—	—	—	—	✓	✓	—	g	*C*
—	✓	—	—	—	—	—	✓	A,C,G,H	—	*C*
—	✓	✓	✓	✓	✓	✓	✓	A,C,P	h,l	*C*
—	—	—	—	—	—	—	✓	A,B,D,G	g,j	*C*
✓	—	—	—	—	—	—	✓	G	d,g,i	*C*
—	✓	—	—	—	—	—	✓	A,C,G,H	—	*C*
✓	—	✓	—	—	—	—	✓	C,G	g,i,m	*C*

Ratings of CD players (continued)

Listed by types; within types, listed in order of overall performance score. Prices are as published in a **May 1987** report.

Brand and model	Price (list/paid)	Bump immunity	Disc-defect immunity	Disc-damage immunity	Fingerprint immunity	Track-locate speed	Time-remaining display	Programming ability	Direct entry, 1 to 99
Changer models									
Pioneer PD-M6	800	◓	◉	◉	○	◒	—	32	✓
Sony CDP-C10	800	◉	◉	◉	○	●	—	20	—
Portable models									
Sony D-7S	350	◉	◉	◉	◓	◒	✓	16	—
Technics SL-XP5	375	◉	◉	◉	○	○	✓	18	—
Toshiba XR-P9	325	◉	◉	◉	○	○	✓	16	✓

Specifications and features

All were judged excellent in: • Frequency response. • Dynamic range. • Stereo separation. • Freedom from distortion and flutter.

All have: • Play, Pause, Stop, Track-select, Open/Close, and Program functions. • Track display. • Display of elapsed time. • Ability to repeat entire disc. • Audible scan.

Key to Advantages

A–Needs no screws to hold laser assembly during transportation, a convenience.
B–Main controls especially convenient.
C–Display more readable than most from above.
D–Nudge on disc drawer closes it and starts play.
E–Has direct entry of tracks 1 through 10 (but to get higher tracks, you must key in a combination—hit 10 three times to get track 30, say). Also has calendar display of programmed tracks.
F–Has direct entry of tracks 1 through 20 (but same limitation as in Advantage E). Also has calendar display of programmed tracks.
G–Has multiple-speed scan.
H–Compatible with external timer that can switch model on (this feature can be disabled with the **Denon** and the **Sony CDP-C10**).
I–Can be programmed to play tracks in random order.
J–Has built-in compressor to limit dynamic range of discs, raising volume of soft passages and reducing volume of loud ones.
K–Displays total program time, handy when recording onto a cassette tape.
L–Headphones included.
M–Remote control can change volume on headphones and line-output jacks.
N–Has rechargeable battery.
O–Autospace feature inserts several seconds of silence between tracks during play; handy when recording with cassette deck that senses blanks between selections.
P–Has Without button for easy programming if you want to skip a track.
Q–Autopause feature pauses at end of every track; useful, perhaps, to make room for commercials in professional use.
R–Autoscan feature samples a few seconds of each track, then moves to next.

● ◓ ○ ◒ ●
Better ← → Worse

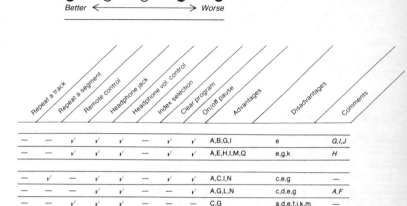

Repeat a track	Repeat a segment	Remote control	Headphone jack	Headphone vol. control	Index selection	Clear program	On/off pause	Advantages	Disadvantages	Comments
—	—	✓	✓	✓	—	✓	✓	A,B,G,I	e	*G,I,J*
—	—	✓	✓	✓	—	✓	✓	A,E,H,I,M,Q	e,g,k	*H*
—	✓	—	✓	✓	—	✓	✓	A,C,I,N	c,e,g	—
—	—	—	✓	✓	—	—	✓	A,G,L,N	c,d,e,g	*A,F*
—	—	✓	✓	✓	—	—	—	C,G	a,d,e,f,i,k,m	—

Key to Disadvantages

a–You must press Scan and Play for audible scan—an impossibility when using remote control.
b–Won't do audible scan unless you press Pause; won't stop scanning till you push another button.
c–Single button gets you Scan (within a track) and Search (track to track), an inconvenience, that may get you one when you want the other.
d–Main controls inconvenient.
e–To start play, you must press Close (for drawer), then Play.
f–Pressing Pause twice by mistake resets disc to beginning of play.
g–On Search, sometimes clipped first note of a track.
h–Cannot scan when programmed.
i–Needs screws to hold laser assembly during transportation, but has no place to store screws, an inconvenience.
j–Drawer sometimes jammed disc in player (but did not damage disc).
k–Has separate Play button to use when programmed, an annoyance.
l–Allows direct entry of track numbers from remote control only.
m–Has slightly poorer dynamic range, judged not very significant.

Keys to Comments

A–Has a jack for remote control when used as part of the same-brand stereo system.
B–Has synchronizing jack for cassette deck.
C–Nudge on disc drawer closes it, but does not start play.
D–Has extra output jacks with additional filtering.
E–Has memory for up to 254 tracks, so can remember order you programmed when you last used specific discs.
F–Has switchable filter to soften harshness from headphones.
G–Has ac outlet.
H–Holds 10 discs.
I–Holds 6 discs.
J–Model discontinued at the time the article was originally published in *Consumer Reports*. The information has been retained here, however, for its use as a guide to buying.
K–According to GE, model was subsequently sold as part of rack system, but at the time of original publication in *Consumer Reports* it may still have been available as a single unit.

conversion of the digital signal into an audio signal. But both types of filter work well.

Indexing. Some models have an indexing feature, which lets you jump to a chosen "index point" within a single track. That, the dealer might say, will let you find a specific passage in a long orchestral movement. Trouble is, the feature is useful only on CDs that incorporate indexing.

Timer. An external timer can turn some models on and start them playing a disc you've previously loaded so that you can have a favorite disc playing as you come in the door.

Day-to-day use

Some players are more convenient than others—but the differences are small. Consider how you start play, for instance. With about half the players you press Open, plop the disc in the drawer that appears, press Close, wait for the drawer to retract, and press Play. The other half take a shortcut: You open the drawer, insert the disc, and press Play; that closes the drawer and begins the music. Auto mode lets you pop in the disc and nudge the drawer shut to start play.

The display on most of the players is readable from above. But some models must be at eye level or close to it to be easily readable.

Configurations

Most CD players are full-sized table models. But the market has been changing. Portable models represent a significant percentage of sales. And changer models are poised to take over much of the market.

Changers hold a number of discs in a magazine (some have a carousel instead). You can program tracks on different discs, and the display shows the disc and track selected.

If you think of changers as being about the size of a bread box (or almost), portables are the size of two slices of bread. That small size is why some people want a portable as much for fitting into a home music system as for portability. Portables can operate on either AC current or battery power (many of them come with rechargeable batteries). But they must be stacked atop other components: Unlike table models, they have controls on top, and a hinged, pop-up cover like that of some conventional turntables.

You pay for miniaturization: Portables cost about as much as low-priced table models and have fewer features than most.

Recommendations

There's no need to spend a lot of money on a CD player. All of them sound superb. Nearly all have the necessary features and even some frills. Spend more money only if you want special features for special needs.

TURNTABLES

There are two main types of turntable: single play and multiplay. The multiplay models, also called "record changers," can play a stack of records automatically. As separately purchased components, however, single-play turntables are far more popular than multiplay models. Most of the best-selling single-play models are automatic to some degree—that is, they will return the arm to its post and stop the platter at the end of play. If the model is fully automatic, it will also start the platter and gently place the stylus on the record, at the touch of a button.

Some turntables are belt-driven. Others have a direct drive. Neither type has any inherent performance advantage. There are other factors far more important in determining the quality of a turntable.

Performance

When Consumers Union electronics engineers test turntables, they measure each model's performance electronically and mechanically.

Flutter. This makes music sound sour or wavery in pitch and can be caused by several different mechanical deficiencies. In recent years, turntables have been very low in flutter.

Rumble. This is a low-pitched noise that comes from vibrations in a turntable's platter-drive mechanism. Rumble is especially noticeable during quiet musical passages played at high volume. You can expect freedom from rumble to be generally superb in mid- and higher-priced turntables.

Tone-arm mass. This is related to the weight of the tone arm, but it's essentially a measure of inertia—the resistance to a change in motion. The higher the mass of the tone arm, the more it will resist.

Inertia can be a problem when the stylus is tracking a warped record. When the stylus reaches the top of a warp, it should immediately change direction and follow the downward path. But with too high a mass, the tone arm may carry the stylus off the record for an instant, which can distort the sound.

The problem can be avoided by using a low-mass tone arm. Straight tone arms are likely to have lower mass than S-shaped ones. And a small headshell, which holds the cartridge, reduces the tone-arm mass further. (A removable headshell makes it easy to install the cartridge you prefer.)

Resistance to vibration. Turntables are usually mounted on some sort of spring system to absorb vibrations from footsteps and other shocks that might make a tone arm skip the grooves. Most models are good at resisting such shocks.

Acoustic feedback. This is an annoying rumble that can occur when the cartridge picks up sound vibrations from a loudspeaker.

Some models are better than others in resisting acoustic feedback. But in any model, you can reduce the problem or avoid it entirely by placing your turntable on a sturdy support, not too close to the loudspeaker.

Speed accuracy. The speed of a turntable's rotation must be very close to the speed at which the record was cut. If it isn't, the pitch and tempo of the record's music won't match the original. Speed accuracy should not be a problem, however.

VTF setting. A good turntable usually has a way to set vertical tracking force (VTF)—the force exerted on the record grooves by the stylus. An incorrect VTF setting could cause distortion and record wear.

On models that use a P-mount cartridge (see section on phono cartridges, page 176), there's no need for tricky adjustments: VTF has been preset to allow good tracking while keeping record wear low.

Tracking error. Many tone arms pivot as the stylus moves across the record. Consequently, the path of the stylus is an arc. However, the original master record is cut on a turntable equipped with an arm that moves straight across the record, with the path of the stylus in a straight line. Because of the difference in stylus paths, the sound is slightly distorted when the record is played with a conventional tone arm. This is called "tracking-error distortion." If the tone arm is properly designed (and most do perform satisfactorily), tracking-error distortion is negligible.

Some tone arms operate differently, using linear-tracking. When playing, the arm moves in a straight line, as does the master disc when the recording is being made. You don't handle the tone arm with these models—just press buttons.

Cable capacitance. This is an electrical characteristic related to the ability to store an electric charge—in this case, in the cables running from the cartridge to the amplifier.

As a practical matter, cable capacitance is rarely a problem. Cartridge performance doesn't change much unless the cable and amplifier capacitances and the cartridge manufacturer's recommended load capacitance are greatly mismatched.

Convenience features

Some turntables are handier than others, usually because of special features.

Automatic operation. Models that qualify as fully automatic perform the entire operation at the touch of a button. Others perform only part of the operation automatically. A fully automatic unit generally has a control that lets you repeat one side of a record as long as you like.

Record-size selection. With a typical fully automatic model, you set a switch to 7 inches (for 45 rpm) or 12 inches (for 33⅓ rpm), which tells the machine where to position the arm.

Some models have a built-in sensing device on the turntable platter that reads the record size and automatically adjusts the tone arm accordingly. If there's no record on the platter and you press the start button, the sensor stops the operation so you won't damage the stylus.

Cueing. This is a standard feature on turntables. By pressing a lever or button, you can slowly lower or raise the tone arm without touching it—a feature that minimizes the possibility of damaging the record or cartridge.

Location of cue control. On some models, the cue control is located outside the dust cover; on others, it's inside—a less convenient location.

Antiskate adjustment. The pivoted tone arm has a natural tendency to "skate"—that is, to push the stylus toward the spindle and against the wall of the record groove—which can cause right-channel distortion. The antiskate device provides a force to counteract this tendency.

To adjust an antiskate device, it's best to follow the manufacturer's instructions.

Speed control. Many models have a knob for making small changes in the turntable speed. This is a useful feature for correcting speed error or otherwise modifying the pitch or tempo. Some models have a more sophisticated mechanism than a manual adjustment device: an automatic device that holds the turntable at precisely the right speed, usually by means of a built-in quartz-crystal oscillator. With a quartz-lock feature, you can't adjust the speed yourself, but it's certainly convenient not to have to worry about variations in turntable speed.

Models with speed adjustments have a strobe indicator that tells you when the turntable is running too fast or too slow. On most models, the strobe can be used for both 33⅓ and 45 rpm. If the strobe shows that the turntable speed is off, you can increase it or decrease it—by approximately 5 percent—by turning the speed-control knob.

Recommendations

It's not hard to find a good turntable. Mid- and high-priced models are commendably free from flutter and rumble.

Under the circumstances, it's possible to base your choice on price or on features that matter to you. You may be attracted by a repeat control or an automatic record-size sensor, for example.

One final note: Read the section on compact disc players, especially if you don't already have a turntable.

PHONO CARTRIDGES

The phono cartridge, a tiny device at the playing end of the turntable's tone arm, holds the stylus (the needle), whose job it is to follow the record grooves. The cartridge converts the resulting vibrations into electrical signals. They go to the amplifier, which uses them to drive the speakers, which reconverts them into vibrations. You hear the vibrations as music. It takes delicate engineering to process the vibrations without distorting them or damaging the record. Except for the loudspeakers, the cartridge is, in fact, the most important determinant of the sound quality you get from records.

The cartridge follows microscopic wiggles in record grooves with a diamond stylus. The stylus is generally attached to a tiny magnet, which also wiggles, near a small coil of wire. The electrical signal thus created drives the amplifier and loudspeakers. If the stylus is kept in constant contact with the grooves, without squashing or otherwise deforming them, the resulting electrical signal ideally should represent an accurate replica of the original recording.

Things can go wrong. If the cartridge stylus rides the grooves with excessive force, it will increase the wear on a record. If the force is too light, the stylus will mistrack, losing contact with the groove. When that occurs, the music you hear is distorted and the record itself may be damaged.

Just about any model you choose will be compatible with virtually all stereo amplifiers and receivers made in the last twenty to twenty-five years or so. Most high-fidelity phono cartridges are the "moving-magnet" or "moving-iron" type, with a fixed coil of wire. A magnetic field generates electrical signals in the coil as it follows the movement of the stylus. The moving-coil type uses the same principle in reverse: It has a stationary magnet and a moving coil.

Usually, a moving-coil cartridge produces a much weaker signal than a moving-magnet one, requiring a special amplifier. But some moving-coil models—the so-called high-output type—produce just enough signal to work with a standard phono input. Some cartridges do without a magnet: They may use a system in which a small nonmagnetic component is flexed by the stylus to produce the signal directly.

Performance

A number of factors are involved in determining performance.

Vertical tracking force. VTF is the downward force that lets the cartridge work. Generally, a low VTF holds down record wear. Almost any reasonably good cartridge will track at 1 to 1.5 grams. If tracking force is too low, the result may be a raspy sound or, at worst, a bump into the next groove.

All turntables (except those designed for P-mount cartridges) have a VTF control, which you should adjust when you install a new cartridge. (VTF and antiskate adjustments are discussed in the section on turntables, page 173.) You should make the appropriate adjustment for your turntable, following the manufacturer's instructions. If your turntable doesn't have an antiskate device, you should increase the recommended VTF by about 25 percent.

If the stylus mistracks, if your records are severely warped, or if you play a lot of specially recorded discs, such as direct-to-disc or digitally mastered records, don't hesitate to increase VTF slightly. This will be easier on your records than mistracking would be.

Frequency response. Your cartridge should be able to handle a wide range of frequencies—from the lowest bass notes to the highest treble—without over- or underemphasizing any. Even many low-cost phono cartridges have a frequency response that is smooth and extended.

Stereo separation. Left-channel and right-channel sounds are recorded on opposite walls of the V-shaped groove. A good cartridge will keep these sounds quite separate. Stereo separation should not be a problem with most phono cartridges. If there does seem to be a problem, the cartridge may not have been properly installed.

Adaptability to tone-arm mass. Cartridge and tone-arm assemblies can be made to resonate like a tuning fork by certain low-frequency vibrations—sometimes enough to interfere with record play. The vibrations are most troublesome above 17 Hz (where recorded sound begins) and below 10 Hz (where inaudible vibrations caused by footsteps, record warp, or other disturbances can cause mistracking).

Some cartridges can team up with either low-mass or high-mass tone arms and still keep their resonance in the safe zone of 10 to 17 Hz. Others are compatible only with lower-mass tone arms. The higher the mass of the tone arm on your turntable, the more important it is to select a cartridge high in tone-arm adaptability.

Occasionally, turntable manufacturers give the range of cartridge mass (in grams) that they consider compatible with a turntable model.

Intermodulation distortion. This has to do with the way the cartridge handles two or more notes played simultaneously. If the cartridge makes one note distort or modulate another, that's intermodulation, or IM, distortion.

High-frequency tracking accuracy. With loud high-frequency sounds, especially near the end of the record, the microscopic wiggles in the groove make hairpin turns, which the stylus may not follow. In that case, the sounds may be muted.

Hum immunity. A cartridge should resist picking up stray magnetic fields, as from the transformers or motors in stereo components themselves. Otherwise, an annoying low-pitched hum may be audible when you play soft music. If a cartridge tends to pick up hum, the problem can often be avoided by placing the turntable a few feet away from the component that is producing the magnetic field.

Features

In recent years, cartridge manufacturers have come up with several useful features.

Plug-in mounting. When you install a conventional cartridge on a tone arm, you may have to carefully position the cartridge on the tone arm's supporting bracket (the headshell), attach the cartridge with tiny screws, and then connect four tiny tone-arm wires to pins in the end of the cartridge.

Cartridges have been designed to avoid that problem. Integrated cartridges combine the headshell and cartridge into a single unit. They can be plugged directly into a tone arm that has a complementary fitting. Not only are the plug-in types easy to install, they also keep the overall tone-arm mass down, which helps when you're playing a record that is a bit warped.

The P-mount cartridge, as these are known, eliminates the need for an attached headshell. The special turntables that use the P-mount come preadjusted for VTF and antiskating, making the task of installing the cartridge a one-step operation.

Record brush. If your cartridge has a small removable brush that sweeps a few grooves ahead of the stylus to remove dust, you should pay strict attention to the manufacturer's recommendations regarding VTF to compensate for the effect of the brush.

Damper-brush combination. A hinged brush that sweeps the record and dampens or cushions the up-and-down movement of the stylus is useful when you're playing a warped record. It also increases the cartridge's compatibility with higher-mass tone arms.

The damper-brush on several models has a white centering mark to guide you when you're cueing a record. You can also lock the brush in the up position if you don't want to use it.

Stylus shape. Almost all styluses are essentially elliptical in shape. The narrow ends of the ellipse touch the walls of the record groove. Other styluses have a modified elliptical shape, which enlarges the area that touches the walls. The modified ellipse doesn't have any significant advantage in tracking or in frequency response.

Installing a cartridge

You may not want to tinker with the delicate workings of a stereo system. In that case, the installation of a conventional cartridge is best left to a hi-fi dealer.

But if you are reasonably deft, you will probably want to tackle the job yourself. Conventional cartridges almost invariably come with the necessary mounting hardware.

Be sure to install the cartridge so that, when viewed head-on, it's perpendicular to the record surface. Seen from the same vantage point, the cartridge and its image reflected in the surface of the record should be vertically aligned. Otherwise, stereo separation may not be up to par. If the cartridge is tilted off the vertical, insert a shim made of cardboard or a few layers of paper under the high side to correct the angle. Follow the turntable manufacturer's instructions regarding proper placement of the cartridge in the tone arm.

Replacing a stylus. You should be able to use a stylus for at least 400 hours. If you keep the stylus and your records very clean, you may be able to use it for a longer time. If you are very cautious, you may want to replace the stylus after 400 to 500 hours of play, either timed or estimated. If you are uncertain, ask an audio technician to examine stylus wear with a microscope. (Check again after 200 hours or more.)

Always replace the stylus if you damage it, or if its cantilever—the thin bar that supports the diamond tip—looks bent out of its original shape. You can replace a stylus at about half the cost of the entire cartridge.

Cleaning a stylus. This is neither difficult nor time-consuming. It's important to keep the stylus free from accumulated dust and dirt, which accelerate record wear and can cause mistracking and distortion.

Clean the stylus with a fine camel's-hair brush lightly moistened with a little rubbing alcohol. Brush lightly from the rear to the front of the cartridge; brushing backwards or sideways could bend the delicate stylus.

Caring for records. The stylus won't need frequent cleanings if your discs are free from dust—and dust-free records last longer. Records can be cleaned with a cloth-pile brush before you play them; audio stores sell such brushes. Keep the turntable's dust cover closed except when changing records, and handle records only by the edges and the label area to prevent perspiration and body oils from attaching dust to the vinyl surfaces. When returning a record to its album, make sure that the opening in the inner sleeve doesn't coincide with the opening in the outer cover and leave the record case wide open for dust to enter. Always store records vertically to reduce the likelihood of warping, and keep records away from direct sources of heat.

Recommendations

You can buy a fine cartridge at a list price of anywhere from less than $50 to more than $200, but list prices may be very misleading. Cartridges are one of the most heavily discounted audio products. Many models sell for less than half the suggested retail price.

In case you are dissatisfied with the sound you get with a low-priced cartridge, or in case you buy a defective model, make sure you have return privileges.

CASSETTE TAPE DECKS

A cassette tape deck is now a standard component in an audio system, taking its place with the receiver, the record player, and the compact disc player.

You can use a cassette tape deck to play commercially recorded tapes, of course, just as you use a record player to listen to records. But a tape deck's real attraction is its ability to record music from your other components. You can, for example, tape FM broadcasts and build a collection of concert performances, make a tape copy of a record, or combine selections from several sources onto a single cassette, creating your own music programs. Once a taped program is in cassette form, it can travel with you, to be played in the car or on a portable cassette player.

Listening

A tape deck should be able to record and play back over the entire audio spectrum, from deep bass through mid-range to the highest treble. It should also be free from distortion and unwanted noise.

Frequency response. This describes how smoothly and uniformly a deck responds to frequencies within the audio range. The response depends largely on how well tapes match a deck's electrical characteristics. One such characteristic is "bias," an inaudible ultrasonic tone generated within the deck to reduce distortion. The amount of bias necessary depends to a great extent on the tape used. A deck provides this bias either automatically, to match the type of tape being used, or by a switch selection.

Tapes are designated Type I, Type II, Type III, and Type IV (metal). Type I describes "normal-bias" tape—the common variety having an iron oxide formulation. Type II includes "high-bias" chromium dioxide formulations and other materials that mimic Type II characteristics. Type III tape is the rarely used "ferrichrome" tape. Type IV is described as "metal" because of the pure metal particles that coat it. Metal tape requires more bias than the others, offers potentially outstanding performance in the treble region, and is probably best used for special applications (see the end of this section).

Tape decks do not perform uniformly well with a particular brand of Type I or Type II tape. To achieve peak performance you should try a few varieties of tape.

Dynamic range. At the quiet end of the range of loudness a deck can handle, background hiss inherent in the tape and the electronic circuits can intrude on the music. The significance of the residual hiss is often described by the "signal-to-noise ratio" (S/N), the proportion of desirable signal, such as music, to the undesirable signal, the background noise or hiss.

At the loud end of the range, overload causes audible distortion. With Dolby B noise reduction (see page 186) mid-band dynamic range may be good. With Dolby C, mid-band performance can improve substantially.

At the high end of the frequency spectrum, tape limitations cause a fall-off of dynamic range. But with Dolby C, treble dynamic range improves, attesting to the effectiveness of that particular form of noise reduction. (Decks equipped with automatic noise-reduction circuitry [ANRS] should perform on a par with units having Dolby B.)

Dolby B is the noise-reduction system most commonly found on prerecorded cassettes. In addition, Dolby B is the system you would ordinarily use in making tapes to play on an automobile tape deck or a portable player. Those machines generally don't have Dolby C capability.

Speed accuracy. This is a measure of how close the tape speed of the cassette deck matches the standard cassette tape speed of 1⅞ inches per second (ips). If a tape moves faster or slower than the standard, the music's pitch and tempo are likely to be off. But speed accuracy is significant only when you play tapes that were recorded on another deck. Tapes recorded on your own deck will be unaffected by speed inaccuracy since they'd be played back at the speed at which they were recorded. Even when playing a tape from another deck, slight inaccuracy is a problem only if you are extremely sensitive to pitch variations.

Headroom. When making a tape, it's desirable to record as strong a signal as possible without actually overloading. In this way you keep noise and tape hiss to a minimum. Recording-level meters on tape decks have reference points that are supposed to tell you where overload is likely to occur. Headroom is the leeway between the reference point and the actual overload point. Since you can't always anticipate the strongest sounds to be encountered in a recording session, some headroom is desirable.

If you have the time to experiment with different recording levels during a recording session, by all means do so. Different kinds of music lend themselves to nonstandard control settings.

Freedom from flutter. Flutter is the consequence of uneven tape motion past the record/playback heads. It produces a wavering in pitch that can make music sound watery, wobbly, sour, or otherwise disagreeable. Flutter can be especially annoying in music with sustained notes played on the piano and on woodwinds. Once a problem with tape decks, flutter is for all practical purposes no longer a concern.

Operating a deck

Because the audio performance of tape decks is quite high, your final choice should be influenced by which features of the deck coincide with the way you expect to use it.

Most models are front-loading, with a hinged door. Stacking a deck under or over other components saves space and is quite acceptable as long as there is adequate ventilation to prevent equipment from overheating.

A deck should let you go from one mode of action to another without first pressing the Stop button. Special circuitry keeps track of the tape motion and acts to prevent tape snarls or breakage.

There should also be automatic shutoff at the end of the tape, a record-level control, a Record-Mute button that permits easy insertion of a four-

or five-second "blank" spot on the tape (useful for separating musical selections), and a compatibility with tape Types I, II, and IV.

Record-level indicators. The record-level indicator is your guide to the strength of the signals you are putting on tape while recording. A "bar graph" indicator is better than the older "twitching needle" variety because it responds almost instantaneously to transient musical peaks, which can ruin an otherwise good recording. The number of illuminated segments varies from one model to another. Although most any will be adequate for home recording, more segments give a more precise indication.

Tape protection. A deck will refuse to record a tape if the plastic record-protection tabs on the cassette shell have been removed. Punching out these tabs can save your recordings from accidental erasure, which is particularly important in decks that don't require pressing two buttons in order to go into the record mode.

Off-tape monitoring. If a deck has three separate heads (record, playback, and erase) you can monitor a recording directly from the tape while the recording is taking place—that is, you can listen to the sound of the tape an instant after a signal has been recorded on it, rather than listening to what the tape is currently recording. Monitoring of this sort is usually unnecessary except for "one-shot" tapings of a critical nature—for example, recording a live performance. But off-tape monitoring is handy for quick adjustments of record-level or bias settings. If the record level is set too low, tape hiss will be heard in the background; if it is set too high, an unpleasant dulling in the treble will usually be the first (and mildest) symptom of tape overload.

Tape-type selection. It's nice to have automatic display of, and accommodation to, the type of tape being used. A deck achieves this by detecting standardized code notches on the cassette shell. Otherwise, you must select the proper tape manually, or a poor-quality recording or playback may result. If you continue to record on older, nonstandardized tapes, automatic accommodation can be a problem. You may want to choose a deck that lets you do this setting manually.

Automatic reverse. Cassette decks with automatic reverse sense the end of the tape and automatically reverse the direction of the tape movement. This lets you listen to both sides of a recorded tape without interruption.

Microphone inputs. If you expect to do live recording, look for a deck with microphone jacks.

Tape counter. An easy-to-read digital display is more desirable than a rather old-fashioned mechanical counter, but counters do differ in certain details that can be important if you take pains with your recording.

Many models feature a "real-time" counter. It directly displays a tape's running time in minutes and seconds. This is useful when recording because you can then determine quite easily whether there is enough tape left to fit the program material to be added.

The counter on some decks is of the "stopwatch" type; it tells you how long the machine has been in the play or record mode. If fast forward or rewind is used on these machines, the counter readouts will no longer be valid. Decks with a real-time counter avoid this problem by calculating tape-spooling electronically in all modes.

Some units have a "time-remaining" option that spares you from having to calculate remaining tape time and subtracts the displayed time from the total time on the cassette. To use the feature, you first "dial in" the playing time of the cassette being used (C-30 or C-90, for example). Other decks feature a kind of time-remaining option that involves shuttling the tape forward to the end and then back to the record spot on the tape. Even when the option works automatically, the extra time consumed is inconvenient.

A tape counter serves another highly useful function: It helps you locate a particular point on the tape quickly. This is called "tape-counter memory." The machine "remembers" the "0000" position of the counter even if the desired marking point is in the middle of the tape. You can rewind the tape to this mark, at which point the cassette deck stops. Most decks can even switch automatically to the play mode at the zero mark. A few machines feature a second memory point that permits them to repeat a portion of the tape over and over.

Timer control. Many models let you make a recording or play a tape under the control of an external timer (an appliance timer, for instance).

In general, the timer-record function should be used with care. If the timer is left in its "record" position with a cassette in place, the cassette may be erased when the deck is turned on.

Automatic search and play. Many decks take advantage of a common occurrence on a tape: the period of silence between music selections. You can program the machine to search ahead for the next such gap and begin playing at that point. Or you can direct the machine to find the previous gap and, thus, the start of the current selection. Some decks go a step further. They permit you to skip ahead by a predetermined number of gaps to a selection well into the tape.

There is, however, a flaw in the use of sound gaps in a tape. Quiet passages in the middle of a musical selection may trigger sensors to stop the tape there rather than at the end of the selection.

Cue and review. Some models provide an old standby method for scanning a tape: Their fast-forward and rewind controls can be partly depressed and so let you hear the "monkey chatter" sound of speeded-up tape, letting you home in on a desired segment of the tape.

FM multiplex filter. A number of decks include a switchable filter to minimize certain kinds of noise when making a stereo recording from an FM radio broadcast. Other decks may include the filter as a permanent part of their circuitry. The effect of the filter on the deck's frequency response is minimal, so there's really no advantage to the switch option. Absence of the switch option may be considered a disadvantage if you are an audio purist who objects to any unnecessary electrical impedimenta in the recording path.

Headphone jacks. A headphone jack on the front panel is rather standard equipment. An added feature would be some means for controlling headphone volume.

Output-level controls. An audio output-level control is useful for equalizing volume between the deck and a tuner or turntable in a component hi-fi system. This control may also serve as a volume control for headphones.

Front panels. Multiple, sophisticated tape-deck functions can be intimidating. The panel design of some models is not as helpful as it could be—for example: function keys labeled with symbols instead of words; a cluster of interconnected boxes labeled with abbreviations. Despite such complexities, however, a careful reading of the instructions and some patience should make even a complex deck reasonably easy to use.

Recommendations

Since good performance is so readily available, your buying decision can be based mainly on the particular mix of features that best matches the way you're likely to use a tape deck.

Thus, for example, if you expect to tape live concerts or record from "super-fi" disks, consider a model with DBX, a noise-reduction system technically superior to Dolby. The DBX circuitry also lets a deck work as a decoder to play DBX-encoded, low-noise discs.

If you intend to tape live performances, consider a deck with three heads. You can then monitor sound the instant it's taped and make any necessary corrections in the controls.

If you like to sit back and just listen, buy a deck that will automatically reverse the tape for virtually uninterrupted recording and playback.

■ Tape noise reduction ■

If you play a blank tape, you hear not silence but a low hissing sound. This "tape noise" is usually inaudible under the sound of recorded music, but it can intrude through soft musical passages. Careful recordists take some pains to avoid tape noise in their recordings, and electronics engineers have developed a number of ways to help them. Prominent among these electronic aids is Dolby circuitry—named after its inventor, Dr. Ray Dolby.

When a tape deck is in the record mode, a Dolby circuit strengthens weak signals as they're being recorded so that the recorded music is louder than the noise. When you play back the same tape, the circuit restores the amplified frequencies to their normal level. Soft passages still sound soft, while the noise of the tape is suppressed.

The earliest Dolby system, type A, was designed for commercial use in recording studios. Then came Dolby B, a system that affects only the treble. It works well, and is used on almost all prerecorded commercial tapes. But Dolby B falls somewhat short of delivering the dynamic range you should expect of a high-quality recording.

Dolby C is noticeably better than Dolby B. It provides much greater noise reduction, even when taping or reproducing very quiet musical passages, and is less likely to dull the sound of musical "highs" than earlier Dolby circuitry.

Dolby C is a common feature on tape decks, but it may not be generally available on portable cassette players or on automobile cassette players. You can record music on a Dolby C tape deck and play it back on a cassette player with Dolby B circuitry; the music will sound slightly bright but otherwise okay.

A later approach to conquering tape noise is called "DBX." DBX is circuitry that electronically compresses the music's dynamic range by half for recording and restores it in playback—a process known as "companding." It is a most effective noise reducer.

When recording with DBX, you may have to change your recording technique, keeping the level of music peaks somewhat below normal on the deck's record-level indicators. Otherwise, music with lots of strong percussion sounds may overload the tape, becoming dull or distorted. DBX recordings are not compatible with audio equipment that lacks DBX circuitry—they can sound shrill and unpleasant when played back without DBX.

Metal tape and DBX. Except for a moderately improved dynamic range in the treble, metal tape offers little to merit its considerably higher price, though it can give exceptional dynamic range when in combination with DBX noise reduction in a tape deck having that system. As a practical matter, however, the combination of metal tape and DBX is unlikely to improve the sound of tapes you make at home. The dynamic range it's possible to record just isn't present in standard discs and FM broadcasts, the musical sources most likely to be recorded. It's present in DBX-encoded records, in compact discs (which require their own player, see page 165), and in live performances.

If you have a tape deck that doesn't have DBX capability, add-on units are available that connect to a deck and receiver.

MID-PRICED STEREO COMPONENTS

The chapter on rack stereo systems (page 191) discusses the easiest way to buy a complete set of hi-fi components. But if you don't need the whole package, or if you'd rather choose the components yourself instead of buying a bundle on a take-it-or-leave-it basis, it's still possible to do so.

Individual components will let you build a system to your liking with the assurance that each element is excellent by itself and will nearly always perform well with the others recommended. Every piece should have a useful complement of features (plus a few frills), the quality to satisfy even critical ears, and enough power to fill a large room with loud music.

A complete system of mid-priced components—receiver, speakers, tape deck, compact disc player, turntable, and cartridge—will list for about $1,700 but may cost more like $1,200 after discounts. For detailed information about individual components, refer to the relevant chapters in this book.

Receivers

Push-button digital tuning lets you preset favorite stations for instant recall. Wireless remote controls let you select the music source, choose a preset station, and adjust volume. The remotes can also operate some CD players, tape decks, and turntables from the same maker (compatible units listed here are the *Sony* turntable and the *Onkyo* cassette deck). The *Onkyo* and the *Sony* provide connections for two tape decks, allowing tape-to-tape copying. There's about 70 watts of power per channel available, adequate for any of the recommended speakers.

The *Onkyo* and *Pioneer* have a switch that provides a "simulated" stereo effect on mono material, and the *Onkyo* has a control that provides a variable bass and treble emphasis. The *Pioneer* has a five-band tone control. The *Sony* has push buttons for all its controls; its volume level is displayed numerically in 1-decibel steps, a barely noticeable increment. That model also has a "surround" switch, which removes the monophonic part of the sound sent to a second pair of speakers, providing an emphasized stereo effect if you place the speakers in the rear or sides of the listening area.

The *Pioneer* has a built-in digital clock with a timer that can be set to turn the receiver (and components connected to its switched outlet) on and off at chosen times. That could be useful for wake-up music, or for recording a radio program using a timer-compatible tape deck.

All three receivers have additional features for the video enthusiast. They provide connections for the video and stereo audio signals from two VCRs or other video sources, and connections to a TV with a video input. They let you edit or copy a tape from one VCR to another. The *Sony* has a second set of input jacks on the front panel for one audio-video input, presumably making it easier to copy from a portable VCR or camcorder. The *Pioneer* and *Sony* even have an alternate selectable cable-antenna input so their FM tuners can benefit from the stereo audio channels carried by many cable-TV systems. (Prices are as published in a November 1987 report.)

Receivers	List Price
Onkyo TX-82	$360
Pioneer VSX-3000	350
Sony STR-AV550	370

Loudspeakers

These three models were the best of the sixteen high-rated speakers in the Ratings on page 152. They scored the highest in accuracy. A wise choice may be made, based on your own listening preference, from among them. Other considerations may influence your selection, such as freedom from bass distortion (of these three, the *Advent* has the edge), or a good discount price. (Prices are as published in a November 1987 report.)

Loudspeakers	List Price
Advent Legacy	$450
Allison CD7	600
Infinity RS4000	470

Cassette tape decks

A cassette tape deck can combine music from several different records or produce tapes for use away from home. Both these models have autoreverse for continuous play of both sides of a tape either once or repetitively. They also have Dolby B and C noise reduction for compatibility with prerecorded tapes and with other tape decks and players.

They have soft-touch buttons for Play, Pause, Fast-Forward, Rewind; record-level controls and meters; and headphone jacks. They're compatible with the three popular tape formulations: "normal" Type I, "high-bias" Type II, and "metal" Type IV. They automatically select the tape type by sensing notches molded into the newer Type II and IV cassettes. That's a convenience, but older Type II or IV cassettes won't record or play correctly. Both models let you fine-tune the recording bias-setting to match that needed for various brands of tape. The *Aiwa* lacks microphone inputs, needed if you want to record live sound.

The *Onkyo* lets you quickly wind the tape to the blank space between selections and can also be set to skip over silent sections—useful for continuous background music with a partially recorded tape. The *Aiwa* lets you start recording or playback with an external AC timer. (Prices are as published in a November 1987 report.)

Cassette Tape Decks	List Price
Aiwa AD-R30	$180
Onkyo TA-R240	265

Compact disc players

These three players are full-featured and have excellent sound quality. They have remote controls for all the major functions. The *Sony* has a calendarlike display to show how many tracks are on a disc and which tracks have yet to be played, as well as an autospace mode that puts a three-second blank space between selections. It also has faster track selection than both the others. The *Pioneer* has a more sophisticated audible-scan function than the others, but no single-track or segment-repeat functions.

All three have the standard features, including play-order programming, disc-repeat, and display of total program time (useful for taping). The *Sony* has a shuffle-play button that lets you play tracks in random order. (Prices are as published in a November 1987 report.)

CD Players	List Price
Denon DCD-1300	$550
Pioneer PD-7030	470
Sony CDP-203	500

Turntables and cartridges

These turntables are relatively free from flutter and rumble, have good speed accuracy, and are convenient to use.

All use P-mount cartridge mounting (also called "T4P"). It makes cartridge attachment a simple plug-in job, assures compatibility with almost any P-mount cartridge, and eliminates the need for adjusting vertical tracking-force (VTF). The VTF is standardized at 1.25 grams: light enough to minimize record wear, yet sufficient to allow good tracking with cartridges of reasonable quality.

All have "linear tracking" design, in which the arm moves in a straight line from edge to center. To select a track, you use Left-Right buttons without touching the tone arm. All allow playing from the beginning of a 7- or 12-inch record automatically, and pick up at the end of play. All have a Repeat switch that lets you play the same record over and over. All have 33- and 45-rpm speeds, a hinged dust cover, a direct-drive platter (no belt), and highly accurate speed regulation.

The *Technics SL-J33* has a sensor that scans the record and memorizes the position of up to eight tracks. You can then select a track by pressing one of a row of buttons on the front of the cabinet, or even program a sequence of tracks.

Both cartridges use P-mount plug-in installation. Both are good performers. The *Shure* has a slight edge in tracking heavily cut test records. And the *Audio-Technica* may sound slightly "bright," due to emphasis in the extreme treble. (Prices are as published in a November 1987 report.)

Turntables	List Price
Sony PS-LX520	$230
Technics SL-J33	approx. 270
Technics SL-L20	approx. 210

Cartridges	List Price
Audio Technica AT231LP	$145
Shure V15-V-P	275

RACK STEREO SYSTEMS

The "rack" is a kind of portable closet stacked with components and sold under a single brand name. In theory, at least, a rack system should make it easy to buy hi-fi gear. The components are supposedly matched to one another in capabilities and quality. You rely on the manufacturer's choices rather than your own intuition or a salesperson's recommendations.

Looking at a row of rack systems in the store, it's hard to tell one from another. There's a sameness to these products that blurs the distinction between the cheapest and most expensive. That similarity challenges the manufacturers to come up with selling points. The route many have chosen is wattage. The typical rack system puts out a lot of power, with 100 watts per stereo channel an unofficial standard.

The biggest attraction of rack systems is also their biggest drawback: One of the components may be inferior. You won't know until you assemble the rack and try it out. Then, if one piece turns out to be subpar, you're stuck. Although you can substitute components in some rack systems, doing so defeats the purpose of buying the system in the first place.

Receiver

In any hi-fi system, the receiver acts as the central connection point for all the components, houses the amplifier they share, and includes the AM/FM tuner. In a good system, the receiver amplifies the signal from the other components and shuttles it to the speakers without introducing any audible distortion of its own.

There may be marginal differences between the best and the worst, but even the worst is likely to be adequate, low in distortion, and with a frequency response good enough to reproduce the music spectrum accurately from deep bass to high treble.

Even the weakest receiver tested by Consumers Union engineers cranked out 73 watts per channel, and the strongest turned out 230 watts. While higher wattage helps the amplifier pump out more volume without distorting the sound, added power soon reaches a point of diminishing returns in home audio gear. There is little practical difference between a 100-watt amplifier and a 200-watt model.

The FM tuners are at least adequate in frequency response, sensitivity to a weak signal, and the ability to pick out the frequency you want in a crowded FM band. All use the now-common quartz digital tuning, so they tune to the center of an incoming station's signal.

AM tuners don't provide real high-fidelity performance.

Loudspeakers

The music you listen to will be only as clear and true as the speakers allow. So it's essential to have speakers that render audible tones accurately, without over- or underemphasizing any particular part of the audio spectrum.

With a receiver's tone control set in the middle position, most speakers are mediocre. Typically, they suffer from booming bass and uneven treble tones. But, since rack systems marry the speakers to a particular receiver, you can give them a second chance by adjusting the receiver's tone controls to try to improve the accuracy of the sound from the speakers. (Most systems have "graphic equalizer" tone controls in lieu of knobs for the treble and bass. An equalizer lets you adjust the sound level at several points along the spectrum, supposedly for more precise control; the improvement over conventional tone controls is minor.)

Cassette deck

The typical rack system includes a dual cassette deck—one that has two "wells" for tape. Both wells can handle playback; one can record, too. That makes it possible to play two cassette sides in sequence—the machine starts the second tape when the first finishes. More significantly, a dual cassette deck lets you make tape-to-tape copies. (Most decks lack a microphone input, needed to make "live" recordings.) But a cassette-to-cassette recording may be audibly degraded to some degree.

Most decks have the standard Dolby B noise-reduction system, which cuts background hiss. A few tape decks also have the latest generation Dolby C, which is even better.

Desirable features include automatic reverse, some means of locating the start of the next selection on a tape (sometimes called "program search"), the ability to start a tape automatically after rewind, and a single switch to select tape type.

CD player

CD players differ from one another primarily in the presence or absence of features, and in how well they can play discs that have been damaged or poorly manufactured.

These CD players have a useful complement of controls, including an audible cue and review function—which allows you to hone in by ear on a particular passage of music—and displays of elapsed time and time remaining on a disc. Other features include: track-sequence programming, allowing you to dictate the order of selections on a disc, and a "shuffle" feature, which lets the player mix the order for you.

Turntable and cartridge

All have at least a two-speed (33 and 45 rpm), single-play turntable with a magnetic cartridge. A few are fully automatic; that is, they start with the touch of a button. The turntable should stop automatically at the end of a record.

Turntable quality tends to be variable. The best ones are component-quality. But the typical turntable may have problems with rumble and flutter. Rumble is audible mechanical noise from such sources as motor or bearings, transmitted through the platter and picked up by the tone arm. Flutter, perceived as a wavery sound, originates in speed variations of the turntable.

The mediocre units have a very flimsy tone arm. Some of them move stiffly enough to contribute to a tracking problem, which distorts the sound.

The cartridge is likely to be barely adequate in most cases, reflecting, perhaps, the diminished importance of LPs in today's audio market.

The racks

The rack itself comes as a kit that must be assembled according to sometimes complicated instructions. When finished, it amounts to a shelving unit about 3½ feet high, with storage space at the bottom for records, tapes, and CDs, and space above for the components. Some models don't provide enough storage space to store LPs upright.

Most racks put the turntable uppermost, with a hinged lid atop the cabinet. Others place the turntable on a slide-out shelf, a slightly less convenient arrangement. The usual recommended placement of the CD player is at the bottom of the stack.

You have to stoop to read and operate the controls on whichever component winds up on the bottom. To compensate for that awkwardness, many systems come with a remote control, which allows you to operate most functions on most components from afar.

Hi-fi components breed unruly tangles of wire and cable. Some rack systems seek to tidy up by connecting the components with neat lengths of multiconductor ribbon wire. Those models also share a single power supply. As a result, you can't upgrade the system with, say, another brand of tape deck or CD player because you can't plug the new component into the system.

Recommendations

A rack system makes sense only if you're in the market for a whole new system, rather than a piecemeal upgrade of components. Because rack components are matched to each other, sometimes literally joined together, it may not be feasible or even possible to switch or upgrade components later on. If you want to put together your own component system by adding a CD player or replacing the speakers, turn to the sections about those components (pages 165 and 147).

Ratings of rack stereo systems

Models are listed in order of estimated quality. For models that don't include a CD player, the model number of the one we added is in parentheses. Prices are as published in a **November 1987** report.

◉ ◓ ○ ◒ ●
Better ⟵ ⟶ *Worse*

					Loudspeakers		Receiver				
Brand and model	Price	Overall system score	Rack height, in.	Speaker height, in.	Overall quality	Deep bass	Overall quality	Convenience & features	FM quality	AM quality	Comments
Technics A640	$1000	◉	38	38	◉	◓	○	○	◒	◒	C,D,L,M,N,S
Technics A630	800	◓	36	36	○	◒	◓	◒	○	◒	D,L,M,N,S
Sears LXI Series Cat. No. 9300[1]	—	◓	43	32	◉	◓	○	○	◓	◉	B,E,G,O,T
JVC GX-3R (XLV-250)	1259[2]	○	36	36	◒	◉	◉	◉	○	◓	A,B,C,O,Q,S
Sony S7700CD	1500	○	43	37	◉	○	◒	◒	◒	○	C,I
Marantz XDX112CD	1300	○	34	34	◓	◉	◉	○	◉	●	G,P,U
Scott RX-300D[1]	1300	○	37	37	◒	◉	◉	◉	◓	◓	A,C,N
Pioneer X4200 (PD-M60)[1]	1500[2]	◒	36	36	◉	○	●	◒	●	○	C,I,T
Sony S4500[1]	1000	◒	44	30	◒	◒	○	◒	○	◒	C,I
Realistic System 601 (CD-2200)[1]	—	◒	40	27	○	◒	○	○	◒	◉	G,O
Onkyo A/V-755[1]	1300	◒	43	33	◒	◓	◒	◒	◒	◒	C,D,K,M,O,Q,S,U
Sharp 5100	630	◒	38	27	○	◒	◓	○	◓	◒	D,F,O

[1] *Model discontinued at the time the article was originally published in* Consumer Reports. *The test information has been retained here, however, for its use as a guide to buying.*

[2] *Price includes CD player added by CU.*

Ratings of rack stereo systems (continued)

Models are listed in order of estimated quality. For models that don't include a CD player, the model number of the one we added is in parentheses. Prices are as published in a **November 1987** report.

					Loudspeakers		Receiver				
Brand and model	Price	Overall system score	Rack height, In.	Speaker height, In.	Overall quality	Deep bass	Overall quality	Convenience & features	FM quality	AM quality	Comments
Sherwood SS-2100RCD	1500	◒	39	39	◒	●	○	⊙	◒	◓	B,C,H,O
Sherwood SS-2050 (CDP-300R)	1300[2]	●	33	33	●	●	◒	●	○	○	F,L
Kenwood Spectrum 56B	1149	●	37	29	◒	◓	◒	⊙	●	◒	A,B,D,O
Fisher 8735D[1]	1200	●	37	36	○	◓	⊙	◓	◓	●	C,T
Realistic System 1000 (CD-2200)[1]	1199[2]	●	45	31	○	●	●	◒	◒	◓	A,B,C,E,K,O
Scott RX111D (950DA)	1100[2]	●	38	27	◒	◓	○	◒	◒	⊙	B,E,F,O,R
Fisher 8700 (AD-922B)[1]	780[2]	●	34	34	●	◒	◓	●	◓	○	F,J

Receiver

All have: • Stereo FM and mono AM tuners with quartz digital tuning and station presets. • FM antenna input(s). • Output-level indicator. • Balance control. • Headphone jack. • FM (wire) and AM (loop) antenna included. • Inputs and function-switching for all components in rack.

Except as noted, all have: • Manually operated station-select feature for locating weak signals. • Autoscan feature that stops at each receivable station. • "Graphic" equalizer with 5 or 6 frequency bands and single set of controls for both channels. • One spare input jack for plugging in auxiliary equipment such as TV sound or second tape deck. • Connections for second (remote) pair of speakers, with independent switching.

A–Has 7-band equalizer, judged more versatile.
B–Has equalizer bypass switch.
C–Has switch to lower sound level temporarily.
D–Tuning beyond one end of dial begins other end—a convenience.
E–Adjusting equalizer requires setting both channels—an inconvenience.
F–Lacks connection for remote speakers.
G–Cannot switch independently between main and remote speakers.

H–Has connection for rear speakers that gives an emphasized stereo effect.
I–Lacks autoscan tuning feature.
J–Loudness compensation on at all times, artificially boosting treble and bass.
K–Audio mute defeated if power is interrupted—can cause loud blast of sound.
L–Lacks unused audio input jacks.
M–Has conventional bass and treble controls.
N–Has bass-boost switch.
O–Has loudness-compensation switch.
P–Volume level suppressed when unit is first switched on.
Q–Has switch to provide pseudostereo effect.
R–Has high-cut filter to reduce noise and high-frequency sounds.
S–Has facility to connect second tape deck.
T–Level display shows levels in 5 ranges.
U–Has audio/video switching facility for VCRs.

Turntable

All have: • Single-play only. • Auto-stop at end of play. • 33- and 45-rpm speeds.
Except as noted, all have: • Cartridge with standard ½-in-mount. • Damped tone-arm cueing control at front of turntable (lowers tone arm softly). • Pivoted tone arm with tracking-force and antiskate adjusters to keep stylus centered in record groove.
A–Has "P-mount" cartridge—easy to replace.
B–Has strobe-lighted variable pitch control, for fine-tuning turntable speed.
C–Has automatic record-size selection.
D–Has automatic start.
E–Record is easy to see while playing.
F–Cueing control inconveniently placed.
G–No antiskate control—could cause mistracking.
H–Has nonstandard cartridge—replacement stylus may be difficult to find.
I–Lacks desirable vertical tracking-force adjustment on tone arm.
J–Tone-arm cueing is undamped—could promote record damage.
K–Has linear-tracking arm with cueing buttons.

CD Player

All have: • Excellent sound quality. • Track-sequence programming ability. • Play, Skip, Pause, and audible Cue/Review. • Digital track-number display.
Except as noted, all have: • Ability to stop play without clearing program. • Elapsed-time display. • Ability to repeat disc program.
A–Has number keypad to select track numbers.
B–Can repeat any segment of track or disc.
C–Can repeat single track without programming.
D–Has multiple-speed Cue/Review.
E–Can display play time remaining on disc.
F–Play can be started with external AC timer.
G–Disc visible during play.
H–Has headphone jack with separate volume control.
I–Has Shuffle-Play button that randomizes order of play.
J–Pushing Play button doesn't close drawer—an annoyance.
K–Lacks elapsed-time display.
L–Cannot repeat disc or program.
M–Pressing Stop button clears program.
N–Cue/Review works in pause only—an inconvenience.
O–Has 6-disc changer with disc-holder cassette.
P–Can insert some silence between tracks.

Ratings of rack stereo systems (continued)

◉ ◓ ○ ◒ ●

Better ⟵⟶ *Worse*

Brand and model	CD player: Overall quality	CD player: Convenience & features	CD player: Disc tracking	CD player: Comments	Cassette deck: Overall quality	Cassette deck: Convenience & features	Cassette deck: Playback quality	Cassette deck: Record/play quality	Cassette deck: Copy quality	Cassette deck: Comments	Turntable: Overall quality	Turntable: Convenience & features	Turntable: Comments	Rack: Overall rack quality	Rack: Comments
Technics A640	◓	◓	◉	D,E,M	◉	◓	◓	◉	◉	E,G,J,L,V	◉	◓	A,E	◉	A,B,C,D,G,H,L,M,W
Technics A630	◓	◓	◓	D,E,M	◉	◓	◉	◉	◉	E,G,J,L,S,V	◉	◓	A,E	◒	A,L,M,Q,W
Sears LXI Series Cat. No. 9300[1]	○	●	◓	G,K,L	●	◒	●	○	◒	A,G,I,R,Z	◉	◒	E,G,H	◒	A,K,V,Z
JVC GX-3R (XLV-250)	◉	○	◉	D	◉	○	◓	◉	◉	G,L,V,Y	◓	◒	I	◓	B,C,D,G,H,I,J,K,L,O,CC
Sony S7700CD	◒	◓	◒	C,E,I,M,P	◓	◓	◓	◉	◉	G,J,M,N,W,Z	●	○	C,D,E,F,H	◉	A,B,C,D,F,G,H,I,J,K,L,S,W,Z
Marantz XDX112CD	◓	◓	◉	C,D	●	◒	●	◒	◒	A,G,R,Z	○	○	A,G	◓	A,B,C,D,E,G,J,L,S,W,AA
Scott RX-300D[1]	◓	◓	◉	B,D,F,M	◓	◉	◓	◓	◓	C,E,G,K,L	◉	◉	A,B,D,E,K	○	A,B,G,H,P,S
Pioneer X4200 (PD-M60)[1]	◒	◉	◒	A,E,H,I,J,O,P	●	◉	●	○	●	C,E,G,L,V	◉	◉	A,D,E	○	A,B,C,G,H,J,S
Sony S4500[1]	◒	◉	◒	C,E,F,I,P	◓	◉	○	◓	◓	C,F,G,L,O	○	◒	E,F,G,I	○	A,K,L,W,Z
Realistic System 601 (CD-2200)[1]	◒	◓	◒	B,D,F,M	◒	○	◒	◒	◒	A,B,L,R,S,X	○	○	E	○	A,B,D,H,X,Z
Onkyo A/V-755[1]	◓	○	◉	B,C	◓	◉	○	◉	◉	D,E,F,G,L,T,V,W,Y	◉	○	E	○	A,B,H
Sharp 5100	○	◒	◓	D,N	◒	●	●	○	●	R,T,Z	●	◒	E,G,H	●	A,T
Sherwood SS-2100RCD	◉	●	◉	D,G,L,M	◒	○	◒	○	○	C,R,V,Z	○	○	D,E,H,K	◉	A,B,C,D,G,H,I,K,BB
Sherwood SS-2050 (CDP-300R)	◉	◒	◉	D,G,L	○	◒	○	○	◓	G,R,V,Z	○	○	E	●	A,R,T

[1] *Model discontinued at the time the article was originally published in* Consumer Reports. *The test information has been retained here, however, for its use as a guide to buying.*

Kenwood Spectrum 56B	○	○	⊙	C	○	◓	◒	◓	○	H,L,O,U	◒	○	A,E,F,G	●	A,Q,W
Fisher 8735D [1]	◒	◒	◒	G,K	○	●	○	◓	◓	R,S,V,Z	◓	○	A,F,G	●	B,E,G,H,J,O,Q,R,S
Realistic System 1000 (CD-2200) [1]	◒	◓	◒	B,D,F,M	○	◓	○	○	◒	A,B,G,L,R,X	⊙	○	E	●	A,Y,Z
Scott RX111D (950DA)	●	⊙	●	B,D,E,F,M	●	●	○	●	●	B,E,L,P,Q	◓	●	E,F,G,I,J	◒	A,K,M,N,U
Fisher 8700 (AD-922B) [1]	◒	◒	◒	G,K	◓	◒	○	⊙	⊙	G,R,V,Z	◓	●	F,G,H	●	A,K,O,R,S

Cassette Deck

All have: • 2 cassette "wells" (only one records) allowing "dubbing" (copying) from one cassette to another. • Record-level indicator. • Mechanical tape-counter. • Auto-stop at end of playback or recording. • Provision for types I, II, and IV tape. • Double-speed dubbing ability.
Except as noted, all have: • Dolby B noise-reduction. • Manual tape-type selector requiring 2 switches to be set. • Ability to mix tape types when dubbing. • Left and right record-level controls, bypassed during dubbing. • Auto-stop at end of Rewind or Fast-Forward. • No microphone inputs or headphone jack. • Ability to start Play or Record with external timer. • Automatic playback of cassette in second well when first is finished.

A–Has mono microphone input.
B–Has stereo microphone inputs.
C–Has Dolby C noise reduction.
D–Has autoreverse on one well.
E–Has audible Cue/Review feature.
F–Has one-button tape-type selector.
G–One well has auto tape-type selector.
H–Both wells have auto tape-type selector.
I–Tape-well light aids in seeing tape position.
J–Can insert 4-second silence between selections.
K–Can locate start of next or previous selections.
L–Has mechanically assisted controls with low-travel keys.
M–Has electrically operated controls with soft-touch keys.
N–Can automatically play after rewinding.
O–Rewind faster than others.
P–Lacks Dolby B noise reduction—has less-effective "DNR" instead.
Q–Has only automatic record-level control.
R–Lacks auto-stop after rewinding.
S–Lacks separate tape-type selector for 2nd well.
T–Rewind time longer than others.
U–Has headphone jack.
V–Single record-level control for both channels.
W–Has feature to mute sound while recording.
X–Record-level active while copying—wrong setting degrades copy.
Y–Lacks two-tape sequential play.
Z–Lacks ability to be started with timer.

Rack

All have: • Wood-grained particle-board cabinet with shelves for all components. • Hinged glass front door with "touch-to-open" latch.
Except as noted, all have: • Cabinet measuring 18 to 20 in. wide and 16 to 19 in. deep. • Casters. • Convenient screwless terminals for speakers.

A–Turntable accessible while user stands.
B–Has wireless remote control.
C–Remote can switch components.
D–Remote can select radio presets.
E–Remote can control scan of presets.
F–Remote can control tape Play, Stop, Fast-Forward, Reverse, Pause.
G–Remote can control CD Play, Stop, Scan, Track-select.
H–Remote can control audio mute button.
I–Remote can control phono.
J–Remote can control TV/VCR functions.
K–Controls and displays judged more convenient than most.
L–Cabinet construction better than average.
M–Easier than average to assemble.
N–Speaker terminals judged inconvenient.
O–Turntable judged difficult to reach.
P–Uses a separate remote for CD player.
Q–Controls and displays judged less convenient than others.
R–Construction judged worse than average.
S–Cabinet judged harder than average to assemble.
T–All units except phono and CD player in one chassis.
U–All units except CD player in one chassis.
V–CD player and FM/AM tuner combined in one unit.
W–Rack lacks casters.
X–Tested with *Radio Shack*'s *Optimus 1000* speakers.
Y–Tested with *Radio Shack*'s *Optimus 600* speakers.
Z–Tape deck uses special cables—cannot be substituted.
AA–Rack wider than others (31 in.).
BB–Has storage rack for 36 compact discs.
CC–Has storage rack for 8 compact discs.

FM ANTENNAS

In cities and their nearby suburbs, where signals are usually strong enough for a low-gain antenna, such as the one that comes with most stereo receivers, reception can be impaired by multipath reflections—signals that bounce off buildings, bridges, and other large objects. Mild multipath distortion can give a slight raspiness to such sounds as the *T* and *S* of speech. Severe multipath can distort sound so badly that listening becomes almost intolerable. A good FM roof antenna is designed to help control these and other reception problems. It has high gain (amplification) to bring in weak signals free from FM hiss. It is also directional, so it can discriminate against the reflected signals that cause multipath.

FM antennas come in various sizes, but the length of the boom is perhaps the most significant dimension in comparing sizes. The boom is the horizontal crossbar that holds an antenna's elements—the metal rods that gather radio signals.

Everything else being equal, the longer the boom, the more elements that can be spaced out along its length. There is also likely to be a correlation between boom length and signal-gathering ability.

Performance

When engineers test an FM roof antenna, they measure its gain and check other important performance characteristics including front-to-back ratio, front-beam width, and back-lobe width, among others.

Gain. This most important characteristic is determined by comparing an antenna's signal-gathering ability with that of a dipole antenna—one that works much like the indoor wire antenna that comes with most FM receivers. Gain is likely to be high with an FM roof antenna.

Front-to-back (F/B) ratio. A directional antenna, by its very nature, should be much more sensitive to signals arriving in front than to those arriving from behind. F/B ratio is the measure of that difference in sensitivity. In practical terms, a high F/B helps reject multipath signals. An FM roof antenna is likely to be at least adequate in F/B ratio.

Front-beam width. This indicates the front angle within which an antenna can best receive signals. Within practical limits, the narrower the angle, the better the antenna can reject unwanted signals.

Back-lobe width. Lobes are teardrop-shaped areas of receptivity for signals arriving from the rear of the antenna. Back-lobe width, like front-beam width, is a factor in resisting multipath. A directional antenna will pick up some signals reaching it from behind, but, as with front-beam width, the narrower the pickup angles, the better the antenna performance.

Flatness. If an FM antenna responds smoothly and evenly from one frequency to another, gain will also be smooth and even across the FM band, from 88 to 108 megahertz. The engineers call this "flat," and most FM roof antennas are at least satisfactory in flatness.

Impedance. An antenna provides its strongest signals to the receiver when the antenna matches the lead-in cable and the receiver in impedance—a complex electrical characteristic measured in ohms. The "standard" impedance for most antennas is 300 ohms.

Installation

Assembling and putting up an FM roof antenna is a fairly straightforward job, but one probably best left to a professional if the roof slopes steeply or the footing is unsure. A roof antenna comes with all the hardware needed to mount it, after assembly, to a mast. However, you'll have to buy the mast and more hardware to mount it firmly somewhere on the roof. Before attaching the antenna to the mast, shift the antenna up and down several feet to test for the best signal.

Install the antenna at least half its length away from such metal objects as pipes and fire escapes, and at least 5 feet away from any TV antenna. As with a TV roof antenna, don't tackle the job unless you're reasonably handy and comfortable on a roof. By all means, work as far away as possible from power lines to avoid the risk of electrocution—and be aware that it's easy to misjudge distance on a rooftop.

Twin-lead or coax. The antenna can be connected to ribbon cable (also called "twin-lead"), which has a 300-ohm impedance.

It's fine to use twin-lead cable if the down-lead from the roof to the receiver is direct and away from electrical wiring, gutters, and other metallic objects. Otherwise, coaxial cable is preferable. (Coax can also lessen spark interference from the electric motors of oil burners, electric shavers, and the like.) With coax, however, you'll also need a balun transformer (about $5 in an electronics supply store) to match the antenna to the cable—and maybe another at the FM receiver if it's not equipped for coax.

Additional accessories. If the coaxial cable is longer than 100 feet, a significant loss of signal strength can occur. To beef up a badly weakened signal, you can add an amplifier ($35 or more) at the antenna.

In areas with especially weak signals, or if the FM stations you want to hear come in from different directions, an antenna rotator could also be useful. It lets you change the antenna's orientation for optimum reception of distant signals coming from different points of the compass.

You might first check the feasibility of stacking two antennas on the same mast, one atop the other. Stacking can effectively double the gain of an antenna.

In addition, it may make sense to protect an antenna from lightning. But only a professional lightning-arrester installation can truly protect the antenna, the stereo equipment, and the house itself.

Obviously, with all possible accessories, an FM antenna installation may well run more than twice the price of the antenna alone. Still, if you have a high-quality FM system that suffers from reception difficulties, you may find the investment worthwhile.

FM indoor antennas

If you have good reception conditions—that is, you're located in a strong-signal area with few obstructions around to cause multipath distortion—you probably can do without the gain and directional qualities of an FM roof antenna. The simple indoor antenna that comes with most stereo receivers should suit you just fine.

Less happily, you may lack good FM reception but are unable to have a roof antenna. It could be too expensive (as much as $100, installed), or it may be impossible to mount an antenna on the roof. In such a situation, an indoor dipole antenna—the standard T-shaped "ribbon" tacked behind a bookcase or stowed under a rug—can provide tolerable reception. A folded dipole is usually a stretch of 300-ohm twin-lead cable fashioned in the shape of a capital *T*, with the top of the *T* (the signal-gathering element) about 5 feet long. You should be aware, however, that a folded dipole suffers from the limitations inherent in any dipole antenna: relatively low sensitivity and little directional discrimination of signals.

Because a dipole is most sensitive to signals arriving broadside, it works best when you can aim the dipole to pick up each station's signal at its best. Even if the signal is strong, multipath interference can be aggravated if a dipole is aimed in the wrong direction.

Rotatable antennas. If you are a critical listener who has no choice except to use an indoor antenna, you may want better reception than a fixed dipole can provide. You would then be best served by an indoor FM antenna like the *Sony AN300 Helical FM Antenna,* $80, the *Technics SHF101 FM Wing Antenna,* $90, or the *Winegard FM4400 FM Antenna,* $70.

These antennas sit on a horizontal surface and have an element that can be rotated for best reception. They must be adjusted whenever you change stations. With the *Sony* and *Technics,* you tune in the desired station on the FM receiver and adjust the antenna's tuning control for peak signal strength. Then you rotate the horizontal element until the signal becomes as clear and noise-free as possible.

The *Winegard* needs only to be rotated. Instead of a peaking-signal control, it has a built-in radio-frequency amplifier to boost incoming FM signals. To power the amp, you plug the antenna into a 120-volt outlet. In a strong-signal area, the *Winegard* is a good choice. With its radio-frequency amplifier, the *Winegard* also looks like the best bet where weak signals are the problem.

Every antenna's performance will vary from station to station and will depend on the antenna's orientation and its position in the room. If multipath distortion is a problem, a rotatable antenna is clearly preferable to a fixed dipole. But if the interference is from motors, car-ignition systems, and other broadband sources of noise, rotatable antennas are no better than an ordinary dipole.

Recommendations

Good FM reception may be hard to get if you live in an area with severe multipath. Unless it's directional, even an FM roof antenna may prove ineffective. Reception often improves with a rotatable indoor model, depending on where you live and what stations you receive. Since these rotatable antennas can be moved around the room, the improvement in reception can be dramatic.

You can try connecting the FM receiver to an outdoor TV antenna. The fact that the TV antenna is up and away from things probably ensures better performance than you'd get with an ordinary indoor FM antenna; such a connection requires a signal splitter. However, most TV antennas, when used for FM, perform no better than a simple dipole antenna installed on the roof, and those that *are* better are still not nearly as good as a directional antenna.

Your best choice, unless your FM reception presents no problems, is a high-gain directional FM roof antenna. A high-gain model is the first choice if you have a weak signal and there's a great deal of trouble with multipath in your area. In a strong-signal area you might experience problems with a high-gain antenna, which may overload a receiver with strong signals. In such a circumstance, choose a model with lower gain.

If you are a city dweller and live near an FM transmitter, a low-gain antenna is least likely to cause overload in a receiver. And a smaller-sized model may be worth considering if roof space is a problem.

Ratings of FM roof antennas

Listed by groups in order of estimated quality; within groups, listed in order of increasing price. Essentially similar models are bracketed and listed in order of increasing price. Prices are as published in a **February 1981** report and updated for **1984** publication.

Excellent ● (dot) · *Very good* ◓ · *Good* ○ · *Fair* ◒ · *Poor* ●

Brand and model	Price	Boom length	Gain	Front-to-back width	Front-beam width	Back-lo… wid…
Channel Master 4408	$83	12 ft.	Excellent	Very good	Excellent	Very good
Antennacraft GFM 10	50	10 ft.	Very good	Very good	Very good	Good
Jerrold QFM9	58	8 ft., 4 in.	Very good	Very good	Very good	Very good
Winegard CH6065 [1]	90	10 ft., 3 in.	Very good	Excellent	Very good	Fair
Archer 15·1636	17	5 ft., 10 in.	Good	Very good	Fair	Very good
Antennacraft GFM6	30	5 ft., 10 in.	Good	Very good	Fair	Very good
Finco FM5 [2]	43	10 ft.	Good	Very good	Very good	Good
Finco FM4G [2]	49	7 ft., 9 in.	Very good	Very good	Good	Fair
Winegard CH6060 [1]	50	5 ft.	Good	Very good	Good	Good
Channel Master 4409	59	7 ft., 3 in.	Good	Good	Very good	Excellent

[1] *Comes with balun transformer for use with 75-ohm coaxial cable.*

[2] **Finco** *brand name was changed to* **Finco-Sonim.** *Model numbers remain the same.*

AUDIO EQUALIZERS

Although a good equalizer can correct for inadequate loudspeakers and other deficiencies of a sound system, you should first consider spending your hi-fi budget on quality components. It's usually a good idea to give priority to upgrading the loudspeakers—the weak point in a typical high-fidelity system. When teamed with a good receiver, a pair of good loudspeakers is capable of reproducing sound quite accurately.

An equalizer is far from being a useless gadget, though. It can dramatically improve the quality of sound in a relatively poor hi-fi system. If you cannot afford a new system with more accurate sound, you could get better results with an equalizer.

If you already own components of high accuracy, there's a different question you should address: Can an equalizer do much for a system that already sounds good? The answer: Perhaps. There are special situations in which even good high-fidelity equipment doesn't produce the expected results. And if you are a musical hobbyist there are interesting things you can do with an equalizer, both in recording live music and in correcting poorly recorded material.

What an equalizer does

An audio equalizer plugs into a music system at the receiver's tape-monitor jacks so it can modify the audio signals before they are fed into the amplifier section, which drives the speakers. From this vantage point, the equalizer is in a good position to improve the sound system.

Compensate for room effects. Even if you could buy a perfect stereo system, you'd still have to contend with some imperfections in sound caused by the size, shape, and furnishings of your listening room. These room effects are virtually certain to impose some acoustic anomalies on the music: too much bass, for example, or weak treble, or unevenness in the mid-band. All hi-fi systems have bass and treble tone controls that help contend with such effects, but in many cases the acoustic anomalies can't be altered with the controls provided.

A good equalizer can help compensate for undesirable room effects, because you can use it to adjust the sound far more precisely than you can with conventional bass and treble tone controls.

Increase accuracy. A good equalizer must be able to adjust the system to respond fully, smoothly, and equally to the entire spectrum of musical sound, from deep organ tones to a triangle's tinkle. Ideally, that response should cover a range of frequencies extending from about 30 Hertz (Hz) in the bass to approximately 14,000 Hz in the treble—the range over which most instruments put out the tones and overtones that give them their special character.

In describing a sound system that can do those things, engineers use the term "flat" response, because an engineer's graph of the response at each frequency suggests a plateau that is flat through the audible frequencies. When response is not flat, the graph will show peaks or valleys resulting from the exaggeration or suppression of certain sound frequencies. Wherever there is a peak, musical notes at those frequencies will sound louder than normal; at a dip, the notes will sound muffled or distant. At either end of the graph, there can be a roll-off, a weakening of response at the low and high ends of the music spectrum.

A good equalizer can (within limits) smooth out the peaks and fill in the valleys by making response flatter, or more accurate. On the other hand, a good equalizer can also make it possible to depart from flatness and enhance music by creating emphasis to suit your personal preference.

Versatility

When setting up an equalizer in a typical home hi-fi system, versatility is the key to obtaining good sound accuracy. In other words, the more adjustments you make in the placement of frequency bands, the more flexibility you can get.

Models with variable bandwidth—and, usually, variable center frequencies—are sometimes called "parametric" by their manufacturers. Other models, often referred to as "graphic" equalizers, have a fixed bandwidth and fixed center frequencies.

Number of bands with width of bands. An equalizer divides the audio spectrum into a number of frequency bands for each stereo channel and provides controls to boost or cut the sound level in each band.

An important design factor is the width of the band—the particular slice of the spectrum that a band-level control can boost or cut. A bandwidth of one-half octave, say, covers six semitones—half an octave on a musical scale. Having a good many bands is the next best thing to having variable width of band.

Average interval in band center frequencies. As a rule, an equalizer's bands are spaced evenly across the audio spectrum. The spectrum covers ten octaves, so the bands of a ten-band model are likely to be spaced at about one-octave intervals between centers. The frequencies of the same note in neighboring octaves vary in a regular two-to-one ratio. (For example, the A above middle C has a frequency of 440 Hz and the A one octave higher has a frequency of 880 Hz.)

Another consideration is where an equalizer's bands are located in the spectrum—where, in other words, each band's center frequency is. With many models, the location of each center frequency is fixed, but some units have center frequencies that can be adjusted up or down the audio spectrum.

Equalizers with the most versatility have bands with adjustable center frequencies, a major asset in tracking the peaks and dips of the response curve.

Performance

There is little doubt that an equalizer can be an effective remedy for irregular bass response. The ability to smooth out the bass is a big plus, because uneven bass response is a common problem with home music systems where reflections from room surfaces and furnishings cause "standing waves"—resonances with fairly narrow peaks and dips in the bass—that are unique to each listening room and change with loudspeaker placement.

An equalizer may improve a poor music system enough to make it sound like a good system. In order to get good results with an equalizer, you have to rely on the tools a manufacturer may supply—a test record, for example, and a sound-level meter.

Because of band overlap among models with fixed width of band, a boost or cut in any band can interact to some degree with the bands adjacent to it. Depending on an equalizer's design, the interaction may result in greater cuts or boosts than you want. A good equalizer will have minimal interaction.

Total harmonic distortion. THD can negatively affect the sound of musical tones and overtones. Although many low-priced equalizers are reasonably free from THD, the perfectionist would probably prefer an audio equalizer that is superior in this respect.

Signal-to-noise ratio. S/N is the proportion of desired signal, such as music, to undesired signal, such as hiss (noise). Perfectionists, again, should probably stick with a model whose S/N is superior.

Features

Equalizers offer a variety of features that contribute to their versatility or convenience.

Band-level controls. These are continuously adjustable controls that let you boost or cut each band by a certain number of decibels. The controls in some models have click-stops over their full range, others just at the center position. The stops make it very easy to set the controls for no equalization.

AC outlets. An AC outlet lets you plug another audio component, such as a tape deck, into the equalizer rather than into a wall outlet. With most models, the outlets are unswitched, or electrically live, as long as the equalizer is plugged in. Outlets that are switched are live only when the equalizer is turned on.

Tape-monitoring provision. Most models have jacks for connecting a tape deck to the equalizer. The jacks substitute for the ones on the receiver that are occupied by the equalizer.

Provision for equalized tape output. This is a switch for routing signals to a tape deck after equalization rather than before. The feature could be helpful for taping certain special effects or to compensate for poorly recorded music when taping. To tape special effects using models without the switch, you must reconnect the equalizer's output to the tape deck's input.

Equalizer bypass switch. This lets you take the equalizer out of the audio circuit by electronic means. The switch makes it unnecessary to disconnect any cables or change any band controls.

Overall gain control. If there's a control for each stereo channel, you can balance the channels after each has been equalized. This feature is somewhat like a conventional volume control. Users tend to set it at the desired point and leave it that way.

Recommendations

Any equalizer should be able to perform the electronic tasks for which it was designed. Therefore, your choice of a unit can be based primarily on the extent of adjustments that a machine can make.

Paradoxically, that doesn't mean you should look for the equalizer with the most buttons. In a system whose sound initially has low accuracy, enormous improvements can be made even with the simpler, less-expensive equalizers. But a system that already produces high-accuracy sound will be improved only slightly, even when using every trick of the most complex equalizers. So before you seriously consider getting an audio equalizer, make a realistic assessment of your music system: Does your receiver have the extra power needed to drive loudspeakers in an equalized system? Will your speakers be able to handle the increased power without overdriving?

When you boost signal levels with an equalizer, the music system needs extra power from the receiver to drive the loudspeakers. Every 1 dB boost in level increases power needs by about 25 percent. If your listening is at relatively conservative loudness levels, that demand for increased power shouldn't pose any problems. But if you use an equalizer to push up signal levels beyond a music system's ability to handle them—to overcompensate for poor bass response in small speakers, for example—you run the risk of overdriving the receiver or speakers.

If you can't say with some assurance that your system can handle the extra power demands imposed by an equalizer, you should keep away from an equalizer until you can upgrade the system.

If power isn't a problem, consider the overall fidelity of the system, which, for most systems, depends primarily on the accuracy of the speakers. If the system's tone quality clearly leaves something to be desired, an equalizer may very well improve the system's overall sound. However, if the speakers are highly accurate—with a room accuracy score in the mid- to upper 80s or higher (see the section on loudspeakers, page 147)—it's doubtful that an equalizer will improve the system's accuracy very much at all.

Another consideration is the bass response. If the bass response in your listening room includes unwelcome peaks and valleys, an equalizer would probably smooth out the system's bass to good effect.

If your system has a lot of room for improvement, it makes sense to choose a less expensive equalizer. It won't match the versatility of some of the more expensive models, but it can still do a creditable job, at an affordable price.

If your system is already a fine one but you hope to increase its accuracy, select an equalizer based on versatility. The more expensive models stand the best chance of improving accuracy even more.

If you want an equalizer for some purpose other than improved accuracy—for special effects, say—the criterion of versatility would also be valid in most instances.

One exception might be the equalizer you want to use as a flexible tone control. For that purpose, you might try a model with fixed bandwidth and ten bands or less.

STEREO HEADPHONES

Headphones can add considerably to the enjoyment of a home audio system. They give you the full benefit of a record's wide dynamic range, from the loudest recorded sounds to the softest. They make the quietest of musical passages clearly audible by reducing or even eliminating outside noise. And they let you listen to loud music without disturbing others.

The stereo effect with headphones can be quite pronounced. Instruments in the middle of the orchestra may seem to be playing in the middle of your head. The violins may sound as though they're coming from far over to the right. Music recorded with multiple microphones, reverberation devices, and other sound-enhancing techniques may seem particularly disconcerting. If you're new to headphones, try them out first to make sure you like the super-stereo effect.

Comfort

Comfort is a key element in the use of headphones. If you don't find them comfortable, they're likely to stay on the shelf. Comfort, of course, is a highly subjective matter, but the style of the headphones can be a critical factor in this important consideration.

Around the ear. Circum-aural types have cushions that enclose ears completely. These cushions fit snugly against the head, isolating you from outside sounds. Around-the-ear models tend to be on the heavy side because their cushioning and sealing are quite elaborate.

On the ear. Supra-aural phones rest against the ears without enclosing them. They let you hear some surrounding sounds, unless the volume is set quite high. (The drawback here, though, is that at high volume levels some music could be audible, and perhaps even annoying, to those around you.) On-the-ear models tend to be small and light. Even smaller and lighter are the headphones designed for walkabout tape players.

Weight and fit. The heaviness of headphones is an important consideration, but for some people it is outweighed by proper fit—headphones that are neither too loose nor too snug. (Headbands are adjustable for fit, but only within limits.)

Some listeners complain that headphones warm their ears and make them perspire uncomfortably. A half-hour audition would let you judge the comfort of a particular set of headphones before you buy them.

Types of headphones. Most headphones are of the type known as dynamic. Each earphone contains a driver that is in effect a tiny loudspeaker. These units plug into the conventional headphone jack of a receiver, amplifier, or a compact stereo system.

Another type is electrostatic headphones, which are more expensive as a class. They reproduce sound by means of a thin membrane with a high-voltage charge. (There is no safety problem; the current they deliver is infinitesimal.)

Electrostatic models don't work with a conventional jack. Instead, they come with a small boxlike adapter. You plug the adapter into a wall outlet and usually connect the adapter to the loudspeaker terminals of the amplifier rather than to its headphone jack. The adapter has a selector switch so you can alternate between headphone and loudspeaker listening. A variation of the electrostatic headphones, one equipped with an electret diaphragm, does not need to be plugged into a wall outlet.

Cord length. With phones plugged into a jack, you're tethered to your audio system by a cord ranging from 7 to almost 14 feet in length. If that seems unduly restrictive, get a headphone extension cord at an electronics supply store. (With electrostatic models, which use nonstandard cords, you can lengthen the wires to the adapter.)

Performance

Accuracy of reproduction is the most reasonable objective standard by which to judge a set of headphones. Impedance, sensitivity, and the bass half-loudness point are also important, but should be secondary considerations.

Accuracy of reproduction. No matter what the design, good headphones, like good loudspeakers, should respond smoothly and uniformly to the full spectrum of musical sound, from the deep bass of the organ and the tuba to the treble sound of the triangle and the cymbal.

The phones should reproduce that broad range of sound without under- or overemphasizing any part of the range. Headphones with that capability are highly accurate—faithful to the original of the sound they reproduce, adding no coloration.

The accuracy of headphones is subtly different from that of loudspeakers. The human head and body affect the way you hear live music or music from loudspeakers, but they have no effect on the sound you hear from headphones because phones funnel music directly into your ears.

Head and body effects are important. By a phenomenon engineers call "acoustic diffraction," these effects boost a certain portion of the treble range of sound at the ear. Unless headphones can supply that boost, the ears won't hear the sound as they would from perfectly accurate loudspeakers.

Impedance. This characteristic, expressed in ohms, is the measure of the electrical load carried by an amplifier. Impedance should range within certain fairly broad limits.

It's unlikely to present any problems by being too low or too high, however, and most headphones should be fine for use with almost any contemporary receiver or amplifier. A few tape decks have a headphone jack that calls for a relatively high-impedance headphone—one with an impedance of 200 ohms or more.

Sensitivity. This helps determine how loud the sound will be at a given volume-control setting. The higher the sensitivity, the louder the sound produced at a given voltage. It's wise to turn down the volume before plugging in headphones that are high in sensitivity, because at high volume levels, you could damage the phones or hurt your ears.

High-sensitivity models are the best ones to use with a tape deck's built-in amplifier, and also as a substitute for the tiny headphones that come with walkabout tape players. (Note that sensitivity has nothing directly to do with quality.)

Stereo headphones can project very loud sound into a listener's ear without distortion. But musically demanding bass passages played very loudly can pose problems with some headphones, especially on-the-ear types.

And, of course, listening at high volume levels for prolonged periods can cause hearing damage; so it's a good idea to condition yourself (and others) to listen to music at moderate levels.

Bass half-loudness point. This is a touchstone for good response in the deep bass frequencies—the frequency at which the loudness of the bass drops to one-half its mean value over the rest of the music spectrum.

The lower the frequency at the drop (known as roll-off), the better. Gradual roll-off usually occurs more at higher bass frequencies with on-the-ear headphones than it does with the around-the-ear models.

Recommendations

Before you buy stereo headphones, decide the type you want and keep comfort near the top of your list of priorities.

Before making your final selection, wear the model for a half-hour or so, if you can, to see if it's comfortable. If discomfort develops over a longer period of listening, find out about exchange privileges.

No matter what type of headphones you choose, avoid extended listening at high volume levels. Such a practice carries a distinct risk of ear damage.

MICROPHONES

Whether plugged into a high-quality cassette deck (most of which have a mike input but include no mike), an open-reel tape deck, or stereo receiver, a microphone is the first link in a chain of electronic signals that can deliver ultimately no better sound than the microphone imposes.

A microphone turns the mechanical energy of sound waves into electrical energy—the reverse of what a loudspeaker does—and passes the energy along for amplification by a sound system.

Because there are a variety of microphones designed for different purposes, no mike will perform optimally in all situations. Microphones that readily pick up the delicate notes of a string quartet, for example, may not be best in spotlighting the bass beat of a rock combo.

Varieties

The choice of a suitable mike is in large part a personal matter: There is no single standard for "good" or "bad" microphones. Making the right choice can be tricky because there are hundreds of models on the market.

Dynamic vs. condenser mikes. A majority of microphones are of the dynamic type, and these have several advantages. As a rule, they're very sturdy and low in cost. They're reasonably sensitive to sound and capable of good fidelity. The diaphragm of a dynamic microphone, however, resonates at frequencies in the middle of the audio spectrum. A well-designed dynamic model can suppress the resonances in order to achieve good fidelity.

Condenser microphones use a different and sometimes more costly design than dynamic mikes and are often used for professional recording. They can offer very high fidelity because their tightly stretched diaphragm resonates at frequencies higher than those in the part of the audio spectrum that is most important for proper reproduction of recorded music.

A condenser model needs a tiny built-in amplifier to adapt its signals so they can be used by a sound system's preamp. Often, a built-in battery is used to supply power for that amplifier. (No built-in amplifier or battery is needed with a dynamic mike.)

Omnidirectional vs. unidirectional. Whether dynamic or condenser, a microphone can be designed to respond in a specific manner to sound reaching it from different directions.

Omnidirectional mikes are likely to be the best all-around choice for home recording. They are designed to pick up sound from all directions—around, above, and below—hence the prefix "omni." An omni mike is especially good at recording a group of performers or speakers. Placed in the midst of a musical ensemble, an omni mike can get an evenly balanced pickup from each instrument; at a roundtable conference, it can record each of the participants. When used in pairs for stereophonic taping, omni mikes serve well for recording a large ensemble, a chorus, or an orchestra.

However, because of their design, omni mikes may pick up unwanted sounds or too much reverberation in some locations. And at high frequencies, some omni models are more sensitive to sound from the front than from the sides and rear.

Unidirectional mikes are more sensitive to sound coming from the front than from the sides and back. Such a mike can help minimize the pickup of background noise—the coughing and fidgeting of an audience, for example, or the reverberations when you record in a big hall or auditorium. Uni mikes are good for pinpointing soloists or a small ensemble within a larger group, but to record a full orchestra you need to move a uni mike back some distance from the orchestra or use several mikes.

Because a uni mike minimizes sounds reaching it from the rear, it's the type to use where surrounding noise may be a problem—recording birdcalls, for example, or interviewing in a crowd, especially if you're unable to get close to a subject. When a uni model is pointed away from the loudspeakers of a public-address system, it will pick up less sound from the speakers and reduce the chance of feedback.

Modular microphones. These consist of a microphone body module that includes the battery for the built-in amplifier as well as a thread-on element called a "capsule." Interchangeable capsules can make an omni into a uni or vice versa.

Performance and features

Here are a few factors to keep in mind when considering the kinds of microphones on the market.

Frequency response. More than any other factor, frequency response determines the overall character of a microphone's sound. Condenser mikes tend to have better frequency response than dynamic models.

Ideally, to reproduce music with maximum fidelity, a microphone should respond to all the frequencies of the audio spectrum, ranging from about 30 Hertz (Hz) in the deep bass to 15,000 Hz in the treble. As a practical matter, though, a frequency response range of 100 to about 10,000 Hz is good for many music applications. To convey speech with near-perfect intelligibility, a microphone's frequency response range can be even narrower—from about 250 to 5,000 Hz.

A microphone that doesn't emphasize or deemphasize any part of the audio spectrum is said to have a "flat" response. The flatter the response, the better it is for serious audiophiles who want to record a wide range of music, from pop to classical, with maximum fidelity. Note, however, that a flat response may not be wanted in every situation: A rock group might want a microphone that exaggerates the bass or punches up the treble, or a thin voice might need the buildup of an accentuated bass. The flattest response isn't necessary for speech.

Impedance. The major significance of impedance, for the typical user of a mike, is its effect on cable length. The lower the impedance, the longer the cable you can use to connect microphone and amplifier without fear of weakening signals excessively in the treble. With a low-impedance model (600 ohms or less), the cable can be several hundred feet long—far longer than you usually need.

Sensitivity. This describes the strength of a microphone's electrical output for a given input of sound. Most mikes are adequately sensitive.

Some may be so sensitive, however, that with loud sound levels they could overload the equipment to which they're connected.

Omnidirectional effectiveness. At frequencies up to about 5,000 Hz, omni mikes are apt to be equally sensitive to sound reaching them from all directions. At higher frequencies, however, some omnis become less sensitive to sounds coming from the sides and rear. They then respond best to sounds reaching them head-on, like a uni mike.

Unidirectional effectiveness (directivity). The degree to which a unidirectional mike can pick up sound from the front while ignoring sounds from the sides and rear depends in part on the frequency of the sounds reaching it. The mike may discriminate against some frequencies strongly, weakly, or not at all. As a rule, unis pick up low-frequency sound from everywhere, including the rear. Their discrimination is strongest in the central part of the audio spectrum, less strong in the treble.

Immunity to wind noise. When you use a microphone outdoors, wind can cause annoying rumble. This is less of a problem with an omni mike than with a uni model, which tends to be more susceptible to wind reaching it from the front than from the sides.

To decrease susceptibility to wind noise, most mikes come with a slip-on windscreen or offer a screen as an optional accessory. (Since an add-on screen is made of stretchable foam plastic, it may also fit a model for which no screen is available.) An add-on screen can reduce noise by 10 to 20 decibels without having a significant effect on frequency response. If no screen is available, you can try changing the mike's position relative to the wind direction.

Immunity to "pop." A windscreen can also soften the unpleasant thump or popping you may hear when words beginning with *P, K,* and other explosive sounds are sung or spoken into a microphone. The closer you get to a mike, the worse the pop effect can be. Thus, with a mike sensitive to pop, it may be difficult to take advantage of being close to the mike. An omni mike is inherently more resistant to pop than uni models. You can count on a windscreen reducing pop by about the same amount as it reduces wind noise.

Immunity to hum. A microphone may pick up a hum of 60 Hz when used near strong magnetic fields produced by AC lines, transformers, electronic equipment, and the like. The dynamic type of mike is inherently more sensitive to magnetic interference than the condenser variety, but even a susceptible one is not likely to present a significant problem.

On/Off switch. This can be a convenience in many ways. For instance, the switch lets you turn off the mike in a public-address system for privacy or when no one is talking into the mike.

Size and weight. These may be important considerations if you require a microphone that is easy to handle, unobtrusive, or readily portable. On the other hand, you may want a hefty mike that has a solid feel.

Recommendations

When choosing a microphone, your selection will be governed by your answers to three basic questions.

Is the mike to be used mainly for reproducing speech? If so, it makes sense to limit your choice to an inexpensive model. You won't need the costly features designed to make a mike effective for recording the wide range of the music spectrum.

Will a sturdy, relatively inexpensive dynamic model meet your needs, or do you require the flat frequency response that typifies some expensive condenser microphones?

Do you want a mike with omnidirectional capabilities or would a unidirectional mike be better for your purposes? An omni has enough versatility to handle roundtable conferences as well as typical home recording needs. A uni mike can zero in on a speaker or a singer while discriminating against sound from other directions.

If you're uncertain of your exact needs, an omnidirectional mike makes the best all-around choice.

OPEN-REEL TAPE DECKS

Cassette decks are much easier to use than open-reel (or reel-to-reel) machines, and are typically much less expensive. Most people probably shouldn't consider buying an open-reel tape deck, but audiophiles who tape live performances still have several reasons to prefer open-reel recording.

Cassette recording has some limitations. It's nearly impossible to edit the narrow enclosed cassette tape, except with painstaking effort. Tape on open reels can easily be edited as unwanted material on the tape can be snipped out, and the tape spliced. The order of performance can even be rearranged on the tape.

Open-reel tapes do a much better job of rendering live music's dynamic range—the span between the loudest and softest sounds—and, unlike cassette decks, they do it with less reliance on noise-reduction systems. This is because the recording track is more than one-and-a-half times as wide as that of a cassette tape, and the speed of open-reel tape is faster.

You don't need tape that has such a wide dynamic range when taping from an FM broadcast or from most recordings; the dynamic range of signals from those sources has already been compressed. But that's not the case with live performances, nor is it the case with the better records or with compact discs.

Finally, open-reel tapes can provide much more recording and playing time than a cassette tape. A C-90 cassette tape, running at its standard speed of 1⅞ inches per second (ips), can record or play without interruption for forty-five minutes per side, ninety minutes altogether. (Longer-playing cassettes are available, but they often lack the C-90's quality.)

Loading a 7-inch reel of standard one-mil tape into an open-reel machine will also give you running time of forty-five minutes. However, a 10½-inch reel loaded with the same tape will give you double that time. Open-reel machines are typically run at a speed of 7½ ips, which takes best advantage of their capabilities, but they also do very well run at 3¾ ips. By recording and playing at 3¾ ips, the running time can be doubled once more to produce up to three hours of uninterrupted performance.

Some machines can automatically reverse the tape in the playback mode. While that also doubles uninterrupted playing time, you won't be able to edit a tape recorded on both sides.

An open-reel deck is designed to plug into a component system that already has loudspeakers and an amplifier. Open-reel tape decks typically have a "four-track" arrangement: They put two stereo programs on a tape, recording one in each tape direction.

A deck's tape-moving machinery includes a supply reel and a take-up reel to hold the tape and, usually, a capstan and a rubber pinch roller to move the tape past the deck's heads in recording, playback, and erase operations. Because tape passes the record head before the playback head, you can monitor what was just recorded to be sure you're getting what you want: The playback head will let you hear the signals a split second after they're imprinted on the tape. Tape monitoring, which can be important to the hobbyist, is possible only with open-reel tape decks that have separate record and playback heads.

You can use the numbers shown on a tape counter to help you find your way rapidly to a particular passage on the tape.

Recommendations

Most people—even those who are hi-fi buffs—can probably do without editing tapes or recording evening-long concerts in their entirety. If so, they are probably better off with a good cassette tape deck than with an open-reel unit, but many audiophiles continue to insist on that combination of sound quality and versatility that characterizes the more expensive open-reel tape decks.

If you do decide to shop for an open-reel tape deck, you can save money with little loss in sound quality by buying a deck that takes 7-inch reels. The capability of playing 10½-inch reels comes with a stiff price jump. But for those who like to edit tape, a higher price tag may be unavoidable: It usually buys more features—an important consideration in tape editing.

COMPACT STEREO SYSTEMS

A compact is a stereo system with a minimum amount of fuss at a minimum price. It's intended for the person who wants to buy a packaged music system that requires the least possible amount of assembly, and not for the hi-fi enthusiast who wants higher sound quality and is looking forward to the challenge of selecting components—a pair of loudspeakers, a stereo receiver, a turntable with a good cartridge, a compact disc player, and a tape deck.

Sound. The sound's not bad with a good compact—though not nearly so good as the sound from a high-quality component system. The sound from a compact stereo will probably suit you only if what you want is music for easy listening, or for a school dormitory, for dancing, for any kind of listening that's not very critical. For serious listening, you'll want something better.

Inadequate speakers are usually the main deficiency with a compact stereo system. There's not likely to be much deep bass at all in a compact's speakers; the mid-bass may be artificially boosted, making the bass sound boomy, and the treble may be weak. Some compacts have a hollow resonant sound, perhaps caused in part by the thin panels of the speaker cabinets.

Most compact systems do not have a very powerful amplifier. Typically, it's barely powerful enough to drive the speakers at listening levels that might be considered loud—for example, some systems can't manage music at party-level volume without distortion.

If you care about music, you would undoubtedly prefer the sound system you'd get spending your hi-fi budget on separate components. Tone quality will be better, overall, that way. The speakers have flatter response in the high frequencies and produce stronger deeper bass. Also, the flutter is lower. Phono tracking is better at lower tracking force (which would save wear and tear on records). The one thing you're *not* likely to get with a compact system is a cassette deck.

Performance and features. If you consider the record player a prime source of hi-fi music, phono performance is a major criterion in deciding whether to get a compact stereo.

The only way to make sure you will be satisfied with the sound of a compact system is to try it yourself. A model whose phono sounds good to you is almost certain to sound equally good when on FM. Your expectations of the tape deck portion of a compact stereo, however, should not be too high. The machine will work, but most likely the performance will not be excellent.

If you do decide a compact system will meet your needs and you're faced with a choice between two models that are about equal in other respects, you might prefer one for a particular feature. For example, some models automatically switch off the whole system—turntable and amplifier—if you turn on the system via the turntable control rather than the receiver's On/Off switch. These models allow you the luxury of dropping off to sleep while listening to records.

Remote-speaker outputs let you connect a second pair of speakers and enjoy a system's music in another room. A headphone jack permits you to plug in headphones for private listening. And you may find it useful to have a tape-output jack with which to send a signal to another tape recorder.

Recommendations

A compact stereo may have a number of sophisticated features and the state-of-the-art look of a component system, but the resemblance is often only skin deep. There's no denying the convenience of a plug-it-in-and-go music system—especially one that's easy on the budget—but you do pay a significant price in the quality of sound you can expect from a compact system. Unless you are severely limited in space, want music mainly for its background effect, or are unwilling to shop for separate components, you are much better off putting together a system of your own.

Portable Tape Recorders/ Players and Radios

A radio provides *casual* listening to music, news, and talk shows; in addition, today's radio can provide a service as well as information. Clock radios can wake you up in the morning. Varying kinds of hobbyists use different types of specialty radios to track weather and scan the public-service communications bands. Shortwave radios keep you in touch with the world. And radios combine with stereo tape player/ recorders in portable units.

Small tape players—with or without a radio—have grown in popularity as some versions have diminished in size. If you want to buy a tape recorder, you need to consider several options. There is the microcassette recorder, whose tiny tape can play for up to two hours. If you want more versatility—for example, the ability to use standard-size cassettes—you should consider a different unit entirely. For less money than you'd pay for a microcassette recorder, you can buy a heavier and bulkier (but still quite portable) cassette recorder. Or you may prefer the company of a walkabout tape player.

WALKABOUT STEREOS

Walkabouts have become smaller, lighter, and fancier. A built-in AM/FM stereo radio is now commonplace. Many models feature such exotica as push-button electronic radio tuning or a graphic equalizer.

Taped music

Most walkabouts are capable of delivering good sound quality. The best are nearly as good as a moderately priced, full-sized cassette deck. Unfortunately, a good deal of that performance potential goes to waste when you actually walk about with one of these players.

Tests conducted by Consumers Union showed that a walkabout's tiny headphones just can't deliver the wide range of sound or frequency response that players are capable of reproducing. Even the best headphones degraded the sound quality significantly. But that doesn't mean the sound through the headphones will be unpleasant. Even mediocre headphones create a "superstereo" effect, which often draws your attention away from defects in the sound. And high-quality headphones for a walkabout are readily available from $30 to $40 and up. Almost every walkabout is capable of providing noticeably better sound quality with higher-quality headphones.

Moving around with a walkabout on your belt may well increase the amount of tape flutter—an unpleasant wavering of the sound caused by fluctuations in tape speed. There will be a lot less flutter when you sit still. Not surprisingly, the units with a snug belt clip jiggle the least and so produce the least flutter when you jog with them.

Some units tested wouldn't hold precisely to the standard speed of 1⅞ inches per second. But the variations aren't likely to be bothersome.

Walkabouts as radios

Overall, the FM radio in a walkabout is just as impressive as the tape player. The performance should be similar to what you get from a good low-cost receiver for a home hi-fi system.

Signal-to-noise. The ratio of signal-to-noise measures the background noise (hiss) detectable behind the sound track. The best players have Dolby noise-reduction circuitry that effectively cuts the hiss without affecting the music.

Sensitivity. This is a key measure of any radio's performance. It tells you how well the radio pulls in weak or distant FM stations.

FM selectivity. This is another key measure of performance. It's especially important if you live in a big city where dozens of stations crowd the FM band. Selectivity is a measure of how well a radio rejects interference from stations adjacent to the one the radio is tuned to.

Strong signal. A nearby FM broadcast antenna can sometimes overload the tuner so that a station appears in more than one place on the dial. Some models have a Local/Distant switch, sometimes labeled "DX," which is intended to reduce strong-signal problems.

Capture ratio. This tells you how well a radio screens out noise from reflected radio signals and from distant stations on the same frequency as the station you've tuned in.

Frequency response. The FM tuner of almost any walkabout should be quite good at reproducing the spectrum of tones.

In general, a walkabout's radio may not be a suitable accompaniment for a couple of laps around the jogging path—not because of the radio's quality but because of the antenna. Walkabouts use the headphone wire as the radio antenna. Moving around may cause the program to fade in and out because the antenna is shifting in relation to the station's transmitter. And you may have to adjust the wire for best reception even when you're sitting down.

Controls and convenience

Clips and straps. Most walkabouts come with a clip that slips over a belt. That's a handy arrangement. A walkabout with a shoulder strap may jiggle around when you run with it. A unit with a belt loop rather than a clip forces you to unbuckle your belt to strap it on.

Control position. For the sake of convenience, it's best to have the tape and tuner controls on top, so you can look down on them when the unit's clipped to your belt. The controls on most units are pretty small and tricky to master.

Tape controls. A walkabout player that can't record has simple tape controls—typically, Play, Fast-Forward, Rewind, and Stop. Some units have no Rewind control. To back the tape up, you have to flip it over and use Fast Forward. That's a very annoying shortcoming.

A number of units also reverse automatically when they reach the end of a tape. It's useful to be able to control whether you want the tape to reverse just once or to do so continuously until you turn the machine off.

Other handy tape features include a light to tell you the tape is running (it can remind you to turn off the player when you remove the headphones) and automatic stop after rewind (not only does it mean one fewer button to press, it also helps preserve the batteries).

Tone and volume controls. More and more walkabouts come with a graphic equalizer instead of conventional bass and treble controls. An equalizer lets you adjust different bands of the music spectrum, but is largely a frill that provides little edge over conventional controls.

Tape-head cleaning. A walkabout's tape heads are best placed at the hinge side of the tape hatch. That makes it easy to reach the heads for cleaning, an important consideration for any tape player.

Batteries. Keeping a walkabout supplied with batteries can be a real headache.

The life of a battery depends on its type and freshness and your tape- and radio-playing habits. Therefore, it's impossible to say how many hours of use a particular player will give on one set of batteries.

Radio battery life. Using the radio consumes much less power than playing a tape does. A good unit can play the radio for several days straight on a single set of batteries.

Tape battery life. This is a vital consideration for a walkabout player. The best units can run all day on a fresh set of batteries; the worst, only about six hours.

A walkabout powered by a solar cell (with a battery as backup) provides the ultimate in battery life—assuming you use the walkabout in a sunny location.

For use indoors, consider an AC converter that plugs into an ordinary outlet. The converter may also serve as a battery charger if you are using nickel-cadmium (nicad) batteries. Since the batteries are out of sight within the machine, however, it's possible to forget which type are in there and use the AC converter with alkaline or carbon/zinc batteries. They won't recharge.

Almost every walkabout has a jack to draw electricity from a DC power source, such as a car's cigarette lighter.

Recommendations

It's possible to play walkabouts loud enough and long enough to endanger your hearing. Most people probably have enough sense to turn down the volume before it becomes dangerously loud. Still, one could learn to live with the volume turned up—especially if the music is blocking out some loud ambient noise such as a lawn mower.

You should also think twice about jogging, cycling, or driving with a walkabout playing loud enough to drown out car horns or sirens. In fact, in many communities it's against the law to wear headphones if you're behind the wheel or jogging down the road.

When you shop, don't take suggested retail prices too seriously. Like most audio gear, walkabouts are heavily discounted.

Ratings of walkabout stereos

Listed by types; within types, listed in order of estimated quality based on performance as tape player, reflected in Overall Tape Score. Differences in score of 8 points or less were judged not very significant. Prices are manufacturer's suggested retail price, as published in a **July 1987** report; + indicates shipping is extra.

				Tape					
	Price	H × W × D, in.	Weight, oz.	Overall tape score	Freq. response, phones	Freq. response, direct	Flutter, moving	Speed accuracy	Signal-to-noise
Tape player/radios									
Aiwa HS-T500	$185	4½ × 3¼ × 1½	16	91	○	◉	○	◉	◉
Sanyo MGR99	130	4½ × 3 × 1½	11	90	◒	◓	◓	◓	◉
Realistic SCP-5	80	5¾ × 3¾ × 1⅞	15	89	◒	◓	○	◒	◉
JVC CX-F7K	160	4½ × 3¼ × 1¼	10	88	◒	◓	◓	◉	◉
Toshiba KT-4066	99	5 × 3½ × 1⅞	14	87	○	○	◓	◒	◉
Sony WM-F41, A Best Buy	45	5¼ × 3¾ × 1¾	11	86	◒	○	◓	◉	◒
Sony WM-F80	160	4⅜ × 3½ × 1½	12	85	○	◓	○	◉	◉
Sony WM-F107	230	4½ × 4⅜ × 1⅝	13	85	◒	◓	◓	◓	◉
J. C. Penney 6090	80	4¾ × 3½ × 1⅝	13	85	◒	◓	○	◉	◉
Panasonic RX-SA77	100	4¾ × 3½ × 1¾	13	84	○	◓	○	◉	◉
Emerson CRS25	50	5½ × 3¾ × 1⅝	12	83	◒	◓	○	◉	○
General Electric 3-5436	48	5¾ × 3¾ × 1¾	13	83	○	◓	◓	◒	●
Quasar GX-3686	97	4¾ × 3½ × 1¾	12	82	◒	○	○	○	◉
Toshiba KT-4046	69	5 × 3½ × 1⅞	13	82	◒	◓	○	○	◉
Sharp JC-786	130	4¾ × 3½ × 1½	11	81	◒	○	◓	◒	○
Sears Cat No. 21169	40+	5⅞ × 3¾ × 1⅝	15	78	●	◒	◓	◒	○
Sharp JC-126	30	5½ × 3⅝ × 1¾	12	75	◒	○	○	○	○
Panasonic RX-SA70	60	5½ × 3¾ × 1¾	14	75	◒	○	○	◉	○
Quasar GX3666	35	5⅞ × 3¾ × 1¾	14	74	◒	○	○	◉	●
Aiwa HS-T200	110	4½ × 3½ × 1¾	13	73	●	○	○	◒	◉
Sanyo MGR67	35	5½ × 3⅝ × 1¾	14	71	◒	○	◒	◉	○
Sansui FX-W51R	120	4¾ × 3½ × 1⅝	13	71	●	◓	◒	◒	●

◉ ◓ ○ ◒ ●

Better ⟵⟶ *Worse*

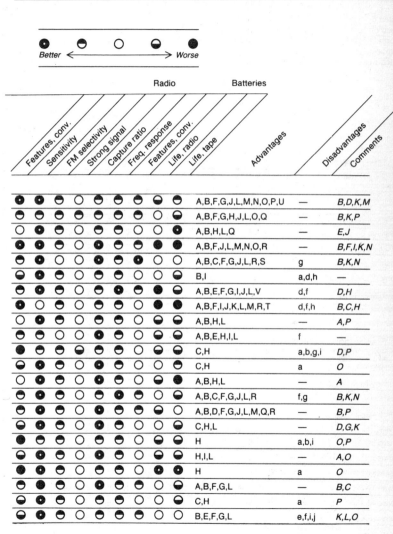

	Radio						Batteries				
Features, conv.	Sensitivity	FM selectivity	Strong signal	Capture ratio	Freq. response	Features, conv.	Life, radio	Life, tape	Advantages	Disadvantages	Comments
◉	◉	◓	○	◓	◓	◓	◒	◓	A,B,F,G,J,L,M,N,O,P,U	—	*B,D,K,M*
◓	◓	◓	◓	◓	◓	◓	○	◒	A,B,F,G,H,J,L,O,Q	—	*B,K,P*
○	◉	◓	○	◓	◓	○	○	◉	A,B,H,L,Q	—	*E,J*
◉	◉	◓	○	◉	◓	◓	●	●	A,B,F,J,L,M,N,O,R	—	*B,F,I,K,N*
◓	◉	○	○	◉	◓	◉	○	○	A,B,C,F,G,J,L,R,S	g	*B,K,N*
◒	◉	◓	○	◓	◓	○	○	◒	B,I	a,d,h	—
◓	◉	◓	○	◓	◉	◓	●	◒	A,B,E,F,G,I,J,L,V	d,f	*D,H*
◉	○	◓	○	◓	◓	○	●	●	A,B,F,I,J,K,L,M,R,T	d,f,h	*B,C,H*
○	◉	◓	○	◓	◓	○	◒	◒	A,B,H,L	—	*A,P*
◓	◓	○	○	◉	◓	○	◒	◒	A,B,E,H,I,L	f	—
●	◓	◓	◒	◓	◓	○	◒	◒	C,H	a,b,g,i	*D,P*
◒	◉	◓	○	◉	◓	○	○	◓	C,H	a	*O*
○	◉	◓	○	◉	◓	○	◒	●	A,B,H,L	—	*A*
◓	◉	◓	○	◓	◉	◓	○	◒	A,B,C,F,G,J,L,R	f,g	*B,K,N*
◓	◉	◓	○	◉	◓	◓	◒	○	A,B,D,F,G,J,L,M,Q,R	—	*B,P*
◒	◉	◓	○	◉	◓	○	○	◒	C,H,L	—	*D,G,K*
●	◓	◓	○	◓	◓	○	◒	◒	H	a,b,i	*O,P*
◒	◉	◓	○	◉	◓	○	◒	◒	H,I,L	—	*A,O*
●	◉	◓	○	◓	◓	○	◉	◉	H	a	*O*
◓	●	◓	○	◉	◓	◓	○	◒	A,B,F,G,L	—	*B,C*
◒	◉	◓	○	◓	◓	○	○	◒	C,H	a	*P*
◒	◉	◓	○	◓	◓	◓	○	○	B,E,F,G,L	e,f,i,j	*K,L,O*

Ratings of walkabout stereos (continued)

Listed by types; within types, listed in order of estimated quality based on performance as tape player, reflected in Overall Tape Score. Differences in score of 8 points or less were judged not very significant. Prices are manufacturer's suggested retail price, as published in a **July 1987** report; + indicates shipping is extra.

					Tape				
	Price	H × W × D, in.	Weight, oz.	Overall tape score	Freq. response, phones	Freq. response, direct	Flutter, moving	Speed accuracy	Signal-to-noise
Tape players									
Aiwa HSG600	150	4¼ × 3½ × 1¼	11	95	○	◓	○	◉	◉
Realistic SCP-19	60	4⅞ × 3¾ × 1⅝	12	88	○	◓	○	◉	◉
Sanyo MGP44	50	4⅜ × 3 × 1¼	10	88	◒	◓	◓	◒	◉
Sony WM-75	110	4½ × 3⅞ × 1¾	12	85	○	◉	◓	◒	◉
Realistic SCP-20	30	4½ × 3½ × 1¾	10	83	○	◓	◓	◓	◒
Sony WM-18	75	5½ × 3½ × 1½	14	77	◒	◓	○	◉	◒
Sanyo MGP 17	25	4¾ × 3⅝ × 1⅝	11	75	○	◓	◒	◒	◒
Panasonic RQ-JA61	22	4⅝ × 3⅝ × 1¾	11	71	◒	◒	◒	◒	○
Sears 21124	20	4¾ × 3⅝ × 1⅝	11	70	◒	○	◒	◉	◒

Specifications and Features

Except as noted, all have: • Headphones included. • Autoreverse. • Clip or strap to attach player to belt. • Rewind button. • FM Mono switch. • 1 headphone jack. *Except as noted, all lack:* • Automatic stop after rewinding.

Key to Advantages

A–Has Dolby B noise-reduction circuitry.
B–Can use Type I, Type II, and Type IV tape.
C–Has 3-band graphic equalizer.
D–Has 4-band graphic equalizer.
E–Has 5-band graphic equalizer.
F–Has tape-running light, a useful reminder that unit has been left on.
G–Light reminds you that radio has been left on.
H–Has FM stereo light, a minor advantage.
I–Has FM Local/Distant switch.
J–Has Autoreverse mode switch for tape.
K–Case is moisture- and dust-resistant.
L–Has switch to change tape direction.
M–Tape stops automatically when it has finished rewinding. (On other units, you must press Stop after rewinding; if you don't, the tape motor will continue to drain the batteries.)
N–Has switch to lock tape controls; prevents unit from being turned on accidentally.
O–Has protective case.
P–Has 2 headphone jacks.
Q–Unplugging headphone stops tape motor.
R–Tape heads easy to clean.
S–Has push-button radio tuning.
T–Has solar-powered battery.
U–Has remote tape switches.
V–Can play sound from TV channels 2–13.

⊙ Better ◓ ○ ◒ ● Worse

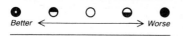

	Radio						Batteries				
Features, conv.	Sensitivity	FM selectivity	Strong signal	Capture ratio	Freq. response	Features, conv.	Life, radio	Life, tape	Advantages	Disadvantages	Comments
⊙	—	—	—	—	—	—	—	○	A,B,E,F,J,L,M,O	f	*B,F,K*
◓	—	—	—	—	—	—	—	◒	A,B,F,L,Q	—	*E,J,K*
◓	—	—	—	—	—	—	—	◒	A,B,F,J,L,O,Q	—	*B,K,P*
⊙	—	—	—	—	—	—	—	◒	A,B,F,J,K,L,P,R	—	*B,H*
◒	—	—	—	—	—	—	—	◒	C,F	a	*E, J*
○	—	—	—	—	—	—	—	◒	B,F,J,L	c	*D,G,K*
◒	—	—	—	—	—	—	—	◒	C	a	*P*
●	—	—	—	—	—	—	—	◒	—	a	—
◒	—	—	—	—	—	—	—	◒	C	a	*P*

Key to Disadvantages

a–Lacks Autoreverse.
b–Lacks Rewind.
c–Has shoulder strap instead of belt holder.
d–Lacks FM Mono switch.
e–Volume control hard to turn.
f–Tape difficult to see.
g–Opening in unit allows dust in tape chamber.
h–Lower tape tension than most; may affect speed accuracy.
i–Slower rewind than most.
j–Fast-forward was slow.

Key to Comments

A–Has loudness control to boost bass and treble.
B–Has tape-door latch.
C–Radio dial on top.
D–Tape controls on top.
E–Has separate volume controls for left and right channels.
F–Belt loop on protective case.
G–Has provision for shoulder strap.
H–Has folding headphones.
I–Comes with rechargeable battery and charger.
J–Ac adapter jack applies voltage to battery.
K–Has tape-direction indicator.
L–Has Cue switch that lets you listen to tape while it fast-forwards; can help you locate a given selection quickly.
M–Music sensor detects silent tape passages; lets you locate a given selection quickly.
N–Autoreverse switch lets you play one side of tape.
O–Pressing Stop button ejects tape.
P–Model discontinued at the time the article was originally published in *Consumer Reports*. The information has been retained here, however, for its use as a guide to buying.

PORTABLE CASSETTE RECORDERS

Portable cassette recorders that use standard-size cassettes are still worth thinking about for recording at meetings, interviews, or family events. One of them is fine for recording thoughts for a diary or for "talking letters." In its play mode, it can supply music on a trip or picnic. The "shoe-box-style" recorder lacks the compactness and very light weight of microcassette recorders and it doesn't have the stereo of a walkabout tape player, but a portable recorder is generally cheaper to buy than either a microcassette or a walkabout.

A recorder is a self-contained unit, with a built-in microphone. It's powered by batteries or from an ordinary wall outlet.

Although the recorder has only a single speaker, it takes the standard stereo cassette. When playing a stereo recording, the recorder merges signals from the two tracks and plays them through the single speaker. When recording, it prints the same signal on both tracks, which results in a monophonic tape.

Recommendations

As audio gear goes, a portable cassette recorder is relatively unsophisticated. Even so, it does a creditable job of recording and playing music for casual listening—and, of course, of recording and playing back speech intelligibly.

There's no extra benefit in using expensive tapes in a portable unit. It lacks the electronic refinements necessary for getting the best out of a tape. Bargain-priced tapes, however, could increase the risk of an exasperating tape tangle or cassette misfit. You probably won't go wrong if you follow a middle course and buy the lowest-priced brand-name Type I (ferric) tape available.

PORTABLE RADIO/CASSETTE RECORDERS

The so-called box is an AM/FM radio, a cassette deck for recording and playback, a pair of built-in loudspeakers, and a pair of built-in microphones. It operates on household current or on disposable batteries. It may also operate on 12 volts DC through an automobile cigarette lighter.

On a picnic, on the porch, or at the beach, a box can provide respectable sound. Although most boxes are only a little larger than a briefcase, big ones may weigh close to 20 pounds. So you'll have to be in pretty good shape if you want to carry the sound very far.

Performance as a cassette deck

Although some models can handle metal (Type IV) tape, you're probably just as well off using conventional ferric (Type I) tape.

Ferric tape costs about half as much and there doesn't seem to be really solid evidence that metal tape is superior with these machines.

Tone quality. A good unit can come close to sounding like a reasonably good sound system—one that consists of a deck, a tuner, an amplifier, and mini-speakers about the same size as those in a box.

Freedom from flutter. This wavering in pitch, which is caused by imperfections in the mechanism of a tape recorder, can occasionally be noticeable. If possible, test a unit you're considering by listening to a tape of piano music. Sustained piano notes are a good test for freedom from flutter.

Speed accuracy. When a cassette tape consistently travels faster or slower than the correct 1⅞ ips, the music's pitch and tempo may be off. Measured by strict high-fidelity standards, some units may wander a bit from the norm—enough, perhaps, for a critical listener to notice. But to an average listener, the inaccuracy would probably be undetectable.

Cassette deck features

A typical box has the usual array of controls: Play, Record, Fast-Forward, Rewind, Pause, Stop, and Eject. You can record from the built-in microphones, from the built-in radio, or from an outside source such as an amplifier, another radio, or another tape deck.

Several cassette deck features are worth noting.

Automatic level control. The ALC evens out the amplification during recording so that loud and soft sounds play back at much the same level. As in a microcassette recorder, ALC is better suited to speech than to music. It may compress the dynamic range of a piece of music too much and make everything sound about equally loud. To preserve the soft and loud passages, you need an On/Off switch to shut off the ALC so you can set your own recording levels.

Noise reducer. A tape usually produces a distinctive background hiss, which you can hear during quiet musical passages. A Dolby noise-reduction system helps to suppress tape hiss in recording and playback.

Sound mixer. With a mixing control and a jack for a monophonic microphone, you can simultaneously record your voice and music from the radio or from an outside source—useful for taping music and narration for a slide show, for example.

Auto stop. In addition to shutting off after recording or playing to a tape's end, a box should shut off automatically at the end of a fast-forward or rewind operation. In this way you don't have to remember to turn the machine off at the end of the tape.

Automatic reverse. With this feature you can record or play the entire tape without interruption—that is, without flipping the cassette over when you're recording or playing. Instead, the tape automatically reverses direction at the end.

Automatic replay. Another convenience lets you rewind and replay all or part of one side of a cassette over and over again. You simply push a couple of buttons that electronically mark the beginning and the end of the segment you want. Or you may be able to automatically replay one side of a cassette by pressing both the rewind and play buttons. The buttons will stay down until the rewinding is finished. Then the machine shifts into play.

Cue/Review. A Cue/Review device lets you fast-forward or rewind the tape without blocking the sound, enabling you to pick out a segment you want to hear. When you're running the tape forward or backward, the sound will be distorted but is usually clear enough for you to identify the passage.

Light-touch control. Piano-key controls are sometimes hard to press. Short-stroke, light-touch controls require practically no effort at all.

Music selection system. If you play a tape that has at least four seconds of silence between recorded segments, some models will sense the silent gaps. If a segment is playing and you want to skip it, you press the right buttons and the machine will fast-forward to the next segment. Or you may be able to skip several segments with the press of a button.

AM filter. A low-level background whistle may sometimes be audible when you record from AM radio. There should be a built-in filter to minimize that interference.

Input jacks are standard with these recorders. "Line-in" or "aux-in" jacks let you record signals from another tape deck, another radio, or some other sound source.

Radio performance

FM performance, more critical than AM, depends on four characteristics: tone quality, sensitivity, selectivity, and image rejection.

Tone quality. With a good FM radio, the music should sound almost as good as a hi-fi system (one with mini-speakers).

FM sensitivity. Using the built-in antenna, sensitivity—the ability to receive weak stations noise-free—is likely to be adequate.

FM selectivity. This is the ability to receive a weak station without interference from strong stations close to it on the dial. Selectivity and sensitivity should be well matched so a radio can sort out all the stations that it is able to receive.

FM image rejection. Most models do an adequate job of rejecting—that is, ignoring—aircraft signals, which can impair FM reception near an airport.

Other features

A box radio-recorder should have a tape counter, a light that indicates when the radio is receiving FM stereo, and connections for external speakers, microphones, and earphones.

Some other features, which vary from model to model, are:

Tone controls. Some models have one control for adjusting the treble and another for the bass—a design that offers flexibility. Others, however, have only a single control for both treble and bass. And still others are equipped with multiband equalizers.

A loudness control boosts bass sounds at low-volume settings. (Your ear tends to be less sensitive to bass sounds played at low volume.)

Meter. Most models have at least one dial-and-needle meter or a light-emitting-diode (LED) meter to help you monitor the recording level. As a rule, the meter also helps tune the radio precisely and indicates the condition of the batteries.

Radio-dial light. If there's dial illumination, it should be push-button operated and stay on only as long as you press the button, to minimize battery drain.

Antenna. Telescopic antennas are either single (monopole) or double (dipole) rods. The antenna tends to be fragile, so look for a model with an antenna that is easy to replace.

Terminals for an external FM antenna make it possible to improve the reception of weak stations with an FM antenna (but they make the set less portable).

Battery operation. Use alkaline batteries for their relatively long life. Remember that the louder you play the radio, the faster the batteries run down. It's wise to remove the batteries from a radio when the set won't be in use for a lengthy period: Batteries occasionally leak and damage could result.

Three-way power. Besides running on batteries, a unit should operate on 120-volt AC power from a wall outlet. With a special cable or adapter, it may be able to draw its power from the cigarette-lighter socket of an automobile or boat. If you do get an adapter from an electronics-supply store, be sure it's the right one. A cable may look fine but fail to work if the polarity is wrong.

Some models can also operate on 240 volts AC, but you have to get an adapter that fits the outlet.

Shortwave band. If there's a provision for shortwave reception, don't expect too much. Chances are the sensitivity will be none too good and the tuning so touchy that you won't be able to locate a station on the dial with accuracy.

Wide stereo. Because the speakers are only about 1½ feet apart at most, there is not very much stereo effect if you're more than, say, 3 feet away from the set.

To improve the stereo effect, many models have a wide stereo option, which makes the speakers seem audibly farther apart than they actually are. It works, but it makes the sound seem harsher than normal. Detachable speakers you can unhook and separate from the set will, of course, increase the stereo effect.

MICROCASSETTE RECORDERS

A microcassette recorder is extremely compact and very lightweight. It is often not much bigger than a pack of cigarettes. The tape is in a tiny cassette about the size of a matchbook. (The cassette can't be played on an ordinary tape deck.)

A microcassette machine typically operates monophonically and is designed primarily to record the voice and to play it back through a built-in speaker or through a plug-in earphone accessory. Because it is no trouble to carry, you can use the machine to tape just about anything at a moment's notice: a classroom lecture, a business discussion, town hall and school board meetings, or professional advice from your doctor or lawyer, among other things. You can use it around the house to leave messages, perhaps, or to tape material from a radio or TV set. Some cassettes are capable of recording as much as three hours of talk. Many units are designed for one-handed operation or are voice-actuated (vox) so they can be put into service unobtrusively.

Performance

An important characteristic of a good voice recorder is its sensitivity to weak sounds. If you expect to use a machine for more than personal note-taking, you will need high sensitivity. The machine should also clearly reproduce the recorded voice sounds.

Intelligibility. You can try out a recorder before you buy it by recording relatively weak sounds in a fairly quiet background. A machine that can't pick up and reproduce such sounds won't perform well under difficult conditions—say, in a large room or lecture hall, where the recorder may have to be some distance from the speaker. (No matter how sensitive your recorder, try to get as close to the speaker as possible when taping lectures or conferences.)

Recording time. All models take a standard microcassette. You record on one side of a microcassette and then flip the cassette over to record on the other side. A microcassette can provide from one-half to three hours of recording time, depending on the type of tape. At a recorder's standard speed (2.4 centimeters per second), an MC-30 tape records for a half-hour, an MC-60 for an hour, and an MC-90 for an hour and a half. A tape records twice as long at the recorder's slow speed (1.2 centimeters per second), but you sacrifice something in sound quality. It's best, then, to reserve the slow speed for recording voices that are reasonably close.

Batteries and power. A microcassette machine generally operates on AA or AAA ("penlight") batteries. You should use the alkaline variety because they will last a lot longer than conventional zinc-carbon batteries. On most models you can save money on batteries by using household current as a power source. Some come with an adapter that plugs into a wall outlet; the rest simply have a jack for an adapter that you buy separately. A few also offer a rechargeable battery pack as an option.

Convenience

Aside from its size, a microcassette recorder is much the same as any portable recorder. It has the usual controls for recording, playing back, stopping the tape, rewinding, and fast-forwarding. Some machines, however, have various distinctions.

Automatic level control (ALC). This special circuitry is designed to even out amplification during recording so that loud and soft voices will be played back at much the same volume. ALC is an asset when recording at meetings because it can boost or tone down voices, depending on how loud the voices are and on how close a speaker is to the microphone. ALC also reduces distortion should a recorder's circuits be overloaded with a stronger signal than it can ordinarily handle. However, ALC may not be useful when you're trying to record weak or distant speech because the circuitry also boosts background noise, possibly to the point where speech becomes difficult to understand.

Size and operation. Some units are small enough to slip readily into a shirt pocket; others might be a tight—or impossible—squeeze. In general, the smaller a unit is, the easier it is to hold and operate with one hand. However, convenient one-handed operation also depends on the placement of the controls, some of which may be a bit close together. Check the machine you're thinking about buying to be sure your fingers fit the controls.

Battery/recording indicator. Most units have an indicator light that comes on only when the machine has been set to record. On some units, the light also gives an indication of the battery life remaining. With a few models, the light blinks during recording to give a very rough gauge of the recording level.

Input sensitivity. Sometimes there's a two-position sensitivity control that you're supposed to set to "high" or "low," depending on whether the sound to be recorded is soft or loud.

Pause control. If brief interruptions hamper a recording session, it's convenient to use a pause control. In this way you can stop and start tape movement without having to touch any other control.

The pause control also makes it easier to set the tape at precisely the point where recording or playback should begin. When the interruption is more than momentary, it's handy if you can lock the control.

Automatic stop. This feature senses that the tape has reached its end and automatically returns the machine to its stop position, disengaging all the tape-motion mechanisms.

It's best if automatic stop works not only in the record and playback modes but in other modes as well; this would prevent the motor from trying to drive a fully wound tape, which would cause wear and tear on the transport mechanism and use up the batteries.

Automatic shutoff. This feature shuts off the motor at the end of a tape running in any mode, which conserves the battery. However, you must still press the Stop button to disengage the recorder's mechanism.

Safety lock. This feature prevents you from working the controls unless there's tape in the unit. Some models let you press the Record button and turn on the recording light even when the machine is empty. In that case, you may think you're recording when you're not.

Cue/Review. This control lets you go from playback to fast-forward (cue) to rewind (review), without having to press the Stop control. This can be a time-saver if you're searching for a particular passage on the tape. With most machines, the sound on the tape is audible but not understandable.

Sometimes there is a slow cue, which permits you to speed up the tape by about 10 to 20 percent. (At that speed, the playback remains understandable.)

Locking controls. Holding down a control during long searches through the tape or lapses in recording can be a nuisance. It should be possible to lock the Fast-Forward, Rewind, and Pause controls.

Tape counter. A counter helps you find a particular passage on tape. And a machine with a memory feature lets you set the recorder to stop the tape at any point you select.

Options. There are usually options to extend operating flexibility. A car battery adapter, for instance, can tap your car's cigarette lighter as a power source.

An earphone affords privacy in listening. An external amplifier-speaker can boost the volume (and perhaps the quality) of the playback. An external microphone or a remote-control switch can prove handy for dictation, or for when you're chairing a meeting. A tie-pin microphone lets you record while on the move, without having to hold the recorder. Various connecting cables are also available.

Some accessories may be interchangeable among brands, while others are not. If you expect to add accessories, be sure to ask your dealer whether they will be compatible with the machine you select.

Recommendations

Which recorder you choose depends to a great extent on how you plan to use it. For example, if all you want is an electronic notepad, you don't have to spend money on exotic features. Even inexpensive ones should perform adequately as a personal note taker.

AUTOMOBILE RECEIVERS/CASSETTE PLAYERS

The technology used in "autosound" components, as in other electronics, has become less costly over the past several years. Many receivers/cassette players offer more power for the dollar than you could buy six years ago. Conveniences and fancy features once found only on the top-of-the-line models—electronic digital tuning, automatic reversal of cassettes, noise reduction systems, for instance—have become commonplace on mid-priced models.

Indeed, even the receiver/cassette player is beginning to feel competition from compact disc players that fit into the dashboard.

However, receivers/cassette players are still the most popular choice as the centerpiece of an autosound system. How to go about choosing one depends on your situation.

If you're interested in a sound system for a new car, you can save yourself a lot of trouble by ordering the better-grade receiver/cassette player that comes with the car you're buying. For years, it was smart to forgo factory-installed sound because you could buy better equipment for less from an autosound specialist. That's no longer always true. Auto manufacturers are offering good quality, price-competitive sound equipment built into their cars.

You may choose to upgrade a factory-installed system by replacing the car manufacturer's loudspeakers with ones of your choice. If you decide to go all "aftermarket," then you should first consider the choices outlined below.

Aftermarket equipment

Some fifty companies sell receivers/cassette players, and many sell dozens of different models. Two factors—size and price—can eliminate a lot of those models from consideration right away.

Fitting your car

The chassis of the receiver/cassette player must fit the cavity that's provided behind the dashboard of your car. As a rule, small cars have less room than big cars.

You may also have the problem of fitting the knob shafts of the receiver/cassette player to the cutouts in the dashboard. Some models simplify this procedure by using an arrangement known as "DIN-mount," a standard agreed upon by the German car companies and increasingly found in other foreign cars as well. The entire DIN-mount component fits into a 2-x-7-inch cutout; depths vary but are usually around 5 inches.

Makers of autosound components generally try to make their products fit into as many car models as possible. Other manufacturers provide an array of installation kits whose faceplates and fittings accommodate most dashboard cutouts. Additionally, many companies design their products so the spacing of the knob shafts can be adjusted.

Some companies have extensive charts in their product literature noting which cars their products will fit; some mail-order autosound dealers provide the same information in their catalogs. Autosound retailers may also be able to help you match the receiver/cassette player to the car.

Prices

Receivers/cassette players carry suggested retail prices from as little as $70 or $80 to well over $1,000. You shouldn't expect much in the way of sound or power from models at the low end of that range. Models at the high end of the range—sporting fancy features, such as remote-control stalks or extra-high-power amplifiers—are more for conspicuous consumers than for ordinary listeners.

Consider three things if you are tempted to buy even a moderately expensive model. First, road noise masks a lot of the low bass and high treble tones whose faithful reproduction you pay extra for. Second, a sound system that wraps you solidly in sound could be dangerous if it keeps you from hearing honking horns, sirens, or a semi bearing down on your tail. And third, in many cities an obviously expensive system makes your car a target for thieves. If you buy equipment beyond a certain price, it would be prudent to figure in the cost of an auto security system.

With most brands, the moderately priced models have a digital quartz tuner. With a few, you get only an old-fashioned dial-type tuner.

Your habits

Another important factor in your choice of receiver/cassette player is what medium you listen to most. If you mostly listen to FM, the considerations outlined below should figure heavily in your choice.

Improved FM

FM became music radio because it can produce better overall sound than AM. When reception is good, FM has a quieter background, free of the whistles and other interferences that often mar AM reception. Perhaps more important, FM was first to offer stereo. However, FM can often sound a lot worse than AM in a car. The medium is not well suited for mobile reception.

Perhaps the most annoying problem—especially for urban and suburban drivers—is interference caused by the FM signal itself. Unlike AM signals, which are stopped only by large objects, such as mountains, FM transmissions bounce off objects as small as buildings and even cars. When there's one strong signal, FM reception may be good. But if there are also strong reflections, the signals arrive at the radio at slightly different times and cause the interference called "multipath distortion."

With a home FM receiver, you can move the antenna a few feet or switch to a directional antenna and usually make the harsh garbling caused by the distortion go away. But you can't drive your car with the position of the antenna in mind. Sometimes you might get a clear signal, sometimes a garbled signal. And in a moving car, the distortion can get a lot worse than simple garbling. As you drive along, the pattern of the FM signal and its reflections constantly change. Multipath can then manifest itself as a rapid, rhythmic fading of the signal, particularly the stereo aspect of the signal. If the signal gets weak enough, the radio produces nothing but noise. The resulting rapid and repeated fluctuation is called "picket-fencing," an obtrusive, pulsing pfft-pfft sound.

How severely multipath distortion disturbs your FM listening depends largely on the potential reflectors offered by the terrain. If you drive only through flat, open spaces and listen to a powerful pop station, you may never hear your radio picket-fence. But if you drive around town, you could encounter a lot of listening trouble, particularly from low-powered stations.

In the last several years, autosound companies have greatly improved the ability of their radios to deal with picket-fencing.

Receivers use three techniques to deal with the pulsing effects of multipath distortion. One is automatically switching to mono from the inherently noisier stereo when the interference causes the signal to weaken. On some receivers, the stereo effect collapses abruptly, only to snap back to full separation when the signal gets stronger. Better are those receivers that gradually blend the right and left channels into mono.

The second technique deals with noise in the treble frequencies, where hissing first becomes noticeable. Many receivers now automatically cut down on the treble volume when the signal weakens ("treble roll-off"). The best receivers do so gradually and unobtrusively. One way of doing that is called "Dynamic Noise Reduction," or DNR, which also works when you play a tape.

Finally, some receivers reduce the entire audio level as the signal weakens, a technique known as "soft limiting." It mutes the bursts of noise that make picket-fencing especially objectionable.

Although those three techniques have gone a long way toward improving the FM reception in cars, they don't conquer the garbling or the worst cases of picket-fencing. Some receivers, called "FM diversity receivers," use two antennas—one on the front of the car and the other on the back. The receiver continually monitors the signal coming from each antenna and rapidly switches back and forth to the one providing the better signal.

Other urban reception problems

Other reception problems that occur mainly in urban or suburban settings include:

Front-end overload. After multipath distortion, this type of interference is perhaps the most annoying. When you're too close to an FM transmitter, which may occur fairly frequently in a metropolitan area, the transmitter's signal can intrude on other signals on the FM band. The front-end electronics that normally screen out unwanted radio frequencies become overloaded (hence the name). The result: buzzing and rasping noise, if not an actual breakthrough of the unwanted signal.

Other radio signals. Radio signals can interfere in several other ways. A strong signal can overwhelm a weak one adjacent to it on the FM band; a receiver's ability to keep the signals separate is measured as "FM selectivity." A receiver might pick up two FM signals being broadcast on the same frequency; its ability to ignore the weaker signal is measured as "FM capture ratio." (Having a good capture ratio also helps a receiver reject some of the signal reflections that cause multipath distortion.)

Electrical-impulse interference. In heavy traffic, the electrical systems of nearby vehicles can cause crackling and popping.

Rural reception problems

An FM receiver can suffer from multipath distortion out in the country, particularly if the terrain is mountainous, but the primary reception problem is likely to be an inability to receive a listenable signal at all. The key test for assessing a receiver's reception far from the radio station is "FM sensitivity"—a measure of the weakest radio signal that still gives noise-free sound.

Picking the features

Performance aside, in order to choose the right model in the manufacturer's line, you need to decide which features are important to you.

Some features affect both FM and cassette performance. Inexpensive receivers/cassette players generally have only one control for adjusting bass and treble. Most useful are the separate bass and treble controls, because you can boost one range of frequencies without reducing the other. A loudness control boosts low bass and high treble—the frequencies most covered up by road and wind noise. A fader adjusts the balance between front and rear speakers.

On many inexpensive receivers with a dial tuner, the only way to change the station is to turn the knob. On the other dial tuners, there are push buttons—*presets*—as well.

Digital tuners are more accurate and stable than dial tuners, but they require more complicated controls. Presets on a digital tuner are electronically set ("programmed"), and on some models you can set as many as twenty. Usually, each preset gives you access to two or more stations, depending on how a band-select switch is placed. To roam around the dial on a digital tuner, you have to use a tuner seek. This control sends the tuner to the next strong signal on the band. On some models, you have to keep pressing the button to move up or down the band. On others, a variation called tuner scan samples successive stations for a few seconds until you stop it.

Digital tuners show the frequency of the station being received with either a liquid-crystal display (LCD) or a lighted-segment display. LCDs don't wash out in bright light; lighted-segment displays show up better at night. Most digital tuners also include a clock.

In addition to checking a unit's features, take a close look at its control layout. Some manufacturers, in providing miniaturized electronic controls, may not have paid enough attention to the size of the human finger and its general unsteadiness when its owner's eyes are on the road.

Selecting features

One feature can directly affect the way in which a cassette player will continue to perform: pinch-roller release. With this feature, the pinch roller, which presses the tape against the capstan during play, automatically disengages when the cassette player is turned off and a tape is inside. Otherwise, leaving the tape in can cause flat spots to develop on the pinch roller, leading to increased flutter. A more convenient variation on this feature is automatic tape eject.

Another important feature is a noise reduction system. Such a system cuts down on tape hiss, which is most noticeable at the treble end of the scale. Dolby B is commonly offered; Dolby C, less so. To get the effect, the tapes you play have to be recorded on a machine that has Dolby.

Some features add a lot of convenience to tape playing. Auto reverse automatically plays the second side of a tape when the first side is finished. Tape search, given such trademarked names as Music Sensor and Electronic Search and Play, stops at each substantial blank portion on a tape during fast-forward or rewind so you can quickly skip from selection to selection.

Recommendations

The first step in deciding which aftermarket model is best for your car is to find out which ones will fit. Installation is not a trivial job, and is usually best left to a professional. The cost of installation can be held down if the equipment is bought from the installer.

If FM performance is important, consider where you drive. Urban drivers need a model good at handling multipath distortion and front-end overload. Rural drivers need a model with good FM sensitivity. Suburbanites need a model good at all three.

If AM performance is important, you don't have to worry unless you live in a fringe reception area. Then consider a model that does well in AM sensitivity.

A receiver/cassette player, of course, is only part of an autosound system. You need at least two other components: speakers and an antenna. They're discussed below. More elaborate components, such as graphic equalizers and power amplifiers, are also available. Attaching them to a receiver/cassette player is much easier if the receiver/cassette player has line-out jacks.

Other parts of the system

To turn a receiver/cassette player into an autosound system, you need an antenna and a pair of speakers.

Antenna. A standard 31-inch whip antenna is better than one built into the windshield, particularly for AM reception, but it's vulnerable to vandals. A power antenna, which automatically retracts when the radio is turned off, should perform about the same as a whip antenna and is less vulnerable. Not all models have an appropriate connection for attaching a power antenna.

Speakers. As in a home system, the choice of speakers for an autosound system can greatly affect the sound. Unfortunately, choosing speakers that will evenly reproduce all the tone levels, or frequencies, in a particular car is extremely difficult.

A car's interior is small, oddly shaped, and full of sound-absorbing upholstery and sound-reflecting glass. The interior can greatly accentuate some frequencies and diminish others. With the right speakers, you can cancel out some of those faults. (A graphic equalizer that allows control over enough frequency bands might be able to do the same thing.) The catch in choosing speakers is that you can't tell how they'll sound in your car unless you can listen to them in a car that's similar to yours.

Reputable speaker manufacturers make their car speakers to much more rugged specifications than indoor speakers. State-of-the-art cone and magnet materials and construction techniques are commonly used.

Car speakers come in lots of shapes, from the common round to the severely flattened oval. So finding one to fit a particular application is usually not a problem. Some speakers have two cones (coaxial); others have three (triaxial).

Speakers mounted in the doors produce sharply separated stereo sound, rather like headphones. Door-mounted speakers should have a waterproof shroud to protect them from rain.

Speakers mounted on the package shelf behind the rear seat give a more blended stereo effect, like what you'd hear in your living room. Such speakers should have a rigid protective grille. Generally, rear speakers produce better bass than door speakers, partly because they are larger and use the trunk space as a baffle or resonator. Ultrathin designs may make it possible to get some of the same effect in a hatchback, with the speakers mounted in the rear pillars.

■ AM radio ■

Although AM radio is scorned by most music lovers, it's still the medium of choice for news, sports, and talk. For drivers who travel between cities and out of any FM station's workable range, an AM station can usually be picked up—AM music can be preferable to no music at all. AM stereo, now becoming more widely available, doesn't necessarily improve the fidelity of AM sound; it adds stereo effect. But it will take radios with improved AM fidelity to capture the promise of AM stereo.

Under most circumstances, the AM performance of any model should be entirely satisfactory.

Ratings of auto receivers/cassette players

Listed in order of estimated overall quality. Differences in overall score of 8 points or less were judged not significant. Prices are as published in a **March 1985** report.

◉ ◓ ○ ◒ ●

Better ⟵⟶ *Worse*

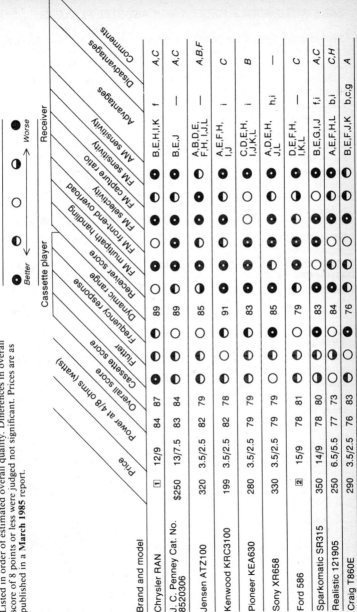

Brand and model	Price	Power at 4/8 ohms (watts)	Overall score	Cassette player: Cassette score	Cassette player: Flutter	Cassette player: Frequency response	Cassette player: Dynamic range	Receiver: Receiver score	Receiver: FM multipath handling	Receiver: FM front-end overload	Receiver: FM selectivity	Receiver: FM capture ratio	Receiver: FM sensitivity	Receiver: AM sensitivity	Advantages	Disadvantages	Comments
Chrysler RAN	[1]	12/9	84	87	◉	◓	◓	89	○	◉	○	◉	◓	◉	B,E,H,I,K	f	*A,C*
J. C. Penney Cat. No. 8520306	$250	13/7.5	83	84	◓	◓	○	89	◓	◉	◉	◉	◓	◉	B,E,J	—	*A,C*
Jensen ATZ100	320	3.5/2.5	82	79	◓	◓	○	85	◒	◉	◓	◉	◉	◓	A,B,D,E, F,H, I,J,L	—	*A,B,F*
Kenwood KRC3100	199	3.5/2.5	82	78	◓	○	◓	91	◉	◉	◓	◉	◓	◉	A,E,F,H, I,J	i	*C*
Pioneer KEA630	280	3.5/2.5	79	79	◓	◓	◓	83	◉	◉	◉	○	○	◉	C,D,E,H, I,J,K,L	i	*B*
Sony XR658	330	3.5/2.5	79	79	○	◓	◉	85	◉	◓	◉	◉	◓	◉	A,D,E,H, J,L	h,i	—
Ford 586	[2]	15/9	78	81	◓	◓	○	79	◓	◉	◉	◓	◒	◉	D,E,F,H, I,K,L	—	*C*
Sparkomatic SR315	350	14/9	78	80	◒	◓	◉	83	◉	○	◉	◉	○	◉	B,E,G,I,J	f,i	*A,C*
Realistic 121905	250	6.5/5.5	77	73	○	◒	○	84	◉	◓	○	◉	◓	◉	A,E,F,H,L	b,i	*C,H*
Craig T860E	290	3.5/2.5	76	83	◓	○	◉	76	◓	◓	○	◉	◓	◓	B,E,F,J,K	b,c,g	*A*

Ratings of auto receivers/cassette players (continued)

Listed in order of estimated overall quality. Differences in overall score of 8 points or less were judged not significant. Prices are as published in a **March 1985** report.

Key: ◉ ◓ ○ ◒ ● — *Better* ⟵⟶ *Worse*

				Cassette player				Receiver									
Brand and model	Price	Power at 4/8 ohms (watts)	Overall score	Cassette score	Flutter	Frequency response	Dynamic range	Receiver score	FM multipath handling	FM front-end overload	FM selectivity	FM capture ratio	FM sensitivity	AM sensitivity	Advantages	Disadvantages	Comments
Hi-Comp HCC1050	230	3.5/2.5	76	82	◉	◓	○	78	○	◉	◉	◉	○	○	C,I,J	a	*B*
Clarion 6900RT	199	4/2.5	74	82	◓	○	◓	76	◉	◓	○	◉	◓	○	E,F	c,i	—
Delco UU7	[3]	10.5/7.5	74	72	◒	◉	○	83	◓	◉	○	◓	◓	◉	E,F,H	a,c	*C*
Concord HPL122R	280	5/3	73	81	◓	◓	◓	76	◒	◉	◓	◉	◓	◉	E,J	a,f	*A,C*
Sanyo FTE25	300	6/5	73	79	○	◓	◉	74	◒	◉	◓	◉	○	◉	B,H,I	—	*A,B*
Alpine 7154	330	3.5/2.5	71	84	◉	◓	◓	74	◉	◓	○	○	○	◓	F,I,L	a,d,e,g	*C*
Sears Cat. No. 50029	200	13.5/8	71	79	○	◓	◉	73	○	◓	◓	◓	○	◉	I,J	g	*B,D*
Blaupunkt Aspen SQR24	320	4/2.5	70	72	◒	○	◓	73	◓	○	◓	◓	○	◓	E,H,I	i	*C,G*
Kraco ETR1089	380	7/5	69	75	◒	◉	◓	71	◒	◉	◉	◉	◒	◉	E	a	*C,D*
Panasonic CQS804	250	3.5/2.5	69	73	◒	◓	◓	73	◉	◓	○	◓	○	◓	E,H	b,c,i	*E*
Audiovox AVX3500	280	7.5/5	60	76	○	◓	◓	58	◒	◓	◉	◓	◒	○	—	c,g	*D*

[1] *1984 factory equipment. Lists for $389 on Turismo/Horizon; $389 on Reliant, Gran Fury, Voyager, and Fifth Avenue; $299 on Laser; $424 on LeBaron, E Class, and New Yorker; and $139 on Executive Sedan/Limousine. Price includes 4 speakers, antenna, and installation. 1985 model has AM stereo and DNR instead of Dolby noise reduction.*

[2] *1984 factory equipment. Lists of $396 on LTD, Thunderbird, Escort, GL, and EXP (Escort and EXP price varies depending on trimline); $310 on Crown Victoria. Price includes 4 speakers, antenna, and installation. 1985 model has tape-search feature.*

[3] *1984 factory equipment. Lists for $277 on 1984 Chevrolet Celebrity or Cavalier. Price includes 4 speakers, antenna, and installation. 1984 models replaced by a later version in 1985 cars.*

Specifications and Features

All have: ● FM stereo indicator light. ● Locking Fast-Forward and Rewind controls. ● Automatic tape stop or play after fast-forwarding or rewinding. ● "Motorola-type" antenna jacks. ● Tone controls.
Except as noted, all have: ● Well-lighted controls.

Key to Advantages

A–Can play tape without radio on.
B–Cassette is easy to insert and eject.
C–Can skip weak stations in seek mode.
D–Seek mode works both up and down band.
E–Tone controls are easy to set.
F–Tone controls are hard to disturb.
G–Has bass-boost switches.
H–Good control layout.
I–Well-lighted controls.
J–Control settings are easy to ckeck.
K–Panel brightness is adjustable with headlight switch.
L–On/Off switch is separate from volume control.

Key to Disadvantages

a–Cassette is difficult to insert or eject.
b–Tuner is susceptible to front-end overload and has no "local" switch to ameliorate problem.
c–Tuner makes noise when changing stations.
d–Has small slide controls, which are difficult to adjust.
e–Tone controls are difficult to set.
f–Tone controls are easy to disturb.
g–Controls poorly laid out.
h–Has poor AVC (automatic volume control), which compensates for varying signal strength on AM.
i–Installation instructions judged sketchy.

Key to Comments

A–Cassette ejects automatically when power is turned off.
B–Has liquid-crystal display with backlight.
C–Has lighted-segment numeric display.
D–Includes device to check wiring-harness installation.
E–Has "ambience" switch, which is claimed to expand the stereo effect.
F–Display lights go on when ignition is on.
G–Has "Automatic Radio Information" feature.
H–Model discontinued at the time the article was originally published in *Consumer Reports.* The information has been retained here, however, for its use as a guide to buying.

Auto receivers/cassette player features

Brand and model	Price	Chassis size (In.)	Shaft spacing code [1]	Power at 4/8 ohms (watts) [2]	Separate bass, treble	Loudness control	Fader	Line-out jacks	Clock	Digital tuner	Presets	Power-antenna lead	Auto reverse	Metal tape	Noise reduction [3]	Tape search	Pinch-roller release
Alpine 7154	$330	2 × 7 × 5¾	A	3.2/2.5	✓	—	✓	✓	✓	✓	12	✓	✓	✓	—	✓	—
Alpine 7155	400	2 × 7 × 5¾	A	6 max.	✓	—	✓	✓	✓	✓	12	✓	✓	✓	B,C	✓	—
Audiovox AVX3000	100	2 × 7 × 5	A,B,C,D	14 max.	—	—	—	—	—	—	0	—	—	—	—	—	—
Audiovox AVX3100	140	2 × 7 × 5	A,B,C,D	14 max.	—	—	✓	—	—	—	0	—	—	—	—	—	—
Audiovox AVX3200	160	2 × 7 × 5	A,B,C,D	14 max.	—	—	✓	—	—	—	0	—	✓	—	—	—	—
Audiovox AVX3300	190	2 × 7 × 5	A,B,C,D	14 max.	—	—	✓	—	—	—	5	—	—	—	—	—	—
Audiovox AVX3400	220	2 × 7 × 5	A,B,C,D	14 max.	—	—	✓	—	—	—	5	—	✓	—	—	—	—
Audiovox AVX3500	280	2 × 7 × 5⅛	A,B,C,D	7.5/5	✓	—	✓	—	—	—	5	—	✓	—	DNR	—	—
Blaupunkt Aspen SQR24	320	2 × 7 × 5⅛	A,B,C,D	4/2.5	✓	—	✓	✓	✓	✓	12	✓	✓	—	DNR	—	✓
Blaupunkt Sacramento	400	2 × 6¼ × 5	A,B,C,D	7.5 max	✓	—	✓	✓	✓	✓	12	✓	✓	✓	DNR	—	✓
Chrysler RAN	[4]	3⅛ × 8⅛ × 5⅞	—	12/9	—	—	—	✓	✓	✓	20	✓	✓	—	B	—	✓
Clarion 6100R	159	2 × 7⅛ × 4⅝	B,C,D	6 max.	✓	—	✓	—	—	—	0	✓	✓	—	—	✓	—
Clarion 6150R	159	1⅞ × 6⅜ × 4⅝	B,C,D	6 max.	✓	—	✓	—	—	—	0	✓	✓	—	—	✓	—
Clarion 6300R	179	2 × 7⅛ × 4⅝	B,C,D	6 max.	✓	—	✓	—	—	—	5	✓	✓	—	—	✓	—
Clarion 6700RT	179	2 × 7⅛ × 4⅝	B,C,D	6 max.	✓	—	✓	✓	—	—	0	✓	✓	✓	B	✓	—
Clarion 6900RT	199	2 × 7⅛ × 4⅝	B,C,D	4/2.5	✓	—	✓	✓	—	—	5	✓	✓	✓	B	✓	—
Clarion 6950RT	199	1⅞ × 6⅜ × 4⅝	B,C,D	6 max.	✓	—	✓	✓	—	—	5	✓	✓	✓	B	✓	—

Concord HPL122R	280	2⅛ × 7⅛ × 4⅜	B,C,D	5/3	✓	✓	✓	✓	✓	✓	10	✓	—	✓	B	—	✓
Concord HPL532 [5]	440	2 × 7⅛ × 4¾	B,C,D	12 max.	✓	✓	✓	✓	✓	✓	10	✓	—	✓	B	—	✓
Craig T760	250	2 × 7 × 5⅛	B,C,D	4.5 max.	✓	✓	✓	✓	—	—	5	✓	✓	✓	DNR	✓	—
Craig T860E	290	2 × 7⅛ × 5¾	A	5/8	✓	✓	✓	✓	—	—	5	✓	✓	✓	DNR	✓	—
Delco UU7	[4]	3¼ × 7⅛ × 5½	—	7/8	✓	—	✓	✓	✓	✓	8	—	—	—	—	—	—
Ford 586	[4]	2¼ × 7⅛ × 6¾	—	15/9	—	—	✓	✓	—	✓	8	—	✓	✓	B	—	✓
Hi-Comp HCC1050	230	2 × 6¼ × 5⅛	A,B,C,D	3.5/2.5	✓	✓	✓	—	✓	✓	12	✓	—	✓	—	—	—
Hi-Comp HC1150	300	2 × 6¼ × 5⅛	A,B,C,D	50 max.	✓	✓	✓	✓	✓	✓	12	✓	✓	✓	DNR	—	✓
Hi-Comp HCC1250	390	2 × 6¼ × 5⅛	A,B,C,D	50 max.	✓	✓	✓	✓	✓	✓	12	✓	✓	✓	B	✓	✓
J.C. Penney Cat. No. 8520306	280	2 × 7 × 5⅜	C,D	13/7.5	✓	✓	✓	—	✓	✓	12	✓	✓	✓	B	✓	✓
Jensen ATZ100	320	1¾ × 7⅛ × 6¼	B,C,D	3.5/2.5	✓	—	✓	✓	✓	✓	12	✓	✓	✓	—	✓	✓
Jensen ATZ200	370	2 × 7⅛ × 6¼	B,C,D	4 max.	✓	✓	✓	✓	✓	✓	12	✓	✓	✓	—	✓	✓
Jensen ATZ300	420	2 × 7¼ × 6⅞	B,C,D	7.5 max.	✓	✓	✓	✓	✓	✓	12	✓	✓	✓	B	✓	✓
Jensen ATZ500	520	2 × 5¼ × 6⅞	B,C,D	18 max.	✓	✓	✓	✓	✓	✓	12	✓	✓	✓	B,C	✓	✓
Kenwood KRC3100	199	2 × 7⅛ × 5	C,D	3.5/2.5	✓	—	✓	✓	—	✓	10	✓	✓	—	K	✓	—
Kraco ETR1081	200	2 × 7 × 5½	B,C,D	8 max.	—	✓	✓	—	✓	✓	10	✓	—	—	—	—	—
Kraco ETR1086	290	2 × 7 × 5⅛	B,C,D	8 max.	—	—	✓	—	✓	✓	10	✓	✓	—	DNR	—	—
Kraco ETR1088	330	2 × 7⅜ × 6	B,C,D	20 max.	✓	✓	✓	—	✓	✓	12	✓	✓	✓	DNR	—	—
Kraco ETR1089	380	2 × 7⅜ × 6	B,C,D	7/5	✓	—	✓	—	✓	✓	12	✓	✓	✓	B	✓	—
Kraco ETR1090	400	2 × 7 × 5½	B,C,D	20 max.	✓	✓	✓	✓	✓	✓	12	✓	✓	✓	B	—	—
Panasonic CQS747	250	1¾ × 7 × 5⅝	A,B,D	20 max.	✓	—	✓	—	—	—	0	✓	—	✓	B	—	—
Panasonic CQS774	230	2⅛ × 6⅜ × 5¼	A,B,D	7.5 max.	✓	—	✓	—	—	—	5	✓	✓	✓	B	—	✓
Panasonic CQS793	290	2⅛ × 7 × 5¼	A,B,D	7.5 max.	✓	—	✓	—	—	—	5	✓	✓	✓	DBX	—	—
Panasonic CQS804	250	2⅛ × 6⅜ × 5¼	A,B,D	3.5/2.5	✓	—	✓	✓	—	—	5	✓	✓	✓	B	—	✓
Pioneer KE7200	350	2 × 7⅛ × 4⅜	B,C,D	3.2 max.	✓	✓	—	✓	✓	✓	15	✓	✓	✓	B	✓	✓

Auto receivers/cassette player features (continued)

Brand and model	Price	Chassis size (in.)	Shaft spacing code [1]	Power at 4/8 ohms (watts) [2]	Separate bass, treble	Loudness control	Fader	Line-out jacks	Clock	Digital tuner	Presets	Power-antenna lead	Auto reverse	Metal tape	Noise reduction [3]	Tape search	Pinch-roller release
Pioneer KEA330	230	2 × 7⅛ × 5⅜	B,C,D	3.2 max.	—	✓	—	—	—	✓	18	✓	—	—	—	—	—
Pioneer KEA430	260	2 × 7⅛ × 5⅜	B,C,D	2.9 max.	—	✓	✓	—	—	✓	18	✓	—	✓	—	—	✓
Pioneer KEA630	310	2 × 7⅛ × 5⅜	B,C,D	3.5/2.5	✓	✓	✓	✓	—	✓	18	✓	✓	✓	—	—	✓
Realistic 121905	250	1¾ × 7⅛ × 6	A,B,C,D	6.5/5.5	✓	✓	✓	—	✓	✓	12	✓	✓	✓	—	—	✓
Sanyo FTE15	200	2 × 7 × 5⅛	A,B,C,D	9.5 max.	—	—	✓	✓	✓	✓	12	✓	✓	✓	—	—	—
Sanyo FTE20	240	2 × 7 × 5⅛	A,B,C,D	9.5 max.	—	✓	✓	✓	✓	✓	12	✓	✓	✓	B	—	✓
Sanyo FTE25	300	2 × 7 × 5⅛	A,B,C,D	6/5	✓	✓	✓	✓	✓	✓	12	✓	✓	✓	B,C	✓	✓
Sears Cat. No. 50028	200	2 × 7 × 5¾	A,B,C,D	14 max.	✓	—	✓	—	✓	✓	12	✓	✓	—	—	—	✓
Sears Cat. No. 50029	250	2 × 7 × 5¾	A,B,C,D	13.5/8	✓	✓	✓	—	✓	✓	12	✓	✓	✓	B	—	✓
Sony XR45B	280	2 × 7⅛ × 6	C,D	4 max.	✓	—	✓	—	—	✓	10	✓	✓	✓	—	—	—
Sony XR65B	330	2 × 7⅛ × 5⅛	C,D	3.5/2.5	✓	—	✓	✓	—	✓	10	✓	✓	✓	B	✓	—
Sparkomatic SR308	230	1⅞ × 7 × 5¼	A,B,C,D	14 max.	✓	✓	✓	—	✓	✓	10	✓	✓	✓	DNR	✓	—
Sparkomatic SR315	350	1⅞ × 7 × 5¼	A,B,C,D	14/9	✓	✓	✓	—	✓	✓	10	✓	✓	✓	B,C, DNR	✓	✓

[1] *Distance between the knob shafts: A = DIN-mount; B = 5⅛ in. (130 mm); C = 5½ in. (140 mm); D = 5¹³⁄₁₆ in. (147 mm).*

[2] *Where two figures given, as measured by CU. Where "maximum" given, as claimed by manufacturer.*

[3] *B = Dolby B; C = Dolby C; DNR = Dynamic Noise Reduction; K = Kenwood's own system; DBX = DBX system.*

[4] *Original-equipment model; see Rating for prices.*

[5] *FM receiver section different from model tested.*

CLOCK RADIOS

Today's clock radios are all electronic. Gone are the mechanical leaves that flip over regularly to signal the passing of time. Numbers in red, green, or blue now signal the time silently.

A few clock radios have shed the cigar-box look and "walnut" finish in favor of trendier shapes and decorator colors.

Some companies have tried to make the clock radio more useful by combining it with other appliances. There is, for example, the clock/radio/telephone combo, which includes just about every appliance you could want on the nightstand. Other combinations include the clock/radio/tape deck and the clock/radio/television.

For the most part, though, people who want a clock radio want just that, without the add-ons. You can buy one for as little as $10 and throw it away when it breaks. (If just the clock goes, the unit can be demoted to the workshop or utility room to live out its life as a radio.)

For $20 to $60 you can expect to get a reliable clock, a good selection of features, and tolerable tone quality.

Features

All clock radios give you drowse time—a button you can press in the morning to gain a six- to eight-minute reprieve before the alarm sounds again. Another basic feature is sleep time; it lets you program the radio to play for a while at night to lull you to sleep.

Battery backup is another common feature; it holds the time and alarm settings in case of a power failure. Typically, the battery is a nine-volt alkaline cell that should last about a year.

The most valuable features include:

Reversible time-setting. This eliminates a common annoyance—controls that only advance the numbers for time and alarm settings. If you miss the mark, you have to keep pressing buttons to go forward until the desired time comes up again. With reversible time-setting, you can move the numbers forward or backward.

Alarm options. Some clock radios will rouse you only with the radio and can be set for only one wake-up time. But some allow you to choose whether the radio or the sound of the alarm will wake you up. Some clock radios give you double alarm settings so that you can set two different wake-up times, choosing the radio or the tone alarm for each time. That's helpful, since sleepers who share a bed don't necessarily have to get up at the same time. Several radios let you adjust the volume of the tone alarms.

Brightness control. Electronic displays are easy to read at night, but they may wash out in bright daylight. An easy-to-read display, however, may be too bright for some people's sleep. Look for a radio with a brightness control. Most models give you a choice of two brightness levels; several offer more.

Readability. For the nearsighted, readability depends on the size of the digits. The typical digit measures about half an inch high. Some models have slightly larger digits, making the display easier to read.

Controls. Typically, these are buttons, sliding switches, touch pads, or combinations thereof. The controls themselves are usually conveniently located.

Tone quality

Clock radios are usually more clock than radio. Their characteristic tinny sound is due to their size, which limits the speaker's reach into the bass frequencies. As a result, you really shouldn't attach too much importance to tone quality when judging the overall value of a clock radio.

Tone quality aside, it's reasonable to expect decent electronic performance from a clock radio.

FM performance. A radio's *sensitivity,* its ability to receive a station without much background noise, is important for receiving weak or distant stations. A radio with good sensitivity should also have good *selectivity,* the ability to receive clearly a weak station next to a strong one on the dial. *Image rejection* indicates the ability to reject interference from aircraft radio transmissions. *Ease of tuning* reflects the radio's ability to tune to a station's true signal, including the effectiveness of the radio's automatic frequency control (AFC) and the absence of "triple tuning" (a signal that comes in at three closely spaced points). If the AFC is too strong, it may tune in a strong station you don't want.

Recommendations

You don't have to spend a lot to get convenient features. But you shouldn't expect much in the way of a radio. You have to pay a lot more for that, and even then, high price is no guarantee of high quality (see page 254).

Ratings of clock radios

Listed in order of estimated quality. Prices are as published in a **September 1986** report; + indicates shipping is extra.

◉ ◓ ○ ◒ ●
Better ⟷ *Worse*

Brand and model	Price	Dimensions, H × W × D (in.)	Overall FM performance	Overall AM performance	FM performance: Sensitivity	FM performance: Selectivity	FM performance: Image rejection	FM performance: Ease of tuning	FM performance: Tone quality	FM performance: Maximum usable loudness	AM performance: Sensitivity	AM performance: Selectivity	Features: Automatic volume control	Features: Interference rejection	Features: Reversible time-setting	Features: Double alarm setting	Features: Selectable alarm sound	Features: No. of brightness levels	Other features	Comments
Soundesign 3789A	40	4 × 12 × 6	◓	○	◓	◉	◉	◉	◒	○	○	◒	●	◉	✓	✓	—	2	B,H,K	f,h
Panasonic RC6360	60	4 × 11 × 6	◓	○	○	◉	◉	●	○	○	◓	◒	◉	◓	✓	✓	—	5	A,B,D, I,L	d,e,g
Realistic Chronomatic-245	48	3 × 11 × 6	◓	○	○	◓	◓	◉	○	◒	◓	○	○	◓	✓	✓	—	2 [1]	B,E,H,J,K,	g
Realistic Chronomatic-248	35	3 × 9 × 5	○	◓	○	◉	◉	◉	●	○	◉	◒	○	◉	✓	✓	—	2	B,E,H,J	d
Sony ICFC30W	55	5 × 10 × 4	○	◓	○	◓	◓	○	◓	○	◓	○	◓	◉	✓	—	✓	2	F	h
Sanyo RM6600	30	3 × 11 × 6	○	◓	○	◉	◉	●	◒	◒	◓	○	◓	◉	✓	✓	—	2	K,L	h
Sony EZ3	60	4 × 10 × 3	○	◓	○	◉	◓	●	◒	○	◓	○	◓	◉	✓	—	✓	2	—	d,f,h,i

Ratings of clock radios (continued)

Listed in order of estimated quality. Prices are as published in a **September 1986** report; + indicates shipping is extra.

⊙ ◓ ○ ◒ ●

Better ⟵⟶ *Worse*

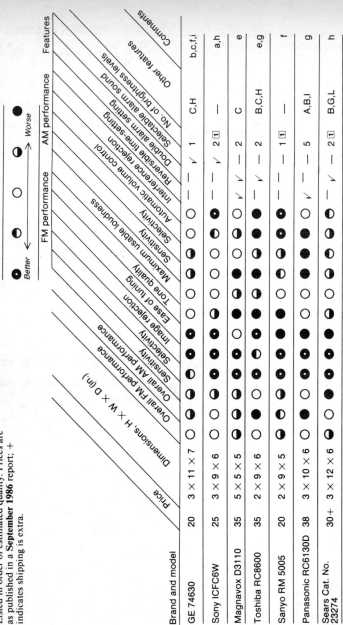

Brand and model	Price	Dimensions, H × W × D (in.)	Overall FM performance	Overall AM performance	Sensitivity	Selectivity	Image rejection	Ease of tuning	Tone quality	Maximum usable loudness	Sensitivity	Selectivity	Automatic volume control	Interference rejection	Reversible time-setting	Double alarm setting	Selectable alarm sound	No. of brightness levels	Other features	Comments
					FM performance						AM performance				Features					
GE 74630	20	3 × 11 × 7	○	◒	◒	◓	⊙	⊙	○	○	◒	◒	○	○	—	—	✓	1	C,H	b,c,f,i
Sony ICFC6W	25	3 × 9 × 6	○	○	◒	⊙	⊙	⊙	◒	○	○	○	◓	⊙	—	—	✓	2 [1]	—	a,h
Magnavox D3110	35	5 × 5 × 5	◒	◒	◒	⊙	⊙	○	●	◒	●	○	◒	○	✓	✓	—	2	C	e
Toshiba RC8600	35	2 × 9 × 6	○	●	○	⊙	◓	⊙	●	◒	●	◒	●	●	—	✓	—	2	B,C,H	e,g
Sanyo RM 5005	20	2 × 9 × 5	◒	◓	◒	⊙	⊙	●	●	○	◓	◒	⊙	⊙	—	—	—	1 [1]	—	f
Panasonic RC6130D	38	3 × 10 × 6	○	●	○	⊙	⊙	●	○	○	●	●	●	○	✓	—	—	5	A,B,I	g
Sears Cat. No. 23274	30+	3 × 12 × 6	◒	○	●	⊙	⊙	●	◒	○	◒	◓	◒	◓	—	✓	—	2 [1]	B,G,L	h

General Electric 74636	27	2 × 11 × 6	◒	●	○	◓	◓	◓	○	◒	●	○	●	○	—	✓	—	1	C,H	b
Sears Cat. No. 23213	20+	3 × 10 × 5	●	○	○	○	◉	◓	●	◒	○	○	○	◉	—	✓	—	2	B	—
Magnavox D3240	22	2 × 8 × 5	●	○	○	◓	◓	◓	●	◒	○	◒	◓	◉	—	—	✓	1	—	—
Emerson Red 5511	20	2 × 9 × 5	●	●	●	○	○	○	◒	○	●	●	●	◓	—	—	—	2	—	c,f

[1] *Lowest setting may be too bright at night.*

Specifications and Features
All have: • *Built-in wire antenna.* • *Drowse period of 6–10 min.* • *Power-failure indicator.*
Except as noted, all have: • *Battery backup for alarm and clock memory.* • *Design to prevent user from resetting time accidentally.* • *Convenient push button or touch plate to reset alarm.* • *Red display digits ½-in. high.* • *Sleep time: Radio-play period of up to 1 hr. before automatic shutoff.*

Key to Other Features
A–Pitch and tempo of tone alarm changes every min. for 5 min.
B–Alarm works during power failure.
C–Tone alarm has adjustable volume level.
D–Displays current time and 1 alarm setting simultaneously.
E–Has earphone jack.
F–Has treble-cut tone control, which can improve tone quality of a noisy transmission.
G–Uses rechargeable battery; unit must be disassembled to replace battery.
H–Has automatic frequency control for FM.
I–Tone alarm can follow radio after 10 min.
J–Has battery-level indicator.
K–Larger-than-average digits.
L–Has illuminated radio dial.

Key to Comments
a–Lacks battery backup.
b–Controls are designed so that time can be reset by accident, a disadvantage.
c–Control to reset alarm is sliding switch.
d–Brightness control and/or time-setting controls located inconveniently on bottom or rear of unit.
e–Only the pointer on radio dial is illuminated.
f–Radio can play for up to 2 hr. before automatic shutoff.
g–Green digits in display.
h–Blue digits in display.
i–Model discontinued at the time the article was originally published in *Consumer Reports.* The text information has been retained here, however, for its use as a guide to buying.

Higher-priced clock radios

You can buy a basic wake-up clock radio for as little as $10. For a few dollars more, you can probably find one with the most important features—reversible time-setting, double alarm, selectable alarm sound, brightness adjustment, and battery backup.

Unfortunately, spending a lot more doesn't guarantee that you get a lot more.

The $99 *Proton 320* is pretty good in sensitivity, tone quality, and maximum usable loudness, but it falls down in AM performance and ease of tuning.

The *Proton* offers basic features, like reversible time-setting, double alarm, selectable alarm sound, battery backup, and automatic brightness adjustment for its display. The radio has a gradual wake-up alarm, which slowly increases in volume until it reaches the level you've set.

For $160, you can get a clock radio that talks—perhaps an advantage for someone with poor eyesight or a child just learning to tell time. The *Panasonic RC6810* has a synthesized voice that volunteers just about all the information you need from a clock radio—the time, the status of the controls and settings, whether a power outage has occurred. If you want a little extra shut-eye, the unit will accommodate, saying, "I'll wake you again in nine minutes." You can set it for a "male" voice for some of the twenty-four hours of the day and "female" for the rest, or choose only one voice around the clock. And you can make the voices shut up altogether.

The *Panasonic*'s overall quality is reasonably good, but not as good as the *KLH200.*

The *KLH 200,* $275, delivers natural-sounding music with very good tone quality, even at high volume.

In appearance, the *KLH* resembles a table radio more than a clock radio. To accommodate stereo sound, the *KLH* comes with a separate speaker that can be placed up to six feet away. It also comes with electronic tuning and buttons that let you preset five FM and five AM stations. Its line-cord antenna provides only adequate sensitivity for reception. Sensitivity improves if you hook the radio's dipole antenna to its external terminals or use a roof-mounted FM antenna.

FOUR KINDS OF SPECIALTY RADIOS

It has been more than sixty-five years since commercial radio broadcasting began filling the airwaves. First came AM (amplitude modulation) broadcasting, which has become so familiar to radio listeners.

Next came the gradual occupation of the shortwave bands, propelled by government use and by enterprising ham (amateur radio) operators. Starting in the late 1930s, the better sound of commercial FM (frequency modulation) broadcasting and the audio elements of television broadcasting began to fill the radio spectrum. Specialty interests—including marine, aviation, police, fire, weather, public service, private enterprise, Citizens Band, and cordless telephone—occupied almost all the remaining gaps. With the advent of VHF and UHF, the traditional radio spectrum neared saturation.

Most of the communications sent out over an enormous range of frequencies can't be captured by ordinary AM/FM radios and TV sets. Hence, the advent of what might be called "specialty radio."

The four types of specialty radios discussed here are tuned more to the broad interests of the general public than to the special focus of such radio buffs as shortwave sophisticates, users of the Citizens Band, or ham operators.

A specialty radio doesn't give hi-fi performance. It tends to be fairly small and can often be used as a portable. Two reception qualities, above all others, determine how well a specialty radio will perform: sensitivity and selectivity.

The two qualities go hand in hand. If a radio has poor sensitivity, it might as well have poor selectivity, too: It probably won't bring in enough stations to create conflicts between strong and weak signals on the dial. If a radio has good sensitivity, it should have good selectivity as well. Otherwise, the radio would receive more stations than it is capable of sorting out.

The radio spectrum

There's a broad range of radio frequencies sent out over the air that can't all be heard over ordinary AM/FM radios, and which specialty radios of one kind or another can tune into.

In general, the use of particular frequencies is regulated by the FCC.

AM transmissions. In addition to broadcasts on the standard AM band, there are AM transmissions that may be available to owners of multiband sets:

- Marine band. These frequencies carry marine weather reports and navigational information by Coast Guard beacons.
- Shortwave bands nos. 1 and 2. These bands carry broadcasts originating from abroad.

FM transmissions. Outside the standard FM band are a large number of other FM transmissions, many of which can be picked up by multiband and scanner radios.

- Public Service VHF low band. Frequencies on this band, whose scope generally covers a county, are used for state and highway police calls, fire departments, taxi and truck dispatching, radio-paging services, and similar local services.
- Aircraft band. This band includes air traffic control, emergency search-and-rescue signals.
- Public Service VHF high band. This band is capable of broadcasting over very large urban areas. Its assigned frequencies cover emergency, fire, weather, and marine communications.
- Public Service UHF band. These frequencies are commonly used for citywide business, such as directing tow trucks, dispatching taxis, and sending service technicians on calls.

MULTIBAND RADIOS

About all that's required for a portable radio to acquire multiband status is that it receive at least one range of frequencies other than the standard FM and AM broadcast bands.

In addition to picking up the Public Service VHF high band, a more versatile unit can receive various parts of broadcast frequencies across the radio spectrum. Such versatility comes at a price. The more expensive sets generally boast broader reception capabilities.

Performance and features

Selectivity is never a problem with TV sound: The FM signals of TV audio frequencies have plenty of space between them. Selectivity on the AM bands, however, affects the reception of Citizens Band and shortwave signals. Good AM selectivity can be especially important on CB, where many stations with different signal strengths are crowded together.

Other points that are worth noting about multiband radios include:

Automatic frequency control. The AFC is usually a desirable feature. In effect, it broadens the dial space over which you can tune an FM station clearly. It's nice to have the option of switching AFC off, though, because at times it may prevent you from picking up a weak station sandwiched between strong ones.

Automatic volume control. AM receivers have a circuit for AVC, which is designed to minimize fluctuations in volume caused by changes in signal strength.

Power sources. Playing at as low a volume level as possible (or using an earphone) prolongs battery life. And though zinc-carbon batteries are relatively inexpensive, they won't last nearly as long as alkaline cells.

Some radios won't run on battery power unless you detach the line cord from the radio chassis. There should be a special compartment for line-cord storage.

To run many radios on 120 volts AC, it isn't enough to plug the line cord into a wall outlet. You usually have to flip a switch to change from DC to AC and back again. Should you forget to switch over, the batteries run down, even though you think the set is using house current. A number of models rule out that possibility by switching automatically from one power to the other when you plug in or unplug the line cord.

Dial convenience. Look for a dial without such drawbacks as widely spaced markings, a lack of the numbers and calibrations needed to locate predetermined frequencies, and parallax (an optical effect that gives you different readings, depending on your viewing angle). Nor should there be only one dial scale for several bands, which gives the dial a cluttered, hard-to-read look.

A minor convenience on many dials is a logging scale, which is just a straight sequence of numbers that runs parallel on the dial to the frequency scale. Although the log numbers mean nothing in themselves, they do help you remember and locate the dial positions of your favorite stations better than do poorly spaced frequency markings.

Most dials are unlighted, so tuning after dark can be awkward if you're outdoors. If, however, your set has a switch-operated dial light, use it sparingly to avoid a substantial drain on the batteries.

Tuning meter. This is a small visual aid to more precise tuning. Its pointer moves upscale as signal strength increases.

Squelch control. A turn of this knob silences the rushing noise you would otherwise hear between transmissions on some FM stations or when you tune between stations. If you set the control too high, though, you may hear nothing at all.

Tone control. Most sets have a single tone control, which affects only the treble. A few have separate controls for treble and bass. Tone quality and intelligibility of speech are generally at their best when you set the control for maximum treble.

Antennas. An internal antenna for AM reception is standard equipment. And so is a built-in telescoping monopole antenna for receiving all the other bands except the Public Service UHF band.

Sets that can pull in the UHF have still another built-in antenna for that band or come with a snap-on UHF loop antenna. Every set capable of receiving the Public Service VHF high band should also be able to pull in the FM weather band, which is located at about 162 megahertz (MHz).

Other features. A multiband radio generally has an earphone or a headphone jack that mutes the loudspeaker. For off-the-air taping, some have a jack that doesn't mute the loudspeaker: You can listen as you record. Or there may even be a microphone input to let you use the radio as a low-power public-address system.

Recommendations

Multiband radios should appeal to you if you want to develop some new listening tastes without forsaking the old. Standard commercial AM and FM broadcasts are the regular fare of multiband radio reception, but more is available, depending on the multiband make and model you decide to buy—portions of at least one of the Public Service bands, some Citizens Band and ham talk, an occasional pilot-to-tower transmission, and a little bit of shortwave as well.

Note that using a multiband to monitor CB channels while you drive may not always be feasible. Portable radios aren't designed to give good reception from inside a moving car, and a driver's hands belong on the wheel, not on a radio's tuning knob. In some localities, it's against the law to use a multiband in a car if the radio can pick up police frequencies.

SCANNER RADIOS

A scanner radio is a receiver that can automatically check the airwaves for signals from stations that are broadcasting at the time. When it picks up such signals, the scanner pauses at the station setting for as long as the signals last—or for as long as you care to listen. When the signals stop for a few seconds, the scanner may automatically tune itself to the next channel that may be broadcasting.

Most scanner radios are designed to receive the VHF and UHF bands that the Federal Communications Commission (FCC) has set aside for Public Service broadcasts. Frequencies within these bands are assigned to police and fire departments, ambulance and taxicab dispatchers, air traffic control, power companies, businesses, radio-paging services, marine weather stations, and the National Weather Service, among others. Although a conventional multiband radio can pick up stations on the Public Service bands, no ordinary multiband can monitor a number of broadcasts automatically, as a scanner does.

Types of scanners

There are two main types of scanners: fixed-crystal models and programmable scanners.

Fixed-crystal models. Many scanner radios are fixed-crystal models in which each crystal brings in a given frequency. Crystal models are relatively cheap to begin with, but the price doesn't include the crystals, which sell for several dollars apiece.

Because you need a separate crystal for each channel, and because each set has eight or ten channels, you can expect to pay quite a bit more than the base price of a fixed-crystal unit to make it fully operable.

To figure out which crystals to buy, you need to know which radio frequencies are worth receiving. Electronic equipment dealers can usually help you ascertain which frequencies are used locally. If they don't know, you can find out from a directory of Public Service radio frequencies, which may be available from your public library. If you tire of one station, you can always receive a new station with a change of crystal, but frequent crystal changes at several dollars each can quickly add to your cost.

Programmable scanners. This type is much more versatile than a fixed-crystal model and much more fun to experiment with. It is, however, quite a bit more expensive (though you don't have to buy crystals for it).

Typically, a programmable scanner has numbered push buttons of the sort found on small calculators. To program the radio for a desired station, you simply enter the station's frequency on the keyboard. Depending on the model and the band received, the frequency shows as a readout display of five, six, or seven digits. You can change the frequency received on a set's channels as fast as you can press the buttons.

Performance and features

An external antenna connection is pretty much standard equipment. Other features vary from one model to another.

Available bands/channels. Frequency ranges commonly extend from about 450 to 512 megahertz (MHz).

Each radio is also apt to bring in the span of frequencies (between 470 and 512 MHz) belonging to part of the UHF television spectrum (TV channels 14 to 20). In some areas, that air space may be allotted to Public Service radio.

Scanner memory. Some programmable models have solid-state memory circuits to retain and recall programmed frequencies. But these models may also need small batteries to power the memory circuits when the radio is unplugged or when the AC power fails. Without standby battery power in these conditions, the scanner's memory might go blank, and you would have to reprogram the radio.

To protect most of the programmable models from memory loss, it's probably enough to fit them with new batteries at the intervals suggested in the instructions or at least once a year.

Search feature. On a programmable model, it's possible to command a search from a given low frequency to a higher frequency—say from 39.02 to 39.98 MHz on the police band. The radio will pause or can be made to stop on reaching an active station within the search limits. (Don't confuse searching with scanning, which is what happens when a radio automatically goes from one preset channel to another.) A search capability makes it easy to find interesting stations that you didn't know about before. When you find such a station, you can program the radio to scan it regularly.

During a search, most scanners take relatively narrow 5-kilohertz steps—5 kilohertz at a time. An occasional model may offer a choice of wider steps. It may also step down or up—not just from a low frequency to a higher one.

Because of its electronic complexity, a programmable scanner will sometimes generate spurious internal signals. These "birdies," as the manufacturers call them, can cause a scanner to stop in mid-search even though no station is being received. Such an unintentional stop is little more than a momentary annoyance; the search can be resumed with the push of a button.

Priority. Every second for about one-tenth of a second, this feature automatically samples one channel you select for priority, while receiving a different channel. The radio quickly switches over completely when a signal comes through on the priority channel. But that split-second sampling gives an annoying staccato sound to speech on the channel you're tuned to.

Recommendations

A scanner radio offers distinctly specialized reception capabilities. Most of them can receive only three FM bands designated as Public Service broadcasts, but within these limitations, a scanner provides custom service. It can be tuned precisely to any of a number of stations. Further, it can be made to monitor preselected stations through periods of transmission and silence alike.

A relatively inexpensive fixed-crystal unit is a reasonable choice if your needs are predominantly practical—for police or volunteer fire department use, for example.

If you want a scanner mainly for amusement, get a programmable model. Expensive as it may be to start with, programmable units can provide a radio hobbyist with an endlessly variable choice of frequencies.

If you do buy a fixed-crystal set, be prepared for problems. If a dealer doesn't stock the crystals you want, you'll have to order them by mail and wait perhaps many weeks for them. In addition, crystals are sometimes faulty when they arrive. The sole recourse in that event is to return the crystals to the supplier and resign yourself to more delays.

WEATHER RADIOS

The United States government's National Weather Service (NWS) operates a nationwide radio network that broadcasts round-the-clock local weather information in the form of continuous taped bulletins that are updated every few hours. The several hundred NWS transmitters are within reach of about 90 percent or more of the population. The NWS broadcasts can be received on single-purpose "weather radios" available from electronic equipment stores or on the weather band of some multiband receivers.

When a hurricane, tornado, heavy snow, or other disaster is imminent, local NWS stations interrupt their regular programming to broadcast a special weather-alert bulletin, describing the problem and advising people what to do. Each station precedes such bulletins with a ten-second alert tone to get the listener's attention.

Performance and features

There's a built-in antenna present in all models—either a monopole antenna, or, rarely, the line cord. An external roof-mounted antenna improves reception, but only a few radios have the necessary connections.

Other features include:

Alert system. Models having this ability respond in various ways to a NWS weather-alert tone. Some turn themselves on to sound the broadcast tone and remain on to receive the bulletins that follow. There may be a light that starts blinking. Some sets placed on alert standby can detect the NWS tone and respond with siren wails of varying durations, sometimes with a blinking light as well. However, you must turn the set on to hear what the emergency is all about.

Station selector. The NWS transmits on seven frequencies from 162.400 MHz to 162.550 MHz. Although only one of these frequencies is on the air in a given area, most sets can receive more than one.

Power. Battery operation is very useful because it enables the radio to warn of a weather emergency even in the absence or failure of AC power.

Despite use of long-life alkaline batteries (the kind you should use), battery life may be surprisingly short. True, you don't operate a weather radio continuously, but even if you played it for only a few minutes several times a day, you'd still probably need to change the batteries every month or so.

Recommendations

The frequencies in one of the Public Service radio bands on which the NWS transmits can be received by many radios intended for consumer use, including multiband and scanner models.

If you live in an area where bad weather can be really bad news, up-to-the-minute information on the local weather may be of vital importance or great interest to you. In that case, look for a weather radio that offers more than simple tuning. The better models also have a weather-alert capability. Even when they're not playing, they can signal the imminence of urgent bulletins.

SHORTWAVE RADIOS

Shortwave frequencies, located between the AM and FM segments of the radio spectrum, provide a source of programming from all over the world. If you become a shortwave listener, you may be able to sharpen your language skills by listening to Brussels or Buenos Aires. Or you will be able to tune to a myriad of English-language broadcasts from Radio Moscow, the BBC World Service, and other stations with points of view that can be quite different from those of domestic TV and radio newscasters. Shortwave signals often travel very far from their point of origin. Some shortwave listeners like to "fish" the airwaves for identifiable signals from a station in, say, Bangkok.

Other programming on the shortwave frequencies known as the International Broadcast Bands includes music, drama, and public service material not usually heard on commercial radio. The shortwave band is also among the areas of greatest activity by amateur radio operators, licensed to communicate on the Amateur Bands of the shortwave spectrum.

Broadcast material on shortwave also includes standard time signals, Citizens Band, and the teletype transmissions sent from in and outside the United States by news agencies, as well as transmissions by the military and by civilian government agencies. The National Bureau of Standards, for example, sends out highly accurate signals for the calibration of clocks and radio equipment.

How shortwave works

Radio signals travel through space in waves, much the way water waves move in the ocean. In radio, the length of a wave—the distance from the crest of one wave to the crest of the next—is one way of designating the segments of the radio spectrum. If the length of a wave is more than 550 meters (about 1,800 feet), the radio transmission is by long wave. Long waves carry mainly commercial aircraft beacon signals and weather signals. Medium radio waves are roughly 550 to 200 meters long and carry standard AM radio broadcasts. Short waves measure less than 187 meters from crest to crest.

Shortwave frequencies. It's customary to designate specific radio broadcasts not by wavelength but by frequency—the number of waves per second that pass a fixed point in space. Standard broadcast frequencies are in kilohertz (kHz), or thousands of waves per second. Shortwave frequencies are measured in megahertz (MHz), or millions of waves per second. Radio stations broadcast on assigned frequencies, and the numbers on a radio dial stand for these frequencies.

Skip signals. Shortwave signals can travel long distances because of "skip," the same phenomenon that lets your AM radio on some evenings pick up stations halfway across the country. Skip can occur as radio signals leave a transmitter, hit the underside of the ionosphere—the electrically charged layers of air in the upper atmosphere—and skip off it, returning to earth at a considerable distance from their point of origin. Skip signals can easily travel overseas. If the skip process repeats itself, the signal can even travel to the other side of the globe. Skipping between radio stations on the ground is especially likely at frequencies below about 30 MHz, which is the upper limit of the shortwave band.

Propagation of signals. The quality of shortwave reception depends on such variables as the year, the time of day, the season, the weather, and cloud cover. Some of these variables have generally unpredictable effects on the propagation of radio signals—the way the signals travel through the air. Others can be reliably predicted. For example, low frequency shortwaves usually propagate better at night, whereas the high frequencies tend to do better in the daytime. Long-term changes, such as those caused by solar activity, also predictably affect propagation. Increased solar activity may allow signals in the 20 to 30 MHz range to travel greater distances than usual. International broadcasters pay close attention to these vagaries of propagation, and often change their frequencies to increase broadcast range.

For a thorough discussion of propagation, you may want to refer to the *Radio Amateur's Handbook,* published annually by the American Radio Relay League. The book is available in many public libraries and in most electronics supply stores, or by mail from the ARRL, Newington, Connecticut 06111. It costs $18 postpaid.

Tuning

Typically, a shortwave radio also receives standard FM and AM broadcasts—useful options in the event you want a change from shortwave. Unlike FM and AM stations, however, shortwave stations are not spaced at regular intervals on the dial. In fact, they often overlap and get in one another's way. (Overlap is common on the Amateur Bands.) Signal strength varies widely, especially in skip transmissions, so a shortwave radio's circuits have to be designed to single out a particular station and pull in its signal with a minimum of interference.

Digital readout. Because each shortwave station occupies only a tiny spot on an ordinary tuning dial, finding a particular station can be nearly impossible without special help. Pinpointing a station is easier with a model that has digital-readout tuning. The readout, an illuminated display, shows the tuned frequency in bright numbers.

Bandswitch. Most models divide the shortwave spectrum into a number of bands—3 to 6 MHz, 6 to 9 MHz, and so on—so you can move quickly from one part of the shortwave spectrum to the other. To tune in a station, you first flip a bandswitch for the range of frequencies you want, then turn the main tuning knob (which works along with the digital display) to select the frequency.

Calibrated bandspread. Other models use calibrated bandspread tuning. It's more cumbersome than digital readout of frequency, but it's almost as accurate. With these radios, you can easily find a signal, but you have to calibrate the bandspread dial with a circuit built into the radio. Otherwise, you won't know for sure what the station's frequency is.

Performance

A number of factors enter into determining shortwave performance.

Selectivity. This is a radio's ability to separate two or more stations that are close together on the dial. With a shortwave set, selectivity depends mainly on "intermediate frequency bandwidth." Narrow bandwidth is an advantage for some kinds of reception because bandwidths wider than about 4 kHz allow noise and interference to intrude.

Dynamic range. This is the difference between the weakest and the strongest signals a radio can receive without a kind of distortion called "overload." In one form of overload, strong signals from a powerful station nearby can crop up at points on the dial where they shouldn't be.

Sensitivity. This is the measure of how well a receiver brings weak signals out of the background noise. When receiving with an outdoor antenna or with a built-in whip antenna, a good shortwave radio should be sensitive enough at the low (3 to 10 MHz) and medium (10 to 20 MHz) frequencies of the shortwave range to receive any signal stronger than the noises that permeate the airwaves. Background noise is weaker at the high (20 to 30 MHz) end of the shortwave range: There, the noise produced by the receiver itself may be an obstacle to clear reception.

Signal-strength indication. A radio with a meter or other indicator makes it easy to determine the relative strengths of the signals being received. As signal strength increases, the meter's pointer (or equivalent) moves across a scale.

Single-sideband capability. Many amateur radio stations (and Citizens Band) use a special form of modulation called "single-sideband." It helps reduce the effects of noise and station-crowding on reception and makes it easier to pick up stations that are weak or very far away. On a conventional radio, single-sideband signals sound scrambled. To make them intelligible, shortwave radios have a beat-frequency oscillator (BFO). On some, the BFO is fixed: You switch it on, then fine-tune until you hear voices with normal pitch. On others, the BFO is variable: You adjust the BFO pitch control for the right sound after you've tuned in the signal. If a radio copes with single-sideband reception as it should, the sounds should be crisp, clear, and relatively undistorted.

Automatic gain control. This keeps loudness more or less uniform despite possible changes in signal strength. If the control acts too quickly, and if signal strength changes rapidly (as it does with many overseas broadcasts), the ebb and flow of background noise will seem to "pump" up and down annoyingly. A control that works too quickly can also cause overload on single-sideband signals. For satisfactory single-sideband reception with some radios, you may have to adjust either a radio-frequency gain control or an antenna-attenuator switch—features designed to reduce the radio's sensitivity.

AM and FM reception. If a radio does well at receiving shortwave, it may be too much to expect superb reception on the regular AM and FM bands. If you can tune them in with reasonable ease and they sound acceptable (considering the loudspeaker size and quality), consider yourself fortunate.

Battery life. A portable that's battery-powered can also be adapted to run on normal house current. For maximum playing time on throwaway batteries, use fresh alkaline cells. If there is a dial light, there should also be a switch to turn it off to prolong battery life.

Features

There are a variety of features that you may find useful.

Two-speed tuning. One speed gets the station indicator across the dial fast; the slower speed allows you to home in on a station. On some radios, the MHz knob permits fast tuning. On others, there's a lever for changing speeds or a knob you push in for slow tuning and pull out for fast.

Radio-frequency preselector. This feature helps to improve signal reception and selectivity. You set the preselector dial close to the desired frequency, tune the radio as usual, and then fine-tune the preselector for the strongest signal. If the preselector doesn't have calibrations, it must be preset by ear.

Antenna trim control. As with a preselector, you adjust this control for the strongest signal after tuning.

Tone controls. Usually there's a switch to reduce the treble, the bass, or both. Most models with FM capability have separate tone controls for treble and bass. A treble control is especially important for shortwave listening because it may reduce background noise.

Other bands. Some models can receive the longwave band below 550 kHz, where you get special service, such as marine radio beacons. Some receive parts of the VHF range from 108 to 148 MHz, above the FM broadcast band. This range includes aircraft communications and amateur stations.

Muting input. If you incorporate a shortwave receiver into a ham station, you have to silence the receiver when transmitting. Some models have a muting input that silences the receiver automatically if it is connected to a transmitter.

Speaker jack and headphones. You may find it useful, in addition to a built-in jack for connection to an external speaker, to have a jack for headphones to enable you to listen to shortwave broadcasts without disturbing others.

Recorder-output jack. You may also want a built-in connection for a tape recorder so you can tape shortwave broadcasts. A number of models have one.

DC operation. For operation while camping or when boating, it's useful to have an input for a 12-volt power source, such as a standard car battery.

Antennas

Quality of shortwave reception can be enhanced by the type of antenna you use. A radio should have a ground connection for use with an outdoor antenna. Some antennas work best when the lead-in wire is coaxial cable. The antenna terminals on the radio you buy should be able to take coaxial cable. (Note that even with the best antenna, shortwave listeners can expect selective fading in long-distance reception. The effect of such fading is to make signal strength fluctuate, with severe distortion, in cycles several seconds long.)

Built-in antenna. A built-in whip antenna is fairly common in shortwave receivers. It should give you generally good reception except inside concrete-and-steel buildings.

Nevertheless, an outdoor antenna is worth having for its ability to pull in remote stations, especially at the high end of the shortwave range.

Outdoor antennas. An outdoor antenna helps to bring in weak or distant signals and, in some cases, to reduce interference from local stations. A simple long-wire antenna works well and can be inexpensive to assemble. The antenna can be made of ordinary copper wire at least 10 meters (about 33 feet) long. You run the wire between your house and a tree or post—not a utility pole. Isolate the wire at each end with a glass or ceramic insulator. Then run a lead-in wire or coaxial cable, connecting the long-wire antenna to the shortwave receiver's antenna terminal. Make the long-wire connection firm, preferably by soldering.

With this kind of antenna, it helps to connect the receiver's ground terminal to a good ground, such as a water pipe. If you want an extra-long antenna, or if the antenna must withstand high winds, you can use the special strong antenna wire sold in ham radio stores—it has a steel core.

To protect radio equipment from electrical surges caused by lightning, ground the antenna lead with a "knife" switch when you're not listening. For protection against direct lightning strikes, hire a professional to do the job.

Other types of antenna. If shortwave reception suffers from local interference, or if you want to listen to a particular band of frequencies, another type of antenna may be more suitable than the long-wire variety. Most of the instructions accompanying radios give advice on the kinds of antenna you can use. A good reference is the *Radio Amateur's Handbook.*

Recommendations

An avid shortwave listener would probably think nothing of spending several thousand dollars for shortwave equipment, but you can get good receivers for much less than that. Locating a store that carries a selection of shortwave radios, however, may require some shopping around.

An important consideration in selecting a shortwave radio is ease of tuning. A good radio shouldn't suffer from backlash—a looseness in the tuning that can make precise adjustment difficult. A radio's calibrations should be accurate as well as easy to read. The best ones can be set and read to within 3 kHz of a desired frequency.

Other Electronic Equipment

AUTO THEFT ALARMS

Car thieves can pick, jimmy, or smash their way through door and ignition locks. They can attempt to start the engine and drive the car off. They can hoist up the car and tow it away. Or they may leave the car and take its tape player, tires, body and engine components, or loose contents.

Their persistence has created a booming industry in auto alarm systems. More and more new cars are coming with alarm systems as original, factory-installed equipment, and alarm systems are now standard equipment on many cars.

In addition, the market for "aftermarket" alarm systems (equipment you buy after you already own the car) is booming.

At the heart of an auto alarm system is an electronic control module. The typical module uses the car's electrical system both for power and for the first line of defense. By tapping into the electrical system, the module senses when a door is opened and the dome light comes on. The device sounds the siren if too much time passes (typically, more than 10 or 15 seconds) before the alarm system is turned off.

If a light comes on when the trunk or hood is opened, then such a system can protect those parts of the car, too. Otherwise, the second line of defense in a typical system is made up of switches at the hood and trunk. Those are designed to set off the siren instantly if the hood or trunk is opened while the system is on. Sometimes switches are installed on the passenger door jambs as well, setting off an immediate alarm if any door except the driver's is opened first.

A third line of defense is a device that senses an attack even if a door is *not* opened—a motion detector or a vibration sensor, for instance. Those sound the alarm if the car is jacked up or moved, or if a window is broken.

Yet another defense is a device that disables the engine. An engine disabler typically interrupts current to the starter or the ignition, making it impossible to start the car. Another type of disabler shuts off the fuel line. That allows the car to be driven, but only for a few blocks.

Many of these aftermarket systems have to be professionally installed. They're generally more comprehensive and often more convenient than systems you install yourself—and always more expensive.

Having a system professionally installed can cause a lot of headaches. Installing a system yourself, on the other hand, requires considerable familiarity with cars and their electrical systems and a lot of squirming around under the dashboard and hood. Figure on spending the better part of a day if you're installing a full system for the first time. Expect extra trouble if you're installing a hood lock.

Because auto-alarm systems are component systems, what you get when you buy a "system" varies. Sometimes you get all the elements as part of a package; sometimes you have to buy some separately. You shouldn't consider a system that doesn't offer an engine disabler at least as an option. While there are many systems that lack one on the market, they don't provide adequate protection.

Convenience

Some features that add to security may go unused—either because they're too much of a hassle or because they make you feel like an intruder yourself. That's why a system's convenience is so important.

Entry/exit. Any system should give you ample time to get in or out of the car before the alarm goes off. Some systems allow too little time for entry—5 seconds to get in is not enough. Adjustable times are usually set by the installer, not the user.

Arming/disarming. Systems with a passive mechanism usually arm when you turn off the ignition and disarm when you turn it on. Because you don't have to take extra steps, they're the easiest kind to use. Best are passive systems that don't start to arm until a door has been opened, and all doors are shut. They don't make you race a timed exit delay. Some systems arm passively but require you to actively disarm them. (Insurance companies in some states give a 5- to 20-percent discount on your comprehensive premium, depending on the type of system.) Active mechanisms include a special key, a numeric keypad into which you punch a code, or a toggle switch. The toggle is less secure than the other types—there are only so many places to hide such a switch within the driver's reach.

Integral disabler. Engine disablers are often an integral part of the system. They're easier to use than disablers that have to be activated separately. A disabler, generally an electronic switch that interrupts current to the starter motor or the ignition system, keeps the car from being started. The installer often decides which system to interrupt. Attaching it to the starter is generally preferred, especially with late-model cars that have a computerized ignition system; attaching the disabler to the ignition can void the warranty on such a car.

Intrusion sensor. Most systems have some way to sense an attack even if the door, hood, or trunk is left shut: a vibration or sound sensor to sense a window being broken; a motion sensor to tell if the car is being moved. A resonance detector can tell if a door has been opened or if a window is broken; ultrasonic sensors do all that, plus sense any motion within the car. Several systems require you to buy sensors separately as options. Their sensitivity can usually be adjusted—important if you are to avoid false alarms from a careless parker, say, or a ball bouncing off the hood.

Hood lock. This typically comes as part of a separate engine disabler and must be actively engaged each time—a nuisance.

LED indicator. Many systems have a little light-emitting diode that shows whether the system is armed or not. LED indicators are much more convenient than systems whose sirens chirp when arming.

Valet mode. Lets you leave your car at a service station or parking garage without revealing the mysteries of the car's security system. It turns the system off—with a switch, a punched-in code, a special key, or a maneuver with the ignition key.

Panic switch. This allows someone in the car to sound the alarm without opening the door.

Backup battery. This is an option with most systems. It supplies power to the alarm should an intruder disable the car battery. A backup battery is good to have, especially if the car's electrical system can be reached from below.

How sensitive?

False alarms diminish the crime-stopping value of alarm systems, particularly in urban areas. The sight of an unmolested car wailing away at curbside is simply too commonplace to galvanize passersby into calling the police. A properly installed and adjusted system, however, should produce no false alarms. The hard part is getting the system properly installed.

To the extent that an alarm draws unwanted attention to a car thief at work, a louder alarm is better than a quieter one, and an alarm that carries on for a while is better than one that warbles briefly then falls silent.

An alarm that blares for at least 30 seconds is probably sufficient to grab attention. Then it should reset itself unless a door or other entry point is open. The alarm shouldn't keep going until it is shut off or drains the car's battery.

Installation

The marketplace is full of put-together component systems sold under little-known brand names. Dealers are accustomed to deciding for themselves what to install.

Leaving the selection of components up to the installers requires faith in their judgment. Apart from that, installation fees can vary widely, and installing the same system is trickier in some cars than in others.

Clearly, it pays to shop around. But price is only one part of the problem. Finding a competent installer may be difficult. You may end up with repeated false alarms and have to reinstall the system.

Recommendations

Electronic auto-alarm systems can't protect a car from a determined assault. But they can buy time or send the thief in search of easier prey.

A system that arms itself passively, when the last door is closed, is a big convenience, since you don't have to race a set exit time. An LED indicator is a plus because it shows whether the system is armed; so is a panic switch, which lets someone inside the car sound the alarm.

Ratings of auto-alarm systems

Listed by types; within types, listed in order of estimated quality. Closely ranked models differed little in quality; models judged equal are bracketed and listed alphabetically. Except as noted: All have sirens that produced between 92 and 96 decibels; all do-it-yourself models lacked minor installation hardware. Prices are as published in an **October 1986** report.

⊙ ◓ ○ ◒ ●
Better ⟷ *Worse*

Brand and model	Price	Convenience	Entry/exit, seconds	Arming/disarming	Integral disabler	Intrusion sensor	Hood lock	LED indicator	Valet mode	Panic switch	Backup battery	Comments
Professionally installed systems												
Crimestopper HP2501	$399	⊙	0–28/[1]	Passive door/ passive	✓	Vibration	—	✓	✓	✓	✓	C,D,E
Clifford System III	550	⊙	30/[1]	Passive door/active remote	✓	Vibration	—	✓	✓	✓	✓	A,B
Alpine 8101	400	◓	0–45/5–45	Passive door/active keypad	✓	Vibration, motion	—	✓	✓	✓	—	E
Maxiguard P-1000; Maxi-Lok L480	370	◓	5–30/60	Passive/passive	—	Motion	✓	—	✓	✓	—	F
Techne Ungo TL1600	485	◓	[2]/32	Passive/active remote	✓	Vibration, motion	—	✓	✓	✓	✓	B,I,Q
Thug Bug Avenger 1001; 0825; 0826	450	○	0–24/40	Passive/active keypad	✓	Motion	—	✓	✓	—	—	A,F
Paragon K6550; KPS9200	550	○	13/65	Passive/passive	—	Optional	—	—	✓	✓	—	—

Ratings of auto-alarm systems (continued)

Listed by types; within types, listed in order of estimated quality. Closely ranked models differed little in quality; models judged equal are bracketed and listed alphabetically. Except as noted: All have sirens that produced between 92 and 96 decibels; all do-it-yourself models lacked minor installation hardware. Prices are as published in an **October 1986** report.

⦿ ◓ ○ ◒ ●
Better ⟵⟶ *Worse*

Brand and model	Price	Convenience	Entry/exit, seconds	Arming/disarming	Integral disabler	Intrusion sensor	Hood lock	LED indicator	Valet mode	Panic switch	Backup battery	Comments
Professionally installed systems												
VSE Digi-Guard VS8200	450	◓	5–30/60–180	Passive/active keypad	✓	Motion	—	✓	✓	—	✓	P
Code Alarm CA1085	325	◓	3–30/3–60	Passive/active keypad	✓	Optional	—	✓	✓	✓	✓	A
Chapman-Lok Generation III System 400	375	◓	11 or 18/60	Active key/active key	✓	Motion	✓	✓	✓	—	—	A,G
Multi Guard MGB II	300	○	0–21/60	Active toggle/active toggle	✓	Vibration	—	—	✓	—	✓	A,G,H,P
Watchdog Trooper[3]; BM707R	260	○	12/55	Passive/passive	—	Vibration	—	—	✓	✓	—	A,G
Anes Pro 900; HL Pro 10	472	◒	15/110	Passive/passive	—	Sound	✓	—	—	—	—	A,K
Pioneer PAS200	324	●	2–25/30	Active toggle/active toggle	✓	Ultrasonic	—	—	✓	—	—	A,F,J,Q
Harrison 7119; 7828	281	◓	17/70	Passive/passive	—	Optional	—	—	✓	✓	—	G,I

Automotive Security Products K400FS	455	○	16/45	Active keypad/active keypad	✓	Optional	—	✓	✓	—	✓	G,L
McDermott AK012; KP-1	467	●	13/30	Active toggle/passive	—	Motion	—	—	✓	—	✓	G,J
Do-it-yourself systems												
Crimestopper CS9502; CS8003	221	◓	0–28/[1]	Passive/passive	—	Vibration	—	✓	✓	—	—	M
VSE Theftrap VS7810; SureStop VS3600	170	○	11/15	Passive/passive	—	Vibration	—	✓	✓	—	—	—
Auto Page MA/07S	165	◓	[2]/46	Passive/active remote	✓	Resonance	—	—	✓	—	—	B,N
Audiovox AA9135; AA7007	140	○	13/60	Passive/passive	—	Ultrasonic	✓	—	✓	✓	—	G
Sears Cat. No. 5980	100	◒	10 or 18/45	Passive/passive	—	Motion	—	—	—	—	—	G,I,P
Anes KD5000	99	●	14/55	Active toggle/active toggle	✓	Motion	—	—	✓	—	—	A,G
Pioneer PAS100	100	●	5/30	Active toggle/active toggle	✓	Vibration	—	—	✓	—	—	G,J
Wolo 612-XP	82	●	15/150	Active toggle/active toggle	✓	Motion	—	—	✓	—	—	O

[1] *Unlimited—arms when last door closes.*

[2] *Unlimited—disarms by remote control.*

[3] *Also sold as do-it-yourself model.*

Key to Comments
A–Hood and trunk switches are an extra-cost option.
B–Has remote transmitter to arm or disarm system from outside car.
C–Needs battery to power memory; inconvenient to replace.
D–Parking lights flash when alarm is tripped.
E–Has override switch that allows car to start should engine disabler malfunction.
F–Headlights flash when alarm is triggered.
G–Alarm stopped too soon, after only one cycle, if entry point was left ajar.
H–Comes with backup battery.
I–Siren emits chirp at start of arming cycle.
J–Siren perceptibly quieter than most.
K–Paging device beeps when alarm is tripped.
L–Has fuel-line cutoff and ignition cutoff as engine disablers.
M–All installation hardware and wire included.
N–Siren emits chirp at end of arming cycle.
O–Once alarm is tripped, it runs until stopped or until battery is drained.
P–Model discontinued at the time the article was originally published in *Consumer Reports*. The information has been retained here, however, for its use as a guide to buying.
Q–Disabler was optional extra.

ELECTRONIC TYPEWRITERS

The electronic typewriter streamlines routine typing chores such as making corrections, starting a new line, and setting margins and tabs. But it can do a lot more: It can check your spelling, file up to a few pages in its memory and print what it has stored on command, and correct typographical errors at the touch of a key.

Compared with the sort of review and revisions you can make on even a modest computer, the word-processing abilities of electronic typewriters are minimal. Some electronic typewriters still just type directly on the paper. Those with a memory store what you type and print line by line or after you've completed and corrected the document. But you are limited to viewing a small "window" of text—typically only 20 characters wide. By contrast, a typical computer monitor shows 2,000 characters at once.

If electronic typewriters are more like typewriters than like word processors, they retain very little of the old typewriter clank. Electronic typewriters have few moving parts. Instead of noisy type bars, a print head darts back and forth across the paper.

The type of element in the print head distinguishes one broad group of electronic typewriter from another. Impact-printing models use an element whose fully formed characters strike the ribbon. Thermal-printing models use an electrically heated grid of tiny pins or rods to transfer the image to paper. Some manufacturers offer both types of machine.

The impact-printing group is by far the largest. The shape of the element is most typically a "daisy wheel," with the type characters on the tips of its "petals." The wheel rotates to position the desired character, and a tiny hammer strikes the tip of the petal against the ribbon and the paper. Cup wheels are similar in function to daisy wheels. The result: clean, crisp type.

Machines that use the thermal print head, by contrast, don't print quite as crisply, since their characters are formed by a matrix of tiny dots. Such machines need paper with a smooth surface—Xerox paper works fine when printing with their ribbon. (They can also print directly on special thermal paper.) On the other hand, thermal-print machines are small, light, and virtually silent. They also run on batteries.

Print quality

Film ribbon has now largely replaced fabric ribbon and allows typewriter printing to be dark and crisply defined. There are no uneven areas or broken characters.

Good print quality is less a certainty with thermal-print models. If the dots aren't dense enough, the characters can be vague and ill-defined. Manufacturers, however, seem to be designing around the problem. The print produced by some thermal models nearly equals that of the typical impact-printing typewriter.

Another measure of print quality is how well the machine types on envelopes and index cards. Most models address envelopes expertly, without any stray marks, scratches, or incomplete characters. Index cards should be no problem at all.

Even good typewriters sometimes have trouble with underlining. Impact models can do a fine job. A few thermal models may make very light or spotty lines.

There isn't much call for carbon copies these days. But if there's no copy machine around or if a form has to be filled out in triplicate, a typewriter that makes copies can come in handy. Most impact models produce at least five readable copies. The thermals cannot make carbons.

Routine tasks

Keyboard design is the most important aspect of a typewriter. The shape and curve of the keys, their proximity to one another, the location of oft-repeated keys, the slope of the keyboard, and the pressure the keys require—all combine to give the machine a certain feel. You really have to try a machine to tell whether it's the one for you. Chances are, however, that you won't like "dead," springless keys or a tightly crowded keyboard.

Another aspect of keyboard feel is the ease and efficiency of using routine keys.

Return. The return shouldn't be too slow-moving. Nor should return keys be hard to reach or too small. Some models have an automatic return, which eliminates the need to hit the return key at the end of each line. On those models, the return is triggered when you type a hyphen or use the spacebar in the "hot zone" preceding the right margin stop. Hot zones typically vary from five to eight spaces. The longer zones require less frequent use of the margin release.

Paper handling. Typewriters move paper in various ways. The easiest assist to inserting paper is a powered paper-injector. A variable-line spacer makes it easy to reinsert and line up a sheet.

Repeating keys. On most models, a special key repeats any letter just typed or function just finished. Other approaches include letting any key that's kept depressed repeat, or letting only certain keys, such as the period, the underline/hyphen, and the *x*, repeat.

Spacing and backspacing. The spacebar and the backspace key should move the print head smoothly and quickly. A half-space key, which makes it possible to fit corrections into tight spaces, is handy, as is the "express backspace," which moves you quickly to the left margin.

Setting margins. Most models provide preset margins when you turn them on. On a couple of models, the preset left margin is awkwardly positioned, way too far to the right. Some machines can also remember any margin settings you make. Some models acknowledge newly set margins with a beep—a very useful feature. The margin-release bell has now also become an electronic warning signal. On some models it is too faint.

Setting tabs. Most machines let you set between six and twenty-four tabs (ten should be plenty). The models that remember margin settings after being turned off also remember their tab settings. And models that beep when you set a margin also beep for new tab settings.

Corrections

Impact-printing models use a "lift-off" correction system, in which a sticky correction ribbon literally lifts the mistake off the paper. The quality of the corrections is generally first-rate. That correcting ability gives impact typewriters a clear advantage over the thermal-printing models, most of which have no means of correcting mistakes once they're on paper; they allow errors to be corrected only on the display.

The latest impact typewriters make fixing mistakes even easier by remembering the keys you've just hit. At the press of a key, the machine recalls what you just typed and snatches it off the paper.

How far back the correction memory remembers depends on the machine. Some have the capacity to remember characters and lines; others to remember a whole page. The auto-correct feature on most models remembers from forty characters to a whole line.

Once you've typed beyond the correction memory, you have to correct mistakes the "old-fashioned" way: You reposition the type head over the mistake, and then, while holding down the Correct key, retype the erroneous character.

One popular variation on the correction feature is a key that erases an entire word. The key, described as "word correct," "word erase," or "word out," works when the print head is positioned over any letter of the offending word.

A handy adjunct to auto correction is a Relocate key. After a correction is made, that key swiftly moves the print head back to where you left off typing.

Machines with displays allow corrections to be made before anything is printed. Word Correct is a special key that erases an entire word (if it's within the correction memory).

Complex functions

Spelling checkers provide reassurance. The electronic dictionaries found in these machines contain from 35,000 to 90,000 words, according to the manufacturers; the higher the model in the manufacturer's line, the more words it contains. (By way of comparison, a standard dictionary contains about 140,000 words, not counting plurals and other variants.)

When you mistype or misspell a word in the typewriter's dictionary, the machine beeps. You look at the paper or the display and make the correction if the word is indeed misspelled. While a machine may recognize plurals and various verb endings as correct, it may not recognize proper names or oddball words you happen to use. So there is usually a custom dictionary, to which you can add your own words (usually, 300 is the maximum).

Some typewriters take this feature one step further. After beeping at a misspelling, they suggest correctly spelled alternative words on the display. You press Return when the desired word is displayed. If you've been using the typewriter mode and the incorrect word is already on the paper, the machine automatically removes the word and types in the correct one. If the new word is longer, the machine will even half-space to fit it.

Some models use spelling checking in conjunction with a document that's been stored in memory, which gives a typewriter's spelling checker much of the power of a word processor's. After a document is typed into memory, you can have the machine search for misspelled words and make revisions before you print out.

The size of the text memories are typically measured in kilobytes (K), or thousands of characters. About 2,000 characters (2K) equal one double-spaced page. Memories range from about half a page of text (1K) to eight pages (16K).

In addition to letting you check a document's spelling, a memory lets you review and revise the document you're working on—as long as it fits within the few pages allowed. The small display screens make seeing what you're doing a bit hard, as do the often-cumbersome means provided for moving back and forth through the text. You'd probably end up printing a draft version, then entering the corrections.

Text memory permits you to store a document—a paragraph you repeatedly use, or your résumé, say—and call it forth at some later date.

Two or three pages, however, would just about exhaust the memory capacity of the typical model with memory. A few models take a small step toward computerhood by letting you add external memory cartridges, which contain anywhere from 4K to 16K of memory.

Models with a memory must contain batteries so you can unplug them without losing any data. Some manufacturers have built-in batteries that recharge themselves whenever the machine's plugged in; others use long-lived lithium cells.

Many machines come with a computer interface built-in; on others the interface is optional and costs an extra $100 to $125. As computer printers, electronic typewriters produce high-quality print. But they are very slow compared with computer printers, taking nearly two to four minutes to print a letter an ordinary dot-matrix printer can turn out in thirty-five seconds.

Features for special needs

Electronic typewriters also come with special format and printing capabilities. For instance, they can automatically center a line, underline as you type, indent a block of text, or justify the lines of type. For special emphasis, they can produce bold type; thermal-printing models can stretch a word out.

Some models offer typists doing tabular matter the convenience of automatic decimal tabulation, which aligns columns on the decimal points. Some can even type vertical lines, allowing you to type boxes around titles and lines between columns.

Nearly all electronic typewriters give you a choice of pitch—that is, the number of characters typed per inch.

Traveling with a typewriter has gotten easier. One reason is that the heavy case has vanished. Only the top gets covered, with either a full or partial lid, which covers just the keyboard.

The thermal models are virtually silent, and most of the impact-printing models make only a pleaşant rat-a-tat.

The main expense of upkeep is ribbon replacement. Some manufacturers still make fabric ribbons, less expensive than film and adequate for casual typing.

All models provide print ribbons in cartridge form, making them easy to snap in and out. Handling the correction ribbon, however, is not so easy.

Aside from ribbon replacement, there's little or no maintenance to these typewriters. Keeping dust out of them is the most important thing you can do; most provide lids or covers for that purpose.

Additional features

Auto underlining works like a shift key. Press it before you type and everything you type will be underlined (some models can also be set to underline only the words, not the spaces). *Bold printing* double types to produce print that's darker than usual. *Enlarged printing* nearly doubles the height and width of characters. With a model that can draw vertical lines, you can make boxes around a title or rules between columns of tabular material. *Auto centering* lets you center a line on the page and sometimes in columns as well. *Decimal tabulation* is another automatic function that lets you stack columns of figures with the decimal points lined up. *Auto return* causes the print head to return whenever you type a hyphen or space in the "hot zone," usually the five to eight spaces before the right-margin stop. A machine with a *variable-line spacer* helps you line up a reinserted sheet or a preprinted form. *Auto text indent* (also called paragraph indent) indents a block of copy. An *On/Off indicator* is useful because the machine is quiet enough to be left on by mistake; it can be a light or a display. Some models can be battery operated.

Recommendations

Impact-printing typewriters tend to be rather big and fairly heavy. Thermal-printing models by nature are much easier to tote around, they're virtually silent, and they can run on batteries. They have some real drawbacks, though: Their on-paper correction systems are cumbersome where they exist at all. They require smooth paper, and they cannot make carbon copies.

Certain features—text and correction memory, a large display, a spelling checker—are likely to drive up the price. Correction memory is the most valuable of those extras. Text memory is most useful if it allows you to enter an entire document, so consider the length of what you generally type if you are choosing among various memory sizes.

Features that seem particularly worthwhile include auto return, word correct, automatic center and underlining, and index and express-backspace.

Regardless of how you narrow down the selection, it's important to try out a typewriter to see if its keyboard suits you. Take advantage of the stores that make this easy, and then price shop.

Ratings of electronic typewriters

Listed by groups as determined by a typist panel. Within groups, listed in order of quality. Bracketed models were judged about equal and are listed alphabetically. Prices are as published in a **November 1987** report.

⦿ ◓ ○ ◒ ●
Better ⟵ ⟶ *Worse*

Brand and model	Price	Weight, lb.	Print system	Visual display	Text memory [3]	Print quality	Keyboard feel	Spacing	Backspacing	Return	Correction ease	Advantages	Disadvantages	Comments
Smith Corona XD8000	$350	18	D	✓	16K	⦿	◓	⦿	⦿	⦿	⦿	C,D,F,I,J,K,N	s,y	*E*
Sears Cat. No. 53051	495	22½	D	✓	10K	⦿	◓	⦿	⦿	⦿	◓	B,C,F,I,K,M,N,O	e,g,v,x,z	*A*
Olivetti CX880	329	20	D	—	—	⦿	◓	⦿	⦿	⦿	◓	C,F,I,K,M,O	e,g,v,x,z	*A*
Olympia XL121	249	15½	D	—	—	⦿	◓	⦿	⦿	⦿	◓	B,C,I,J,M	e,g,	—
Royal Alpha 605	273	17	D	—	—	⦿	◓	⦿	◓	⦿	⦿	C,D,J,L	—	—
Panasonic RKT36	270	14	C	✓	8K	⦿	◓	⦿	⦿	⦿	◓	B,D,F,I,L,N	h,o,t,v,y,z	*F*
Olivetti CX440	199	16	D	—	—	⦿	◓	⦿	⦿	⦿	◓	B,C,I,J,M	e,g	—
Swintec Student	319	16	D	—	—	⦿	◓	⦿	⦿	⦿	◓	B,C,I,J,M	e,g	—
Smith Corona SC110	198	16½	D	—	—	⦿	◓	⦿	⦿	⦿	⦿	C,H,J	a,g,i,n,s,y	*C,E*
Sharp PA3120	239	12	D	✓	8K	⦿	◓	⦿	⦿	⦿	○	B,D,F,P	i,o,q	*E*
Panasonic KXR200	180	11	D	✓	1.3K	⦿	◓	◓	○	◓	◒	B,C,D,E,F,I,J,L,N,O,P	k,o,v	*E*

Brother AX33	320	13	D	1	5K	◉	○	◉	◉	◓	○	A,B,F,I,J,M,N,P	g,i,o	—
Swintec 1146CM	479	20	D	—	—	◉	◓	◉	◉	◉	◓	C,I,M,O	g,k,v,x,z	*A*
Canon S16S	299	12½	D	✓	5.5k	◉	◓	◉	◉	◓	◒	B,D,E,F,J,K,M,N	b,k,s,t,v,w,x	*E*
IBM Actionwriter I	479	24	D	—	—	◉	◓	◓	◓	◉	◓	C,D,I,J,K,M,N,O	m,p,r,t,u,v,z	—
Sharp PA3100	162	11½	D	—	—	◉	◓	◉	◉	◉	○	B,D,P	i,o,q,r	*E*
Royal Alpha 110	150	16½	D	—	—	◉	◓	◉	◉	◉	◉	C,D,J,L	i,j,k,o,q	—
Brother AX15	159	13	D	—	5K	◉	○	◉	◉	◓	○	A,F,I,J,M,P	g,i,o	—
Panasonic KXW60TH	210	8	T	✓	6K	◓	◓	◓	◓	◓	●	C,D,E,F,G,L,M,N,O	b,k,o,p,q,t,v	*D*
Sears Cat. No. 53002	165	11	D	—	—	◓	○	◉	◉	◉	◉	C,J	j,t,y	—
Smith Corona SL80	150	11	D	—	—	◓	○	◉	◉	◉	◉	C,J	j,t,y	—
Silver Reed EZ20	159	13½	D	—	—	◉	◓	◓	◓	◉	◓	C,D,J,L,P	o,p,v,aa	—
Casio CW500	249	8	D	—	—	◉	○	◓	◓	◉	◒	B,C,N	d,h,j,n,o,q,r,s,v,x,y	*E*
Olympia Carrera	175	15½	D	—	—	◉	◒	◉	◉	◓	◓	C,I,M	c,k,p	—
Canon S58	150	15½	D	—	—	◉	○	◉	◉	◉	◒	B,G,I,J,M	c,g,j,k,q,r,t	—
AT&T 6500	300	17	D	✓	—	◉	○	◓	◓	◓	○	B,D,L,N	f,m,z	—
Sharp PA1050	179	6	T	✓	6K	◓	◓	◓	○	◓	—	D,F,G,L,M,N	c,d,m,q,t,u,v,bb	*E*
Canon Typestar 4	157	5½	T	✓	—	◓	○	◉	◓	◓	—	B,E,F,L	g,k,n,p,t,v,w	*B,E*
Brother EP43	130	5½	T	✓	—	○	○	○	◓	◓	—	G,M,N	b,c,g,i,k,l,o,q,t,v,bb	—

Ratings of electronic typewriters (continued)

[1] *Instead of showing type, display shows status of various functions.*

[2] *D = Daisy Wheel; C = Cup Wheel; T = Thermal*

[3] *K = 1,000 characters*

Specifications and Features

Except as noted, all have: • Lift-off correction system. • Platen that can accept paper 11-in. wide. • Linespacing capability of 1, 1½, or 2 lines. • Repeat key that repeats all characters and functions. • At least 10 adjustable tabs. • Pica and elite pitch, produced by the same printing element. • Full or partial lid and carrying handle.

All thermal models: • Must use special, smooth paper. • Are much quieter than electronic models. • Cannot correct characters already printed (except for *Panasonic*). • Operate on batteries and (except for *Sharp*) come with AC adapter.

Key to Advantages

A–Can correct one whole line without having to hold down key throughout.
B–Auto return can be set to ignore space or hyphen in "hot zone."
C–Beeps when margins or tabs are changed.
D–Margin signal sounds 8 or more spaces before right margin stop.
E–Switches to low power draw if left plugged in for extended period.
F–Retains margin and tab settings when machine is turned off.
G–Very portable because so light in weight.
H–Very comfortable handle.
I–Has paper release or insertion lever to raise paper bail.
J–Shuts off when ribbon cover is raised.
K–Has powered paper-injector lever (*Canon S165* and *Olympia Carrera* have injector key).
L–Has characters for typing French, Spanish, and German.
M–Has characters for technical use.
N–Can automatically type copy with uniform right, instead of left, margin.
O–Correction ribbon judged fairly easy to replace.
P–Daisy wheel judged very easy to change.

Key to Disadvantages

a–Had trouble typing in corner of envelopes.
b–Produced relatively uneven underlining.
c–Slowed down typing speed of some panelists.
d–Envelopes hard to insert (and cards, with *Sharp PA1050*).
e–Sometimes bent or ripped corners of paper, envelopes, and cards.
f–Preset left margin judged much too far to right.
g–Margin signal sounds too late, only 5 or 6 spaces before right margin stop.
h–Corrected letters showed traces of ink (or, with *Panasonic RKT36*, left deep indentation in paper).
i–Lid judged somewhat difficult to remove or install.
j–Only 6 or 8 adjustable tabs.
k–Margin warning signal judged too quiet.
l–Margin set, tab set, tab clear, and tab keys inconveniently small.
m–Judged noisiest of all models.
n–Lacks paper support.
o–Lacks impression control.
p–Paper shifted position when platen rolled up and down.
q–Difficult to line up reinserted sheet for additional typing.
r–Gives no indication when some functions are activated.
s–Character being typed is partially obscured by print head.
t–Lacks paper guide.
u–Lacks platen knob.
v–Cord cannot be stored in housing or lid.
w–Light enough to carry easily, but lacks protective lid and carrying handle.
x–Printing ribbon judged somewhat difficult to replace.
y–Correction ribbon judged difficult to replace.
z–Daisy wheel or cup wheel judged somewhat difficult to replace
aa–Reverse index (to move paper up) judged extremely slow and jerky.
bb–Print head doesn't move to type (or backspace) until screen is filled (or emptied).

Key to Comments

A–Not designed to be portable.
B–Optional case is available.
C–Words cannot be added to spelling checker's dictionary.
D–Instructions say bond paper can be used; results were poor.
E–Model discontinued at the time the article was originally published in *Consumer Reports*. The information has been retained here, however, for its use as a guide to buying.

Features of electronic typewriters

						Correction features						Special print			
Brand and model	Price	Print System	Text memory[1]	Modes[2]		Capacity	On paper	Word correct	Half space	Spelling checker[1]	Paper width, in.	Auto underlining	Enlarged print	Vertical lines	Auto centering
AT&T															
6500	$300	D	—	T,J	✓	80 ch.	—	—	—	12	✓	—	—	✓	
Brother															
EP43	200	T	—	T,L	—	—	—	—	—	9 ½	✓	✓	—	✓	
AX15	300	D	5K	T	✓	1 line	✓	✓	—	12	—	—	—	—	
AX20	400	D	5K	T	✓	1 line	✓	✓	60K	12	—	—	—	—	
AX33	550	D	5K	T	✓	1 line	✓	✓	60K	12	✓	—	—	✓	
Canon															
Typestar 4	220	T	—	T,L	—	—	—	—	—	9 ½	✓	✓	—	✓	
Typestar 5	240	T	—	T,L	—	—	—	—	—	9 ½	✓	✓	—	✓	
Typestar 7	350	T	6K+	T,L,J	—	—	—	—	—	9 ½	✓	✓	—	✓	
S58	250	D	—	T	✓	31 ch.	—	✓	—	14	✓	—	—	—	
S68S	400	D	6K+	T,L,J	✓	1 line	—	✓	90K	14	✓	—	—	✓	
S16S	500	D	5.5K−	T,L,J	✓	1 line	—	✓	90K	9 ½	✓	—	—	✓	
Casio															
CW500	250	D	—	T	✓	1 page	—	—	—	9 ½	✓	—	✓	✓	
IBM															
Actionwriter I	545	D	—	T	✓	2 lines	—	✓	—	13	✓	—	—	✓	
Olivetti															
CX440	359	D	—	T	✓	40 ch.	—	✓	—	13	✓	—	—	✓	
CX880	499	D	—	T	✓	1 line	—	✓	—	15	✓	—	—	✓	
Olympia															
Carrera	249	D	—	T	✓	24 ch.	—	✓	—	13	—	—	✓	—	
XL120	299	D	—	T	✓	40 ch.	—	✓	—	13	✓	—	—	✓	
XL121	329	D	—	T	✓	40 ch.	—	✓	80K	13	✓	—	—	✓	
Report Electronic	399	D	—	T	✓	46 ch.	—	—	—	14	—	—	—	—	

Panasonic														
KXW50TH	220	T	—	T,L,J	✓	2 lines	—	—	—	10	✓	✓	—	✓
KXR200	240	D	1.3K	T,L,J	✓	1 line	—	✓	—	11	✓	—	✓	✓
KXW60TH	270	T	6K+	T,L,J	✓	2 lines	—	—	—	10	✓	✓	—	✓
RKT36	370	C	8K	T,L,J	✓	1 line	✓	✓	86K	12	✓	—	—	✓
RXT40D	420	C	7.2K	T,L	✓	1 line	—	✓	—	12	✓	—	—	✓
RKT45	500	C	8.1K	T,L,J	✓	1 line	✓	✓	86K	12	✓	—	—	✓
Royal Alpha														
110	280	D	—	T	✓	40 ch.	✓	—	—	13	—	—	—	—
115	370	D	—	T	✓	40 ch.	✓	—	80K	13	—	—	—	—
605	470	D	—	T	✓	1 line	✓	—	80K	13	✓	—	—	✓
Sears														
53002	198	D	—	T	✓	1 line	✓	—	—	12	—	—	—	✓
53015	245	D	—	T	✓	1 line	—	✓	50K	13	—	—	—	—
53035	295	D	—	T	✓	1 line	—	✓	80K	13	✓	—	—	—
53041	345	D	—	T	✓	40 ch.	—	—	—	13	✓	—	—	✓
53051	495	D	10K	T,L,J	✓	300 ch.	—	✓	80K	15	✓	—	—	✓
53061	595	D	32K	T,L,J	✓	300 ch.	—	✓	80K	15	✓	—	—	✓
Sharp														
PA3100	240	D	—	T	✓	65 ch.	—	✓	—	12	✓	—	—	✓
PA3120	300	D	8K	T,L,J	✓	65 ch.	—	✓	—	12	✓	—	—	✓
PA1050	280	T	6K	T,L,J	—	—	—	—	—	9 ½	✓	✓	—	✓
Silver Reed														
EZ20	179	D	—	T	✓	40 ch.	—	✓	—	13	—	—	—	—
Smith Corona														
SL80	199	D	—	T	✓	1 line	✓	✓	—	12	—	—	—	✓
XL1000	229	D	—	T	✓	1 line	✓	✓	—	12		—	—	✓
SC110	289	D	—	T	✓	1 line	✓	✓	35K	13	✓	—	—	✓
XE5200	299	D	5K	T	✓	1 line	✓	✓	50K	13	✓	—	—	✓
XD8000	559	D	16K+	T,L	✓	1 page	✓	✓	75K	13	✓	—	—	✓
XD8500	619	D	16K+	T,L	✓	1 page	✓	✓	96K	13	✓	✓	✓	✓
Swintec														
Student	319	D	—	T	✓	40 ch.	—	✓	—	13	✓	—	—	✓
1146CM	479	D	—	T	✓	46 ch.	—	✓	—	15	—	—	—	—

[1] *One K = 1,000 characters*

[2] *T = Typewriter; L = Line-by-line; J = Justified (straight margins)*

HOME BURGLAR ALARMS

Given a choice between a home with an alarm system and one without, a burglar, and particularly an amateur, is likely to choose the one without. Police statistics from one New York City suburb bear out this fact. The police found that 90 percent of the burglaries in a six-year period occurred in homes without an alarm system. In the homes that had alarm systems and *were* burglarized, the system didn't fully protect the house or wasn't turned on.

A burglar alarm diminishes your freedom in your own home. You and your family must remember that the alarm system is there and change your habits to suit its peculiarities. You (and the neighbors) also have to be tolerant of false alarms. If the system is too intrusive on your routine, chances are you'll abandon it. Of all the alarm systems installed in American homes, it's likely that only half are used regularly.

If you make your house hard to get into and follow commonsense security precautions, most common thieves will be discouraged. But ordinary measures may not be enough for people who live in a neighborhood where burglary is fairly common or for those who are often away from home.

Types of systems

There are many install-it-yourself burglar alarms available. Because different houses have very different needs, there's no one "best" alarm system.

The skill required to install a system ranges from a little to a lot. The least expensive devices cost around $50; a more complex system costs upward of $200, depending on the layout of your home. This may sound expensive, but it's low compared with the cost of professionally installed systems, which typically start at $1,000.

If you want a comprehensive alarm system, a professional installation might be well worth the expense. Most of what you're paying for is labor, because it takes a lot of skill to install such a system. That skill can help eliminate the many false alarms caused by guesswork and faulty workmanship. A reputable installer will thoroughly test the equipment and instruct you on how to use it. A professional can also make the system quite inconspicuous, as well as provide a "silent" alarm to the company's office.

System components

The basic elements of an alarm system, whether installed by a professional or by you, are the same. Magnetic or electronic sensors detect an intrusion; the sensors alert the master control, which then triggers the alarm. Some systems run on household current, with batteries as a backup in case of a power failure; others run on batteries alone.

In the simplest systems, all those functions are performed by a single device the size of a toaster. In more complex systems, the work is divided among several components, which are usually connected with wires.

Sensors. There are two main types of sensor. Perimeter sensors guard the outside of your house—the windows and doors. Interior sensors detect movement or intrusion within specific areas inside.

The main advantage of using a system based on perimeter sensors is that it sounds an alarm before an intruder has gotten inside or very far inside. The main disadvantage is that such a system is hard to install.

A system that uses an interior sensor is easier to install. With most, you simply plug in a single device, point the built-in sensor toward the space to be protected, and do a bit of testing to adjust its sensitivity. The single sensor of a "space protector system," as these units are called, can often guard more area than one perimeter sensor. However, these systems have disadvantages, too: They don't sound an alarm until the intruder is already inside; many can be easily set off by innocent things, including pets; and they restrict your own movements within the house when they're turned on.

You can combine both types of sensor in the same system. In a two-story house with a basement and relatively inaccessible second-story windows, for example, you might use a perimeter system on the ground floor and interior sensors to guard the stairwells. In a studio apartment, you might use just one perimeter sensor on the door and a space-protector system to guard the rest of the room.

Wiring. Perimeter sensors always are, and interior sensors sometimes are, connected to a separate control unit, either with wires or with small radio transmitters.

Most security experts consider a hard-wired system the more reliable type. Its components are analogous to the switches, lights, wires, and fuse box of a household's main electrical system. You usually buy the components of a wired alarm system part by part, as dictated by the layout of your home. Installation requires drilling holes in walls, snaking the wire through walls and floors or stapling it to baseboard moldings, and making lots of electrical connections. (The wiring conducts very low-voltage electricity and poses no shock hazard.)

In a large house, the wiring for such a system can quickly become extensive and complicated, especially if you're trying to conceal it. In such cases, a "wireless" system using radio transmitters may be easier to install.

Wireless systems aren't completely wireless, but they do away with the long runs of wire from the sensors to the control unit. One or more sensors are attached to a battery-powered radio transmitter. When a sensor is disturbed, the transmitter sends a brief, coded signal to a receiver in the control unit, which triggers the alarm.

A wireless system is likely to be more costly than a comparable wired system. A typical wireless transmitter costs $30 to $40, about five times the cost of a typical wired magnetic sensor. You also have to check the batteries in the transmitters regularly (some transmitters have test lights to make that easier).

Wireless systems once suffered from false alarms caused by other radio devices, a problem now lessened by the use of a coded signal unique to the system. Still, professionals rarely use wireless systems, largely because more can go wrong with sophisticated transmitters than with simple wires. Transmitters are apt to fail eventually. Since replacement parts are hard to get, you'd probably have to replace the entire transmitter.

Control unit. A separate control unit is a necessary part of any perimeter system. In fact, what you buy with most perimeter systems is just the control unit, with perhaps a token sensor and a siren thrown in. Additional sensors cost about $4 to $5 each.

In a perimeter system, the control unit is the brain of the system. The unit is generally plugged into a wall outlet and holds the backup batteries. Most also contain the system's On/Off switch and perhaps various indicator lights.

When a wired perimeter system is turned on, the control unit monitors the presence of electrical current through the sensors and the connecting wires. In a wireless system, the control unit monitors the radio receiver.

Systems commonly operate "normally closed": Electrical current normally flows continuously through the system. If the flow is interrupted—by a door being opened or a wire being cut—the alarm is triggered. Closing the door or reconnecting the wire won't silence the alarm; the control unit must also be reset.

Most systems can also work "normally open": Current flows only when a sensor is disturbed. Normally open systems are less reliable than normally closed systems. A broken wire, which sounds the alarm in a normally closed system, renders a normally open system inoperative—and you wouldn't know it unless you tested each sensor regularly.

Some interior sensors, like perimeter sensors, are designed to be hooked up to a master control panel. Most interior sensors, however, are part of complete or near-complete space-protector systems, with sensor, control unit, and alarm all in one box. Some can even function as the control unit for a wired perimeter system.

Switches. Easy access to the system's controls is important, since you'll probably have to turn it on and off several times a day. And the easier the system is to use, the more likely you are to use it. The switch can be a simple toggle switch, a key-operated switch, or a keypad that requires a code to be punched in. A toggle switch is not very desirable, since a thief could operate it as easily as you could. A keypad code saves you from carrying another key but can be less secure than a key switch if a lot of people need to know the code.

Most systems have an entry and exit delay time (typically ten to thirty seconds) for you to get in and out without setting off the alarm. Some delays are fixed; others are adjustable. More sophisticated systems can restrict the delay to selected doors; the alarm sounds instantly if any other protected opening is breached.

Obviously, a short delay means the On/Off switch can't be very far from the door. If you want to hide the control unit, many systems allow separate "control stations" to be installed, either indoors or outdoors. With a control station outside the front door, you have no need for exit and enter delays; you turn the system on and off from outside. With another station by your bed, you can wait until you've retired for the night to arm the system. Some control stations are as simple as a light switch; others require a key or a code and have indicators to show the state of the system.

Most space-protector systems have controls that may be difficult to reach. Often, making the device unobtrusive makes it even harder to reach the controls. For instance, an ultrasonic device may be disguised as a book, but its usefulness is diminished if the controls are at the rear and therefore difficult to reach if the device is sitting on a bookshelf with other books.

The actual alarm. Before you decide what kind of alarm to use, decide whether you want your system to scare off an intruder or to summon assistance. Some systems can be arranged to do both.

Many systems contain a built-in siren or other noisemaker. Such an alarm may be fine for scaring a thief away, but a noisemaker sounding inside your house may not be audible outside, so it wouldn't be very effective at alerting a neighbor. In addition, a built-in system allows the intruder to quickly smash it or otherwise disable it.

Better systems give you the choice of adding a separate alarm that can be placed indoors or outdoors, depending on what you intend your system to do. The alarm can be a horn, siren, bell, or other noisemaker, or it can be lights that suddenly illuminate a dark room or yard.

One important feature of an audible alarm is the ability for it to silence itself. An audible alarm doesn't have to be on for very long to scare away an intruder. If it sounds for more than fifteen minutes or so, you're likely to irritate neighbors and the police. It might also cost you a stiff fine. Many municipalities now require that all audible alarms silence themselves.

Most systems have an automatic silencing provision, shutting up after five or ten minutes. Some space-protector systems both silence the alarm and rearm the system for continued protection. In addition, some systems have an indicator to show whether the alarm has gone off while you were away.

If you want a system to summon assistance, however, there are more direct ways than with an audible alarm. Sophisticated systems can be hooked up directly to the police station. Many police departments, however, don't allow residential alarms to be directly connected because of the high rate of false alarms, which is where an alarm-monitoring service comes in. A monitoring service will check the validity of an alarm (generally by calling your home and requiring a prearranged code word as a response) before it calls the police. Some services will monitor equipment you've installed yourself. More commonly, these services are arranged by the companies that install alarm systems. Fees are typically $20 to $25 a month.

An alternative to a monitoring service is an automatic telephone dialer, which will forward a recorded emergency message to telephone numbers you designate (assuming the telephone line is intact). Many systems have a suitable fitting (a "relay contact") for attaching an automatic dialer. If your local police don't allow such calls, you could have the message sent to the office, a neighbor, or a friend.

Other functions. With many systems, you can attach a "panic button"—a button that sounds the alarm when you press it. Some systems even come with panic buttons built in.

A comprehensive security system can protect against more than intruders. Smoke and heat detectors are commonly connected to a perimeter system and can be connected to an automatic dialer that makes the emergency call. Other types of sensor are also available—temperature sensors and moisture detectors, for instance, to protect against burst or frozen water pipes.

Recommendations

Living with an alarm system is, in a way, living under siege. You may find that you can't live that way, so before you spend the money for an expensive system, carefully consider whether you really need the added security it promises. If you decide to install an alarm system, first look at your home as a burglar would—that's how a security consultant would start.

No single alarm or system can possibly suit every requirement. A rambling colonial obviously needs a very different system from a studio apartment on the eighth floor. The exterior siren that would summon help in a neighborly neighborhood would be fairly useless on an isolated house in the country. Pets can preclude using certain types of sensors (ultrasonic and microwave motion detectors, for instance). Whatever system you choose, make it as simple as possible.

Your choice also depends on whether your main intention is to protect yourself while you're at home or to protect your property while you're away. To safeguard your possessions, scare tactics may be sufficient. To protect yourself, you'd probably add some provision for summoning help.

The best and most reliable type of alarm system has as its heart a wired perimeter system. But because such a system can be difficult to install—especially if your home is large or if you want to hide the wires—installation is probably best left to a professional.

Before signing a contract for an alarm system, shop around and get written estimates from several companies. Contracts for security systems range from outright purchases to lease-maintenance agreements; understand what you're getting. Warranties also vary; ninety days is typical. Check a company's reputation as you would for any home-improvement contractor. Ask for references from its customers. Consumer-protection agencies often keep track of complaints filed against specific firms and can warn you away from charlatans.

Do-it-yourself installation of a perimeter system is a project only for those with a good grasp of carpentry and electronics.

Even if you install it yourself, a perimeter system can be fairly expensive. Before you spend the money and time, it might be wise to determine if a space-protector system is enough.

Of the various types of interior sensors, infrared sensors are passive, merely sensing changes in heat instead of sending out waves. Consequently, they are less prone to false alarms than ultrasonic or microwave sensors, as long as they are not pointed at objects that change temperature rapidly (radiators, air-conditioners, and the like) and aren't in direct sunlight. You can aim sophisticated infrared sensors very precisely, which permits the system to ignore the lower half of the room, where pets might set it off.

Once an alarm system is installed and functioning, be sure everyone in your family knows how to work it. And don't throw away any decals that come with the system. They do have a deterrent value when conspicuously displayed outside.

■ Burglar alarm summary ■

An alarm system can use just one type of sensor or a combination of types. Perimeter sensors protect doors and windows. Interior sensors detect intrusion in certain areas within the house. All perimeter sensors and some interior sensors must be connected to a separate control unit, usually with wires. Many interior sensors are built right into their control units and function as complete space-protector systems.

Perimeter sensors

How it works	Things to consider
Magnetic switch	
Two-part plastic switch and magnet assembly mounted on door and frame or window and frame; electrical current interrupted when door or window is opened.	Separate switch must be installed for every protected opening. Windows can't be opened for ventilation unless system is turned off or second magnet is installed. Do-it-yourself installations are hard to conceal.
Window foil	
Strips of foil stuck on window glass; electrical current passing through foil is broken when glass or foil is broken.	Glass must be broken for alarm to sound. Difficult to install; not recommended for do-it-yourselfers. Prone to false alarms. Impossible to conceal.
Glass-breakage sensor	
Placed on window; sensitive to vibration or sound when glass vibrates or is broken.	Wind or innocent noise can set off alarm if it's not properly adjusted.

How it works	Things to consider
Wired window screen	
Fine wire woven into fiberglass window screen sounds alarm if screen is cut; magnetic switch sounds alarm if screen is removed.	Separate screen must be installed for every protected window.

Interior sensors

How it works	Things to consider
Passive infrared	
Measures changes in infrared (heat) radiation; senses body heat of person entering protected area.	Sensors can be aimed above a certain height so pets won't set off alarm. Innocent heat sources—radiators, air-conditioners, sunlight—can cause false alarms.
Ultrasonic	
Transmits and receives high-frequency sound waves; movement in area distorts wave patterns.	Susceptible to false alarms caused by air currents, moving drapes, very small pets, and the like. Changes in temperature and humidity affect sensitivity, so system must be periodically readjusted.
Microwave	
Transmits and receives high-frequency radio waves; movement in area distorts wave patterns.	Radio waves pass through plaster and glass, so one sensor can cover several rooms. Sensor needs careful placement to avoid false alarms. Metal objects and mirrors can reflect waves.
Pressure mat	
Thin metal strips in mat used under carpet or rug sound alarm when stepped on.	Covers small, specific area, such as doorway or stair landing.

How it works	Things to consider
Photoelectric beam	
Beam of light projected between two points; anything passing through beam trips alarm.	If seen, easy to bypass.

SMOKE DETECTORS

Your best protection against fire is early warning. Smokc is a natural warning, but you may not see or smell it early enough. And, depending on the type of fire, you may have only minutes or even seconds to reach safety. Most people aren't willing to take that risk. A majority of homes now have at least one smoke detector.

Do smoke detectors really make a difference? Are they worth the frustration of nuisance alarms? Absolutely. In homes with detectors, deaths and injuries are substantially lower and property loss is substantially less than in unprotected homes.

According to a study by the National Fire Protection Association, more than 90 percent of fire fatalities and more than 80 percent of fire injuries in a recent year occurred in homes that apparently had no detectors. Residential property loss in unprotected homes accounted for 87 percent of that total.

Homes with detectors are obviously safer. Yet they still account for nearly 10 percent of the fatalities, 20 percent of the injuries, and 13 percent of the property damage. Some of those detectors, it is known, failed to operate. One possible reason: Their batteries were dead or missing.

A detector with dead batteries is worthless. So is one whose batteries have been removed to avoid nuisance alarms. Therefore, you should pay special attention to such features as a low-battery signal and a temporary-disabling mechanism that would encourage you to keep the smoke detector operating.

Detector types

The least expensive smoke detector is the ionization type. Photoelectric detectors cost more to buy and may not be as easy to find. Smoke detectors that combine the two types are also available.

Ionization smoke detectors rely on a small amount of radioactive material to let the air between two electrodes conduct a constant electric current. When smoke particles disrupt the current, electronic circuitry sets off the detector's alarm.

Ionization detectors have had a reputation for responding quickly to the small, invisible smoke particles from fast-burning fires. The photoelectric models' forte was detecting the larger, clearly visible, particles of slow-burning fires. A combination unit should combine the best of both designs.

How speedy a warning?

Consumers Union measured speed of response to smoke from four different burning materials. They simulated fast fires by igniting shredded paper, wood strips, and polystyrene. They simulated slow, smoldering fires (typically caused by a cigarette dropped on a mattress or upholstered furniture, one of the most common types of home fire) by placing wood strips on a controlled-temperature hot plate.

The most recent tests affirm that ionization detectors are the first to detect smoke from the test fires, but only by three or four seconds. Manufacturers have apparently succeeded in making the photoelectrics more responsive to the smaller particles of smoke from a fast fire.

The tests using slow fires held no surprises. As in the past, the photoelectric models sounded off first by a substantial margin.

On every detector, the alarm is appropriately shrill and unpleasant, usually a high-pitched pulsating sound.

Consumers Union tested to determine how loud the alarms would sound when the batteries were weak. CU tested each detector with batteries that were running low, then triggered the alarm and measured the sound level.

The alarms remained assertive with weak batteries, sufficiently loud to awaken most sleepers if the alarm is near the bedroom. Heavy sleepers, hard-of-hearing individuals, and sleepers whose detector is positioned at some distance from the bedroom should choose a particularly loud model.

Nuisance alarms

Smoke detectors can't discriminate between the smoke from a dinner cooking or that from a house burning down. An alarm that sounds when there's no danger is a nuisance. (The problem of nuisance alarms may be exaggerated. According to a 1982 survey by Market Facts Inc., a survey research firm, more than twice as many nonowners as owners believe that smoke detectors go off too easily.)

Nuisance alarms can lead owners to disable their smoke detectors permanently, or to remove the batteries "temporarily" and then forget to replace them. In either case, the detectors provide no protection.

Some detectors may become cranky because they're in the wrong place. Avoid placing a detector near a kitchen or close to a fireplace or woodstove. Garages and furnace rooms are also poor locations; car exhaust or a small back draft from your furnace can trigger a detector's alarm. Placing a detector near a bathroom can lead to trouble: Droplets of moisture inside the chamber have the same effect as smoke particles.

Nuisance alarms may also sound because dirt has accumulated inside the detector's chambers. The air that constantly drifts through a smoke detector leaves behind dust and grime, which may eventually accumulate to the point where the unit triggers spontaneously or ceases to operate.

To keep your smoke detector operating properly, vacuum it yearly, cleaning with a vacuum wand. On a detector with a fixed cover, run the wand across the openings. If the cover of the detector is removable, gently vacuum the sensor chambers.

Another cause of nuisance alarms is insects. If your smoke detector is in their flight path, you'll know about it soon enough. Unfortunately, there's no fix you can make on an existing detector that wouldn't reduce the unit's sensitivity. As of March 1986, however, Underwriters Laboratories require all smoke detectors to have a bug screen or equivalent protection.

Batteries

The power source for the vast majority of detectors used in homes is a single 9-volt battery. After a year or so, the battery will need replacing. Replacements cost less than $2 and are simple to install.

To keep a smoke detector ready, you must know when the battery is failing. The detector should signal that fact with a gentle, periodic beep. How long a detector continues to beep a warning depends on the model. Some beep for at least seven days; others give you at least thirty days' warning that the battery is dying. A thirty-day warning is better—just in case the battery starts failing while you're on vacation.

A number of smoke detectors give a second signal of battery distress. Those models have a pilot light that normally blinks slowly. When the battery is dying, dead, or removed, that pilot light goes out.

A few models utilize a second battery, which is used to power a small lamp. When the alarm sounds, the lamp goes on, illuminating a small area under the detector for at least four minutes. At night this could be a distinct advantage.

Test buttons

Every detector should have a test button to let you check if the unit is functioning. The button, for models made after November 1979, is a complete check. It indicates that the battery is live and the sounding device working, and it simulates a fixed amount of smoke to check the unit's smoke-sensing capabilities. However, clogged air slots, which may not let real smoke enter, would not be discovered.

Use experience

A Consumers Union Annual Questionnaire included a number of questions about smoke detectors. More than one-third of the respondents said they owned one or more detectors—a total of 90,000 units.

The detectors proved quite reliable: Only 2.5 percent had required repair. The respondents had experienced a total of 1,879 fires, and their smoke detectors sounded an alarm in 96.6 percent of those fires. In 956 such incidents, the respondents believed, the alarms helped prevent or reduce injury or death; in 1,133 cases, they helped reduce or prevent property damage.

The sixty-four units that didn't respond to a real fire may not all have been defective, in truth. In the absence of periodic cleaning, they may have gotten too begrimed to work. The batteries may have been faulty, or even missing. Weekly testing would quickly reveal such faults. The survey revealed that 67 percent of the units were checked less than once a month.

The respondents' worst problem by far appeared to be nuisance alarms. Almost one-third of the detectors had been tripped by innocuous smoke from the kitchen or fireplace (at that, though, the average was only three times a year).

What about radiation?

Most ionization smoke detectors contain a tiny amount of americium 241, a radioactive material. In sufficiently large doses, radiation from any source, including the sun, can do biological damage. Thus, the very presence of americium in the home may make some people nervous.

In the judgment of Consumers Union engineers, the hazards associated with americium in smoke detectors are very small. However, questions regarding the acceptability of any level of risk from radioactive materials keep coming up.

In trying to answer these questions, Consumers Union's staff reviewed a number of scientific studies undertaken since 1977. They looked not only at the possible exposure of people to radiation from smoke detectors in the home, but also at exposure of workers to americium during the manufacture of the devices. They also considered possibilities of accidents or fire conditions that could release the material, and at long-term environmental consequences of the disposal of millions of the radioactive sources in discarded smoke detectors.

The available information led them to make the following estimates: If there were no ionization smoke detectors, some 2,590,000 persons would die of cancer over the next seven years. Assuming every household in the country installed an ionization smoke detector, that number might grow to 2,590,001.

The added risk is equivalent to the additional radiation risk that would result from moving from your present home to one 16 feet higher in elevation, and hence 16 feet nearer the sun.

Where to put detectors

Ideally, smoke detectors should be on every level of a house, in every bedroom, in every hallway, and even in the basement. How many detectors you end up with will probably be determined by the size of your pocketbook. Other factors to consider are the size of your house, the number of rooms, and whether or not anyone in your family smokes.

If you can afford one smoke detector, make it a photoelectric unit. It may cost twice as much as an ionization model, but it's worth the extra price and shopping effort. The photoelectric models tested by Consumers Union responded to slow fires considerably faster than the ionization units.

The best location for a single detector is just outside the bedroom.

If you install a multiple-detector system, choose one ionization detector and make the rest of the units photoelectric. Place one outside the bedroom and one in the living area, but not too close to the kitchen or a fireplace. Additional detectors can go inside the bedroom of any smoker or in the basement (at some distance from the furnace).

All detectors can easily be mounted on a ceiling or a wall. A good spot is the middle of a ceiling. Nearly as good is someplace on the ceiling off to one side but no closer than 6 inches to a corner.

Recommendations

The best protection against fire is a smoke detector that responds quickly to all kinds of fires. Photoelectric and combination units do that best.

If you can, it's best to buy more than one detector. For a multiple-detector system, one detector should be an ionization or combination unit; the rest should be photoelectrics. An ionization unit's fast response to a fast-developing fire makes it good supplemental protection. Furthermore, since no two fires are alike, it's possible that a fire could produce a relatively transparent smoke that could greatly accentuate the difference in response times between the two types of detector.

LOW-COST CALCULATORS

Four things have made the useful and ubiquitous calculator small and light: integrated circuits etched with ever-increasing complexity on silicon chips; the liquid crystal display (LCD), which requires only the tiniest trickle of electricity to show segmented black numerals floating on a silvery field; the solar cell for power from a lamp bulb's light and the button cell battery, also used in cameras and watches.

Basic calculators, now packed like razor blades in blister packs, have become a product sold in drugstores and supermarkets.

Expensive models

Sometimes the complications of the more expensive models are playful. The calculator has now been crossed with the radio, the music synthesizer, and the video game.

Calculators that keep time. The difference between a watch and a calculator is becoming more a matter of whether you can wear the device. Just as some watches now incorporate calculators, many calculators now keep time—with multiple alarms, stop watches, 200-year calendars, and electronic tunes. Some calculators have even been taught to talk, doing their figuring in a synthesized voice.

Printing calculators. These have dropped dramatically in price and size. There are even some as small as some basic calculators. The more advanced printing calculators come with keys for letters as well as numbers—a further incursion by the calculator into territory once occupied by the scratch pad.

Solar cells. Calculators powered by solar cells, often sold with the slogan "never needs batteries," are energized by sunlight or room light. However, the button cell batteries commonly used in a basic calculator cost only about $2 and last for a year or two of heavy calculating, and you may well lose your calculator or want a new one before you've used up even one set of batteries. The advantage of powering your calculations with solar cells is really just a novelty.

Calculators for special needs. Of course, many of the more complex calculators on the market are simply capable of more complex calculations. Scientific calculators are indispensable for any of the sciences or for mathematics at or above the level of trigonometry; business calculators can figure discount rates, and statistical calculators will do standard deviations. The distinction between programmable calculators and hand-held computers grows increasingly fuzzy. For $100 or so, you can now hold in your hand the computing power that once would have filled a large room.

The basic calculator

For the simple arithmetic of checkbook balancing or dividing up restaurant checks, the basic $10 calculator suffices.

Calling a calculator cheap is no insult. With this product, you pay no premium for style—there are plenty of thin, sleek machines in handsome brushed-metal cases. If a small package appeals to you, you can get a calculator no bigger than a credit card. Some are in a vertical format, while others are arranged horizontally.

Standard features on calculators in this price range are a single memory, the ability to use a number again and again as a constant in a calculation, and an automatic shutoff to conserve the battery. In addition to the keys for the four basic functions of addition, subtraction, multiplication, and division, keys for percent and square root are common. Some basic calculators have even more.

Switching on and off. Some calculators go on at the flick of a switch, usually on the face of the calculator. Others go on with the press of a key. Because these keys are vulnerable to accidental pressure while you're patting your pocket or searching in your handbag, many calculators are designed so that the On key is nestled safely in a little valley or protected by a plastic ridge.

Sequence of functions. Calculators of this class perform the four basic functions in the same sequence as you would say them. Two plus two equals four, for instance, is entered as 2 + 2 =, and the display shows 4. A clear key (C) erases the previous calculation, and a clear entry key (CE) lets you correct a miskeyed entry. Sometimes the two clearing functions are combined in one key.

Other operations. There are other operations that aren't quite as straightforward. While you'd say "20 percent of 80," most calculators require you to push 80 × 20% to get the answer. On some calculators, you have to push the = key, too.

Percent keys. These are idiosyncratic in other ways. Some calculators wait for you when you're adding a tip to a restaurant check: Tap in 25 + 15%, and the display shows that the tip is 3.75. When you press =, the total, 28.75 shows up. Some calculators show 28.75 right away, and to find out what the tip is, you have to subtract the amount of the check or do another percentage calculation: 25 × 15%. This idiosyncrasy also shows when you're subtracting percentages, as in figuring discounts.

Square root keys. These are common on calculators, perhaps because figuring a square root by hand is a difficult and tedious arithmetic chore. But apart from that rare problem involving the hypotenuse of a right triangle, finding a square root isn't a demand of everyday life. Square root keys work in reverse of the way you'd say it: To find the square root of a number, you enter the number and then push the √ key.

Change-sign key. Less common on a basic calculator, but more useful, is a change sign: the ± key. The change sign key makes it easy to enter a negative number (as opposed to subtracting one number from another) or to add a series of numbers—say, the various amounts of outstanding checks—and then subtract the total from another number, such as the balance on a bank statement. Again, the key has to be pressed after the number whose sign you want to change.

Additional keys. If there's a reciprocal key (1/X), you can easily convert cents per kilowatt-hour into kilowatt-hours per cent, or hours per mile into miles per hour. The key for the constant pi (π) could be handy if you needed to calculate the diameter of a tree from its circumference or wanted to bake a cake in a round pan instead of a square pan and needed to compare areas. A parenthesis key looks like (), and with it you can do calculations such as 4 + 3 × 2, which equals 14, in a different sequence: 4 + (3 × 2), which equals 10. An exponential key (X^n) is more versatile than a simple squaring key (X^2), though neither is used much in day-to-day arithmetic.

Using the constant. Most cheap calculators allow you to use a number over and over as a constant in the four basic arithmetic functions. If you wanted to find the average monthly cost of several annual expenses, for instance, you would key in the first item and divide it by 12: 1243.38 ÷ 12 =. For the rest of the items, you would key in only the annual dollar figure and the = sign (448.23 = 2926.62 =) to get the average monthly cost.

On some calculators, the constant is always the second number entered. On others, the constant is the second number in addition, subtraction, and division, but it's the first number in multiplication. On some you have to key in the constant first, strike the function key twice, then key in the variable number. This latter method is somewhat confusing, especially in division or subtraction, but you eventually get used to it. A few calculators allow use of a constant only with multiplication and division.

Error signal. Basic calculators typically have room for eight digits in their display. If you exceed eight digits by multiplying 52,444 by 6,877, for example, some calculators reject the calculation completely and display a row of Es for error. Other calculators show a modest E and the number 3.6065738. Moving the decimal point eight places to the right gives an approximate answer to the problem. You have to clear the calculator before the next operation, however. Other calculations that produce an error signal are invalid operations, such as dividing a number by zero or taking the square root of a negative number.

Basic memory

Some calculators on the market have lots of memory—that is, they can store and recall many numbers, formulas, or even names and telephone numbers. Basic calculators have only one eight-digit memory—room for one number up to eight digits long. Typically, you can add to it, subtract from it, or recall it for use.

Using memory. An example of memory's use is keeping a running tab on the groceries in your shopping cart. After figuring the price per ounce of various brands of tuna to see which is the best buy, you can calculate the price of the six cans you've chosen and add it to the calculator's memory with the M + key. Then, you can change your mind and put back two cans, subtracting their cost with the M − key. When you need to know if you can afford a can of olive oil this week, you can recall (without erasing) the total that's in memory with the RM key. And when you go to the supermarket the next time, you will be able to start fresh by clearing the calculator's memory with the CM key.

Four-key memory vs. three-key. Many basic calculators have what's called a four-key memory: M +, M −, RM, and CM. Other calculators have a three-key memory, which combines the RM and CM functions on one key (one press for RM, two for CM). Three-key memories are somewhat less useful than four-key memories because you have to recall the number in memory before you can clear the memory. The number in memory will then appear on the display, replacing any number you may have there.

Continuous memory. A calculator with memory generally shows a small M in the display when something is in memory. There are a few calculators with the useful feature of continuous memory. With them, what's in memory remains in the calculator even when the calculator is turned off.

The keys

How fast you can make entries on a calculator depends partly on the size of its keys, and that's largely a factor of the size of the calculator. A big model may sit firmly on a desktop and have keys nearly as large as those of an office adding machine. A small calculator or credit-card-sized calculators are strictly one-finger or pencil-eraser models.

Key feedback. Even a very small calculator, however, can be used quickly and without fear of miskeying if you get some sort of positive indication that you've made an entry when you've pressed a key. Watching the display while you're entering the numbers is one way to make sure a number is correctly entered, but that's difficult and tiring when you're going fast. Better is a calculator that tells your sense of touch or your ears that a key has been successfully pressed.

The keys on many calculators provide tactile feedback. Before a key is pushed all the way to the bottom, you feel a little bump. That bump, which sometimes makes a faint clicking sound, signals that an entry has been made. Some calculators, particularly the very small thin ones, use a beep to tell you an entry has been made. (The tone can usually be turned off if you're working in a room where it might disturb others.)

Tone vs. touch. Feedback is an important advantage in a calculator. Tone feedback is a strong indication that a key has been pressed, but most models that beep provide no indication when the tone is turned off. Touch feedback can be quite strong, or it can be so weak as to be practically imperceptible.

Key feel. Keys on different calculators can feel very different from one another—but unlike key feedback, key feel is a matter of personal preference. People with broad fingertips may like a calculator whose keys have a heavy feel: They tend to feel insecure with a model whose keys require less effort to push. People with small fingertips may dislike calculators with heavy keys, preferring models with a medium or light touch.

Other key points. Other aspects of a calculator's design may or may not be to your liking. On most calculators, when you key in a series of digits, one digit makes way for the next by simply moving to the left. But on some, the entire display goes blank when any key is depressed. Such a calculator seems to respond slowly, and someone making rapid entries would see nothing on the display until a pause.

Keyboards are laid out in various ways. On some calculators, the numeral keys or the clear keys are a different color from the other keys, which may help you find your way around the keyboard. Some calculators have light-colored numbers on dark keys and some dark on light. Some have the number above the key.

Look at a few models in the store to see which you find easiest to read. Some calculators have tiny keys with a fair amount of surrounding space. Other models look more crowded because they have bigger keys and less space, though the actual distance between the centers of the keys is the same.

Calculator count

LCD models with up-to-date circuitry should run a long time on one set of batteries. A calculator should operate for more than a thousand hours, or about three hours a day for a year, before it will need new batteries. In fact, the batteries may expire from normal aging before they actually run down.

From the design of many calculators, it's clear that the manufacturers don't expect you to change batteries frequently. Most have no battery compartment as such. To get at the batteries, you must unscrew the back of the calculator.

Button cell. The typical battery used in the slim LCD models is the 1.5-volt button cell, which comes in an alkaline or a silver-oxide version. Manufacturers usually recommend the appropriate size of each type. Silver-oxide batteries last longer but cost two or three times as much as alkaline batteries and don't outlive the alkaline batteries long enough to repay their extra expense.

Lithium batteries. Some calculators use the longer-lived lithium battery. Because the shelf-life of a lithium battery is about five years, the battery in these calculators may never need to be changed.

Automatic shutoff. Many models have an indicator in the display—usually a small dot—that fades when the battery begins to run down. To further conserve the already long life of their batteries, LCD models usually turn the display off automatically if no key is touched for five minutes or so. To reactivate the display, you simply press a key.

Recommendations

When you buy a low-cost calculator—one selling for less than $15, perhaps less than $10—you should expect several fairly sophisticated features.

- A good basic calculator should have a liquid crystal display, keys that by touch or by tone let you know when an entry has been made, and an automatic shutoff provision.
- It should run for at least 1,000 hours on one set of batteries, which are usually included.
- Also common are such functions as a four-key memory and a constant capability in addition, subtraction, multiplication, and division.

CORDLESS TELEPHONES

A cordless phone's base station plugs into your phone line. The handset sends radio signals to the base station and receives signals from it so you can take or make calls practically anywhere in your house or outdoors.

Early models, those built before government rules affecting cordless phones changed in October 1984, often had serious problems. Connections crackled or buzzed with noise. Neighbors competed for the same airwaves because the government had allotted only five cordless phone channels for the whole country. Phone pirates cruised neighborhoods with cordless handsets, making unauthorized calls over any responsive base station within range. Worse, some owners suffered hearing loss when ringers built into certain handsets unexpectedly blasted them at close range with noise exceeding 125 decibels. There were even reports of batteries exploding in cordless handsets, shooting debris in all directions.

Improved technology, design changes, and additional cordless channels have now solved or minimized those problems.

Distance from house base

How far from the phone's base can you expect to roam with a cordless handset?

The older cordless phones used the utility company's power lines to transmit their signal, which had the effect of limiting their range; the farther you moved from a power line, the weaker the signal. The latest phones aren't limited in that way.

A cordless phone has two ranges. *Speaking* range is the farthest handset-to-base distance a model reaches before static makes the connection unintelligible. *Ringing* range is the maximum distance from the base at which an incoming call can ring the phone's handset. With some phones, the speaking range is greater than the ringing range. On others, the reverse is true.

A cordless phone's range can be cut significantly by metal siding, chain-link fences, screened porches, power lines, or metal plumbing. Even interposing your body between base and handset can reduce the range. For best results, position the base station near a window, extend the antennas fully, and remember that a clear line of sight works best. Inside a typical house, the range of a cordless phone will be about one-fifth the distances given in the Ratings, or less. But a range of even 50 feet is useful for many households.

Sound quality

When you're ready to dial a call or answer one, you typically flip a switch on the phone's handset from Standby to Talk. (Standby keeps the handset ready for incoming calls.) A cordless phone's main drawbacks soon become clear.

Even the best cordless units won't match the clarity of a good conventional telephone, whether for listening or talking. Cordless phones generally transmit voices a little more clearly than they receive them. That means you may have more trouble hearing the person at the other end than he or she will have hearing you.

Loudness, though, is not a problem: Incoming voices come through quite strongly. Most models have a volume control.

Background noise. All phones generate some background noise. Even the best cordless phone will be noticeably noisier than a good corded phone. Radio-frequency (RF) noise from refrigerators, vacuum cleaners, TV sets, computers, fluorescent lights, and the like add to the problem. You can expect to hear hisses, whistles, sputters, and other sounds.

If electrical noise is especially troublesome in your house, you may be able to improve matters with an RF filter. They're sold at electronics-supply shops for about $7. You just plug the filter into a wall outlet and plug the noise-generating light or appliance into the filter. You may have to experiment to find the offending appliance.

Ringers. Like many modern phones, cordless phones have an electronic ringer that tweets, chirps, or warbles. To prevent the ringer from sounding off directly into your ear, the newer models have their ringer positioned away from the earpiece, on the phone's mouthpiece or on its backside. The most robust ringers are as loud as regular desk phones. The weakest sound roughly one-fourth that loud. Heard at the earpiece, the ringer on some phones is quite loud, but not raucous enough to damage your hearing.

Privacy

The Federal Communications Commission has designated ten channels for cordless phones—twice the number available just a few years ago. Each channel is actually a pair of FM frequencies: The base unit transmits on a frequency in the 46 megahertz (MHz) band; the handset, on a frequency in the 49 MHz band.

Channel selection. The typical cordless phone allows you only one channel—the one the manufacturer built into the unit. (That channel number is usually printed on a sticker on the handset or the base.) If your neighbor also has a cordless phone on the same channel, you'll have what amounts to a party line: You won't be able to use your unit while your neighbor is talking. Worse, you'll find yourselves eavesdropping on one another. So it's a good idea before you purchase your phone to find out, if you can, which channels are being used nearby. You may, however, be able to exchange the phone for one with a different channel.

A few models let you use one of two channels. But even better are phones that offer multichannel use. Some let you choose any of the ten channels by setting switches on the base and handset.

Since cordless phones do broadcast over the air, be aware that other people can and do listen in, as some instruction manuals warn. People with cordless phones, FM scanners, and walkie-talkies can eavesdrop. Even an ordinary FM radio may be able to pick up a nearby cordless conversation. Since eavesdropping is so simple, you should never give out credit-card numbers or other personal information on a cordless call.

Digital security. Eavesdropping is one thing, but outright piracy is quite another. Fortunately, the new technology used in some phones severely limits the ability of phone pirates to make unauthorized calls through your base station.

The digital security system in these phones works something like that in an automatic garage-door opener. A complex series of signals moves between base and handset whenever you try to make or answer a call. If the base station "recognizes" the handset's code, it unlocks itself so the call can go through. Another handset, even if tuned to the correct channel, would almost certainly be unable to use that base unit. It could be used to listen in on a call, however.

Some models come preset with only one security code. Others let you select one code from hundreds or thousands of possible settings, either by flipping small switches in the base and handset or by entering a digital code through their keypad. For added security, most of the phones cannot be dialed by another handset whenever their own handset is in the base station's battery charger.

Phantom operation. Most cordless models are susceptible to "false ringing." Electrical noise activates their ringers, seemingly at random, when no one's calling. Some phones are also capable of "phantom dialing," conjuring up a number on their own.

Batteries

Rechargeable nickel-cadmium batteries (nicads) power the handset. When the batteries are properly charged, they'll generally let you talk for several hours straight. Increased static is often a sign that batteries need recharging. Most phones also have an indicator that's supposed to warn you when the batteries are weak. You can't always rely on the warning, however.

Nicad batteries are finicky. When you first get a cordless phone, you may have to charge it a few times for ten or twenty hours at a stretch before the handset holds a full charge. Further, nicads have a "memory." They seem to know how much of their power has been consumed between chargings. If you charge the batteries too often, they will expect frequent recharging and so will quickly use up their charge. That's why cordless-phone manufacturers recommend "deep discharging" batteries every two to three months. Leave the handset in the Talk position until the low-battery light comes on, then charge the batteries as if they were new. Be sure to disconnect the base station from the phone line. If you don't, you may not be able to get incoming calls on your other phones.

Nicad batteries should last through several hundred chargings, but they will eventually need to be replaced. Some phones let you replace dead cells quite easily. The simplest has two slip-in battery packs—one in the handset and the other in the base. The battery in the base gets charged. When the handset's power weakens, you simply swap battery packs.

Many phones lack easy-to-change batteries. You have to take the phone in for servicing when the batteries go. That's not very convenient.

Features

Some cordless phones are loaded with extras. Here's a rundown on the handier features:

Pulse or tone dialing. This feature is becoming standard fare on modern telephones, cordless or not. It allows you to use pulse dialing for most calls and switch to tone dialing only when you need it (for electronic banking, say, or to access an alternative long-distance carrier). That way, you avoid paying a surcharge for a tone line. Some phones give very short tone beeps, perhaps too brief to activate answering machines or computers. A few phones also offer the option of speedier pulse dialing, at twenty pulses per second instead of the standard ten. The quick pulses work in many areas.

Memory dialing. Many phones can be programmed to dial frequently called numbers at the push of two or three buttons. Often, the memory feature has a Pause button to insert several seconds of silence into a dialing sequence. That's useful when you have to wait for a switchboard or a long-distance computer to respond with a dial tone. Models vary in how many phone numbers can be retained and in how many digits each number may contain. Memory phones usually have a card index to help you recall the codes for the numbers.

Extra keypad. A few models have a second keypad on the base station. Both bases can be used as a speakerphone, so the extra keypad lets you call from the base, even if someone has taken the handset away. The keypad on many models has a Mute button to silence the microphone; the person at the other end can't hear a conversation you have with someone else in the room.

Paging and intercom. All models have one-way or two-way paging buttons. One-way lets someone at the base station signal whoever's carrying the handset. Two-way paging lets either person beep the other. Generally, models can also be used as an intercom, so the person at the base can talk to the one with the handset—without also hearing a dial tone in the background. The most convenient of the "true intercom" phones use a speakerphone built into the base station as part of the intercom; others require you to plug in a second phone.

Auto standby. Normally, you hang up a cordless phone by shifting the handset's Standby/Talk switch to Standby. (Leaving it in Talk keeps the phone off the hook.) With auto standby, though, you can hang up even when the handset is turned on—just put the handset in the base's charger. The handset will ring with incoming calls as long as it's sitting in the base.

Recommendations

Cordless phones don't deliver the voice quality you may be used to with a conventional phone. Background noise can also be intrusive on a cordless phone. The batteries can be finicky and can quit unexpectedly. Most cordless phones won't work at all in a blackout, since they need household electricity. Other folks may be eavesdropping. Repairs may be expensive and time-consuming.

But a lot of people feel that a cordless phone's convenience outweighs its shortcomings. For people with big houses, for parents at home with young children needing constant supervision, and for those who have trouble walking, a cordless phone may be a boon. The phones also make easy-to-install extensions for rooms without telephone wiring. Further, new technology and additional channels have improved cordless telephones.

You can get a feel for how a cordless phone sounds only by trying one—something you really should do before buying it. If you decide a cordless just isn't worth the trouble, you may want to consider a new corded phone or an answering machine. (See pages 317 and 326.)

Ratings of cordless phones

Listed in order of estimated quality, based on performance and features. Except where separated by bold rules, closely ranked models differed little in quality. Prices are as published in a **November 1986** report.

◉ ◓ ○ ◒ ●
Better ⟵⟶ *Worse*

Brand and model	Price	Handset dimensions & weight		Speech quality: Talking	Speech quality: Listening	Speech quality: Background noise	Ring loudness	Speaking
Freedom Phone FF 1700	$180	8½×2⅝×2⅛ in.	13 oz.	◓	◓	◓	◓	1500 ft.
Cobra CP46OS	235	8⅜×2½×2½	13	○	○	◓	◓	1400
Uniden EX 4102, A Best Buy	130	8¼×2¼×2¼	12	◓	○	◓	◓	810
Sony SPP100	190	7 5/68×2×2¼	14	◓	○	◓	◓	750
Freedom Phone FF 750	130	8½×2¼×2¼	12	○	○	◓	◓	1500
GE 29660	160	8⅝×2¼×1½	12	○	◓	◓	◓	690
Radio Shack ET410	180	8¼×2⅝×1⅞	13	◓	○	○	◉	660
AT&T Nomad 8000	200	7⅜×2½×1⅞	11	○	◒	◓	◓	960
Panasonic KXT3831	140	8¼×2½×2	12	◓	○	◓	◉	990
AT&T Nomad 4400	140	7⅝×2⅝×1⅞	11	○	○	◓	◓	870
ITT PC 1700	170	7¼×2¾×2¼	10	○	◒	◒	◉	1020
Sears 34986	130	8⅛×2⅜×2⅛	9	○	◒	◒	◓	540
Radio Shack ET415	160	5⅞×2⅜×1⅜	6	◓	◒	◓	○	690
Cobra CP302S	120	9⅛×2⅜×2¼	12	○	○	◓	◉	510
Panasonic KXT3821	120	8¼×2½×2	12	◓	○	◓	◉	660
Uniden XE300	100	8×2¼×2½	11	○	○	◓	◓	750
Webcor Float 600	165	7⅞×2⅜×2	12	◓	○	◓	◓	540

[1] Checked models do not have background dial tone annoyance.

Maximum range			Features					
Ringing	Built-in channels	Security settings	Memory dialing	Paging	Intercom [1]	Advantages	Disadvantages	Comments
1500 ft.	10	512	9/16	2	✓	C,D,G,H,J	e,g	*A,B,E,I,J*
1400	2	512	9/16	2	✓	C,D	g,j	*G,I,J,K*
870	2	256	—	2	✓	C,D,H,J,K	e,g,h,j	*C,F,K*
600	10	1	—	1	—	A,B,C,I,K,L	e,h,i,j,n,r	*B,H*
1500	1	1024	—	1	—	H	—	*B,E,J,K*
1500	10	256	—	1	—	C,G	c,e,h,j	—
960	1	10,000	32/28	2	✓	C,F,G,H	e,h,j	*I,J*
780	1	1	9/16	2	✓	D,E,H,J	q	*A,E,L*
1500	2	64	9/16	1	—	G	b,o	*G*
1050	1	1	—	2	✓	H,J	—	*C,E*
540	1	256	9/16	2	✓	K	b,e,h,j	*B,C,D,G,I,J,L*
1500	10	256	—	2	✓	D,H,J	b,g,i,r	*B,D*
1500	1	3	—	1	—	C	e,i,j,n	*C,H,J*
1268	1	32	—	1	—	C	e,j	*B,G,J,K*
1020	1	16	—	1	—	G	b,o	*G*
1050	1	1	—	1	—	—	c,i,j,l,m	*C,E,J*
480	1	10,000	2/15	1	—	D,I	c,e,g,h,j,m,p	*J*

Ratings of cordless phones (continued)

Listed in order of estimated quality, based on performance and features. Except where separated by bold rules, closely ranked models differed little in quality. Prices are as published in a **November 1986** report.

Brand and model	Price	Handset dimensions & weight		Speec qualit				
				Talking	Listening	Background noise	Ring loudness	Speaking
Record-A-Call Cat. 400	125	8½×1⅞×2⅜	11	○	◒	◓	○	570
Record-A-Call Cat. 300	100	8½×2⅛×2⅜	11	○	◒	○	○	540
Audiovox AT14	110	8¼×2¼×2½	11	◒	○	○	◓	750
Webcor 584	75	8⅞×2¼×2⅛	11	○	◓	◓	◓	420

Specifications and Features

All performed well in tests of speaking and listening loudness.

All have: • Low Battery indicator. Charge lamp. • Keypad in handset.

Except as noted, all have: • Ability to redial last number tried at press of 1 or 2 buttons. • Battery that cannot be replaced by user. • In-use light. • Switch for pulse or tone dialing. • Additional security provision: Dialing circuit locks when handset is in charging cradle; normally, no calls can be made through base station. • Extra phone jack on base, • Volume control on handset. • Antenna and phone cord that can be replaced by user. • Base unit that can be wall-mounted, if desired.

Key to Advantages

A–Automatically selects first unused channel.
B–Comes with extra nickel-cadmium battery in base unit's charger. Batteries are very easy to interchange.
C–Battery can be replaced by user.
D–Rings at base unit as well as at handset.
E–Base has switch to adjust ring volume.
F–Handset has switch to adjust ring volume.
G–Has Pause button for memory dialing or last-number redial.
H–Has Mute button.
I–Handset antenna is shorter and more flexible than most; less prone to break.
J–Base has speakerphone.
K–Switch lets you select 10 or 20 pulses per second for pulse dialing.
L–Will continue to function during power blackout.

Maximum range	Ringing	Built-in channels	Security settings	Memory dialing	Paging	Intercom [1]	Advantages	Disadvantages	Comments
				Features					
480	1	3	9/18	2	✓	D,J		c,d,f,g,i,m	*B,C,I,J,K*
480	1	3	9/18	1	—	—		c,d,f,i,m	*B,C,J,K,L*
660	1	1	—	1	—	—		b,c,i,j,m,r	*B,G,J*
480	1	1	—	1	—	—		a,b,c,i,j,k,m,r	*B,J*

Key to Disadvantages

a–Had more "false ringing" than most.
b–Loads telephone line more than others.
c–Handset lacks volume control.
d–Volume control for handset is located on base unit.
e–Short dialing tones may not work with some answering machines or computers.
f–On/Off switch is hard to reach.
g–Ringer in base unit can't be switched off.
h–Dialing is possible even when handset is in charging cradle.
i–Base unit lacks extra phone jack.
j–Phone cord can't be unplugged from base unit.
k–Lacks last-number redial.
l–Base unit lacks in-use light.
m–Handset lacks Flash/Cancel switch.
n–Lacks auto-standby feature.
o–Operating the Talk switch may inadvertently turn ringer off.
p–Lights on handset buttons are hard to see under bright light.
q–Lacks tone-dialing capability.
r–Antenna cannot be replaced by user.

Key to Comments

A–Has extra keypad for dialing from base unit.
B–Has "brick"-type AC power adapter.
C–Base unit has no power indicator.
D–Lacks intercom in-use indicator.
E–Handset lacks Talk indicator light.
F–In-use light hidden under handset in cradle.
G–Talk switch located on side of handset.
H–Cannot be wall-mounted.
I–Base unit has Hold feature.
J–Handset has separate On/Off switch.
K–Handset can be charged away from base unit.
L–According to the company, this model had been discontinued at the time the article was originally published in *Consumer Reports*. The information has been retained here, however, for its use as a guide to buying.

Comparing Speaking and Ringing Range

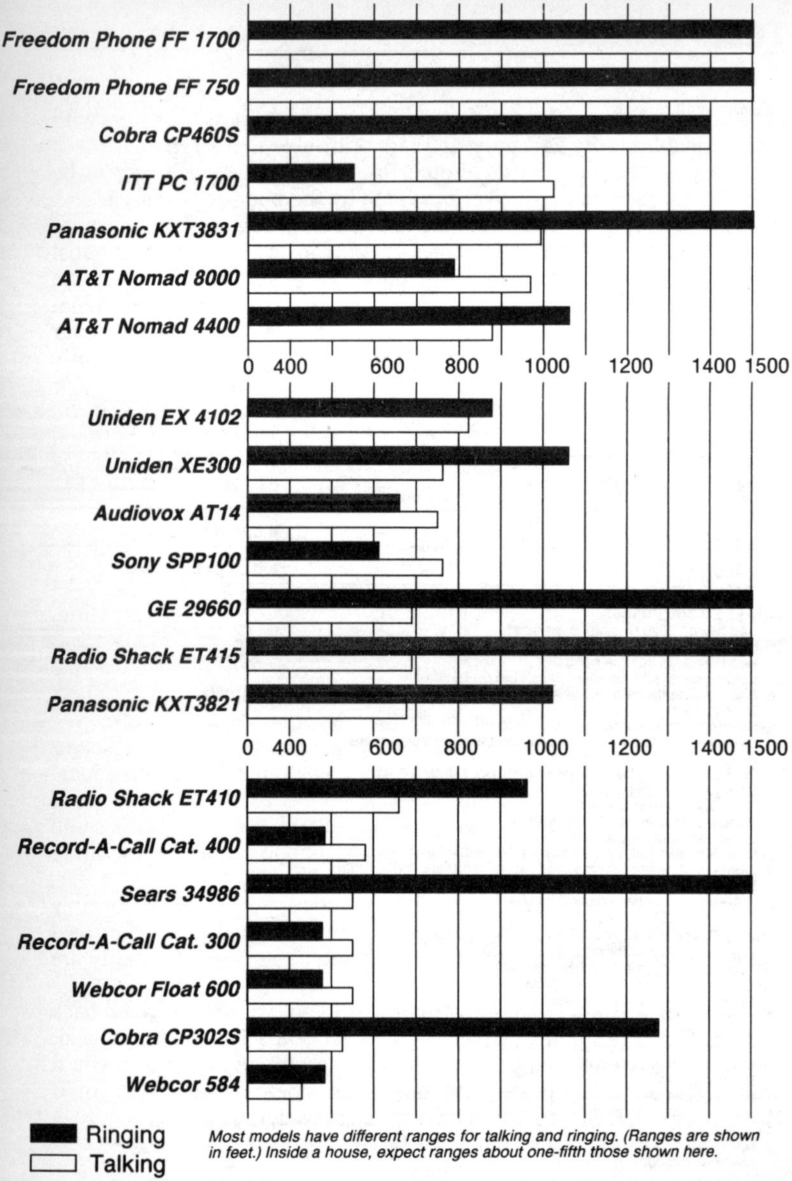

Most models have different ranges for talking and ringing. (Ranges are shown in feet.) Inside a house, expect ranges about one-fifth those shown here.

TELEPHONES

It wasn't very long ago that all telephones seemed to transmit voices with commendable clarity and provide years of trouble-free service.

Today, a telephone is just another basic appliance, no longer to be taken for granted and no longer provided by the phone company.

Buying a telephone presents some unaccustomed choices: You need to know, if possible, which of the scores of companies makes the best, most durable phone. What happens if the phone breaks? Should you buy the cheapest one around and throw it away when it stops working? What about frills? Telephones can now do tricks the old Bell phone never dreamed of. Many models can remember a dozen or more frequently dialed phone numbers and let you call them at the touch of a button or two. Some will time your calls, let you know when another extension is in use or off the hook, or even doggedly redial busy numbers. Such phones are becoming so commonplace and inexpensive that the no-frills basic telephone could eventually become a relic.

Loudness

The cleverest of features cannot make up for weak or muffled reception. With new long-distance carriers and satellite transmission, loudness and clarity are more crucial than ever: You never know how good a line you'll get when you dial a call or answer one. If you get a fuzzy connection, a poor phone can only make matters worse.

Listening volume. The loudness you hear at a phone's receiver depends, in part, on the length of wire between your home and the local phone company's switching station—the more wire, the softer the connection will sound. Using the main phone and an extension at the same time reduces volume further. But a good telephone can overcome some of that falloff in loudness.

Voice quality. Many of the new phones are capable of outperforming the traditional AT&T desk phone—no doubt due to the improved microphones on some newer models. Differences in speech quality are slight.

Sidetone. When you talk on a phone, some of your voice is fed back through the instrument's earpiece—you hear yourself speak. Long ago, engineers discovered that such "sidetone" is essential: Without it, you'd probably raise your voice. With too much sidetone, however, you'd lower your voice. Most phones have adequate sidetone.

Hearing-aid compatibility. The receivers of some telephones generate a magnetic field strong enough to "drive" many hearing aids electrically. People with a hearing problem should try a phone with their hearing aid before buying.

Clicks or tones

Dials are out and push buttons are in. Only a handful of the most basic phones are still available with a rotary dial. Most phones have push buttons. A few have a flat "membrane" keypad like some microwave ovens.

But even though a phone has buttons—a feature usually associated with Touch Tone dialing—you don't need to use it as a tone dialer. The phone should switch easily between tone and pulse dialing.

Pulse dialing is the system rotary phones use. On newer phones, the buttons can trigger a sequence of electric pulses that sound like the clicks of an old-fashioned dial swinging back into place. New phones dial faster: You can punch the buttons at your own pace without waiting for the clicks to stop. The phone "remembers" which button you've pushed and catches up.

Some phones offer even greater speed. They can switch from the standard speed of ten clicks a second to one twice as fast. The faster clicks work in many areas and are a nice compromise if you want speedier dialing without paying a monthly premium for a tone line.

A phone that can be switched from pulse to tone dialing is very handy if your phone company hasn't yet converted to the simplified long-distance dialing known as "Equal Access."

Tone dialing is significantly faster than the fastest pulse telephone. But there may be a problem: Some phones emit only a very brief tone, even if the key is held down continuously. The short tones dial calls well, but they may not last long enough for such uses as electronic banking from home, or calling home to retrieve messages from a telephone-answering machine. If you intend to use a tone-dial phone for banking or other computer applications, be sure the store will take the phone back if it doesn't work.

Voltage surges

Electrical storms often cause brief power surges that can wreak havoc on telephone equipment, particularly on the microchip components of modern phones. Typically, surge suppressors installed by the local phone company protect the line where it enters your home. The protectors work well with older phones, which are made largely of electromechanical parts. But the electronic circuitry in modern phones is more easily damaged.

Ringing

Compared with phones of yesteryear, modern telephones often have a feeble ring. The old-fashioned but vigorous bell ringer has given way to tweets, chirps, and warbles.

To measure each ringer's volume, Consumers Union's testers turned it up to its loudest position and took readings with a sound-level meter. The loudest ringer on a new phone is roughly equivalent to the old electromechanical bell in a standard desk telephone; the weakest ring is less than half that loud.

Many electronic ringers are rather high-pitched, and their sound may not carry as well as a bell's. People with hearing loss in the upper frequencies may have trouble with such ringers. You should audition a phone before you buy to be sure you like the sound of its ring.

If you must have a bell ringer, you can get one as an add-on. You plug the bell box into the phone jack on the wall and the phone into the box.

REN values. When you hook up a new telephone, your local phone company may ask you for two registration numbers: The unit's Federal Communications Commission registration number, and the Ringer Equivalence Number. The REN is a measure of the current the instrument's ringer draws. The REN is important if you have several extension phones because the RENs for all your phones should not add up to more than 5. If they do, the total current required will exceed the amount supplied on a standard phone line, and the phones might not ring at all. For most phones the REN is about 1, but some are considerably more, and others are less.

All the numbers

New phones excel at remembering frequently dialed phone numbers. Memory capacity ranges from just nine digits to more than thirty. You'd need thirty digits in a phone number mainly for complicated long-distance calling. Thirty digits could hold an entire dialing sequence: the local phone number of a long-distance carrier's computer, the required access codes, and, finally, the area code and local number.

Phones that can't store lots of digits can handle such calls if you split the dialing sequence into two or even three parts, storing the long-distance company's number and access code under one memory button and the individual's area code and local number under another. You'd press the first button to initiate the call and the second to complete it—a procedure called "chain dialing."

Typically, you can lengthen the pause by hitting the button more than once, but each tap uses up one digit of the phone's memory capacity.

Some phones can even remember whether a number was programmed to be dialed in pulse or tone, and dial it in the proper mode. A few phones go further than that, allowing you to switch the mode ("mixed mode") within a stored phone number.

One-touch dialing. Phones with this feature are easiest to use once they are programmed with numbers. One-touch dialing means there's a separate, dedicated key for each phone number in memory. Labels near the memory keys help you recall which button gets which number.

Several phones offer one-touch dialing for only three numbers. They're labeled as emergency numbers, but you can use them for any phone number you want. On those phones, other memory dialing can't be done with only one press of a button.

On phones without one-touch memory, you recall a number from memory with the same keys you use to dial. Typically, you press a special Memory button, then the one- or two-digit code used for the telephone number you want.

Last-number redial. This common feature lets you press the appropriate keys, and the phone repeats the last number dialed—a good arrangement for coping with busy signals. An even better system is automatic redialing. On phones with this feature, the unit repeatedly tries busy numbers about once a minute. When not trying the number, the phone hangs itself up to let incoming calls through.

Memory backup. Most phones take their electric power from the telephone line and thus might lose their memory if ever unplugged. Fortunately, nearly all have a battery backup for just that contingency. (All the phones let you dial manually, even if the batteries are dead.)

Hang-ups. Some phones have their switch-hook on the handset, or a Flash key to let you hang up without putting the receiver back on its base. If you're making several calls in a row or must rouse the operator, those features can be helpful.

Mute and Hold. Most phones have these buttons. Holding down Mute disconnects the microphone so the person on the other end cannot hear side conversations. A Hold button works similarly but needn't be held down continuously.

In-use light. Some phones have a little light that glows when an extension is lifted or off the hook. Others have a light to warn you when they haven't been properly hung up.

Shopping for a phone

The store should allow you to work a phone's controls, hear its ring, and listen to voice quality on a call. Some retailers—typically phone stores, but others as well—have "central office simulators," devices to put a phone through all its paces. Others are unable or unwilling even to ring a phone for customers. Shop somewhere else. When you find a cooperative store, here are things to check:

Design and controls. The phone should feel comforable to hold and to dial. The handset is particularly important. Many people prefer a traditional handset—the familiar bowed receiver, narrow in the middle and flared at the ends. That shape may be more comfortable than other designs if you want to prop the receiver between head and shoulder for hands-free talking.

Be aware, too, of clumsy controls. Some keyboards are too cramped to use easily. And some phones have only numbered keys—they lack the familiar set of letters alongside the numbers. That may seem unimportant, but just try dialing 1-800-4-PHONES, say, on a phone without letters. All the phones we tested have numbers and letters.

The disconnect button on some handsets is too close to chin level—you could accidentally cut yourself off as you talk.

Instructions. Well-written instructions are essential, since it's not always obvious how a phone's features work.

Recommendations

Some of the fancy features are nice to have, but it's more important for a phone to deliver loud, clear speech to people at both ends of the line. A phone should also be sturdy enough to withstand most man-made or natural disasters that befall it.

As for features, only you can decide just how sophisticated a telephone you need. If you intend to make many calls to the same people or if you work at home, you'll probably appreciate memory dialing. If simplified long-distance calling hasn't yet reached your area, you'll likely relish the ease of "chain dialing": Once programmed, a couple of buttons will get you through the most convoluted dialing sequence to practically anywhere. And the ability to switch between pulse- and tone-dialing makes a phone even more useful. You need not take on the added expense of a tone line to use an alternative long-distance carrier; you can use a cheaper pulse line and switch between pulse and tone as required.

No matter what phone you buy, try it out first—play with the buttons, cradle the receiver, listen to the ring. If you can, plug in the phone so you can judge its sound quality. Some stores are equipped to let you do that.

Ratings of telephones

Listed in order of estimated quality. Differences between closely ranked models were slight. Prices are as published in a **May 1986** report; + indicates shipping is extra.

⊙ ◓ ○ ◒ ●

Better ⟷ *Worse*

Brand and model	Price	Style [1]	Loudness: Listening	Loudness: Speaking	Speech quality	Ring loudness	Hearing-aid compatibility	Resistance to abuse [2]: Drop test	Resistance to abuse [2]: Voltage surge	Resistance to abuse [2]: Static electricity	REN [3]	Memory capacity, numbers/digits	Advantages [1]	Disadvantages	Comments
AT&T Touch-a-Matic 1600	$120	C	◓	◓	◓	⊙	○	⊙	✓	✓	0.4	15/11	A,B,C,K,L	a,b	*B*
ITT Medallion 2480	70	C	◓	○	○	⊙	●	⊙	✓	✓	0.5	10/15	E,F,K,M	f	*I,J*
Uniden EX-1020	80	C	○	◒	◓	⊙	◒	○	✓	✓	0.7	13/25	A,B,E,K,	b,f	*B,K*
GE 2-9260	50	T	◓	○	○	○	◒	⊙	✓	—	0.9	12/15	B,D,E,F,	e,m	*C,F,J*
Panasonic VA-8080	55	T	◒	○	○	◓	●	⊙	✓	✓	1.0	13/16	B,D,H,I,O	b	*C,D,K*
AT&T Ranger 600	50	T	◓	⊙	○	○	◓	⊙	✓	✓	0.7	9/16	D,N	d,e,f,k,l	*C,E*

Radio Shack Duofone-170	70	C	○	○	○	○	◒	◉	✓	✓	1.0	33/30	B,C,K,	b,n	*J*
Soundesign 7364	70	C	◓	◓	◓	○	○	◉	✓	✓	0.9	13/16	B	c,e,f,h,n	*K*
GE 2-9275	70	C	◒	●	○	◓	◓	◓	✓	✓	0.8	16/22	A,B,K,O	b,h	*J*
GTE Ultrastyle 300	50	T	○	○	◓	○	●	◓	✓	✓	1.0	9/16	D,I	e,f	*C,F*
Webcor ZIP-757	60	C	◓	○	◓	○	●	○	—	✓	1.5	16/16	G,M,N	f,h,i	*H,K*
Sanyo MD 3100	79	C	◓	●	◓	○	●	○	✓	✓	0.6	9/18	J	h	*K*
Panasonic KX-T2320	65	C	○	◒	◓	○	◒	◓	—	✓	1.0	33/30	B,C,K,	b,n	*J*
Sears Cat. No. 34401	30+	T	○	◒	◓	◒	○	◓	—	✓	0.6	9/18	D	b	*C*
Cobra ST-410	55	T	○	◒	◓	○	○	◓	✓	—	0.8	12/16	D	a,e,k	*C,F,G*
Conair Prima PR1001	40	D	○	◓	○	○	◒	○	—	—	1.3	9/18	K	a,b	—
Telequest Grand Prix Plus 123	80	D	◒	●	◓	○	◒	◒	—	✓	0.7	10/15	K	f,g,j	*A,J*

1 *C = console model. T = **Trimline**-style model, with controls on handset.*
*D = desk/table model, with controls on base (**Conair** is size of **Trimline**-style phones; **Telequest** is larger).*

2 *✓ indicates that the phone passed the test with no damage;—indicates that the phone sustained damage.*

3 *Ringer equivalence number.*

Specifications and Features

All have: • Last-number redial. • Switch for pulse or tone dialing. • Speed-dialing memory. • Chain-dialing capability (ability to recall two numbers from memory in sequence). • Volume and On/Off switch for ringer.
Except as noted, all: • Have pause feature; can be set to wait for computer tone or switchboard line when dialing automatically. • Have Mute/Hold button. • Can be used on desk or wall-mounted.

Ratings of telephones (continued)

Key to Advantages

A–One-touch dialing for all numbers in memory.
B–Has 3 keys for one-touch dialing emergency calls.
C–Has low-battery indicator.
D–Has disconnect/flash button on handset.
E–Has 2 speeds for pulse dialing: 10 and 20 pulses/sec.
F–Last number dialed can be stored in permanent memory.
G–Can automatically redial a number 15 times when busy signal is detected.
H–Has volume control for earpiece.
I–Has in-use light on back of handset.
J–Very compact phone.
K–Handset is of traditional design; may be more comfortable than some newer designs.
L–Has convenient digital display for number dialed, low battery, ringer off, clock time, date, and call timer.
M–Directory/index concealed for privacy.
N–Pause feature more convenient than most.
O–Somewhat louder reception than most, an advantage for those hard of hearing when background is noisy.

Key to Disadvantages

a–Lacks Pause control.
b–Lacks Mute/Hold button.
c–Battery can't be replaced by user.
d–Phone must be turned on and off by hand.
e–Last-number redial requires press of more than one button.
f–Dialing tones are too short to work with some telephone-answering devices and other computer entry equipment.
g–Keyboard less convenient than others.
h–Telephone wire cannot be unplugged from phone's base.
i–Phone must be plugged into ac outlet.
j–Poor battery cover; batteries fall out easily.
k–Retains memory for shorter time than most when unplugged from phone line.
l–Phone may possibly ring in ear. Sound may be jarring, but judged unlikely to damage hearing.
m–User cannot hear tones when dialing.
n–Sidetone somewhat deficient under worst-case conditions.

Key to Comments

A–Phone can't be wall-mounted.
B–Membrane keypad (only for memory keys on *Uniden*).
C–Push buttons on handset.
D–Push buttons on back of handset.
E–Privacy feature allows user to erase last number dialed.
F–Has lamp on keyboard side of handset.
G–Ringer can be switched between bell and electronic chirp.
H–Has listen-only loudspeaker.
I–Has memo pad on phone.
J–Memory recalls dialing mode (pulse or tone) for each number; allows "mixed mode" dialing.
K–Model discontinued at the time the article was originally published in *Consumer Reports*. The information has been retained here, however, for its use as a guide to buying.

■ Memory dialing ■

If you're content with the phones you already have but are tempted by the convenience of memory dialing, you don't necessarily need a new phone to get that amenity.

Many local phone companies offer a "Speed Calling" option that lets any phone—tone or pulse—mimic a memory dialer. You program frequently called numbers—up to fifteen digits each—into the company's central office computer, associating each number with a digit or two on your phone's dial. To reach those numbers, you simply dial their codes, and the distant computer does the rest of the work.

On pulse lines, the call takes a few seconds to complete because the computer waits to see if you're going to dial an entire number manually. On tone lines, pressing the # key speeds things along. The system is a cinch to reprogram, too. No need to call the phone company—you do it right from your phone.

Speed Calling usually costs a few dollars per month for options with eight- or thirty-number memory. There may also be a one-time connection charge.

Is it worth it? Only if you have several bare-bones phones and want memory dialing on *all* of them. Otherwise, it probably makes better sense to spend, say, $90 for a respectable memory phone than to pay the phone company up to about $100 a year for Speed Calling. If you buy the phone, you can take it with you—memory and all—if you move.

■ Hands-off dialing ■

Chances are, you dial your phone in the time-honored way—with your index finger. There are a few phones that may change all that; they dial by the sound of your voice.

You program the phones by reciting your list of names into the mouthpiece and punching in the corresponding phone numbers on the keypad. Then, to reach one of the people on your list, you speak the name into the phone. A built-in microcomputer analyzes your voice, matches it against voice-prints of the names in memory, and retrieves and rings the right number. These phones performed with impressive speed and accuracy.

At a price exceeding $200—several times what a conventional phone costs—these voice-dialers are likely to appeal mainly to the gadget-happy consumer or to people with a handicap that limits the use of their hands.

The *Voice Dialer 1000,* by Innovative Devices (1333 Lawrence Expwy., Santa Clara, CA 95051), holds up to 100 names, and two phone numbers for each.

The phone is simple to program and to use, in part because its computer-synthesized voice prompts, instructs, and repeats names to verify calls. The voice can be quickly run through an index of names and numbers. At times, though, the clipped, robotlike voice garbles names.

The *Command Dialer II VRT-1150,* by Audec Corp. (299 Market St., Saddle Brook, NJ 07662), holds just sixteen spoken names and numbers, which may be dialed by voice or with dedicated buttons on the phone's console. As a voice-dialer, the phone is somewhat cumbersome to program and use; you have to speak names and push buttons in a hard-to-remember sequence.

Full names work best, with near-perfect accuracy—the phones are capable of dialing without error at least 98 percent of the time. With one-word names, the phones' accuracy drops to as low as 70 percent for the *Command Dialer* and 90 percent for the *Voice Dialer,* depending on the speaker. Full names, however, consume more memory: The *Voice Dialer* holds only about 80 such entries.

Both models work on tone or pulse phone lines. And both must be plugged into house current. The *Voice Dialer* has a battery backup to keep it going for a couple of hours in case of a power failure. The other phone won't work in a blackout. The *Voice Dialer*'s electronics may also interfere with the picture on a nearby TV set.

Since every voice is unique, a voice-dialer programmed by one caller may not work for anyone else. So each user should program the phone separately. The *Voice Dialer*'s large memory can accommodate several renditions of the same name and number; the *Command Dialer,* with its skimpier memory, can't hold many variants.

TELEPHONE ANSWERING MACHINES

Some callers don't like answering machines. But these machines can be a real convenience, as their popularity attests.

Callers and users alike will find the latest answering machines friendlier and more sophisticated. They give callers ample time to leave a message. Their recording and playback circuitry is "smart" enough to overlook callers who hang up, while making it easy for you to review the messages you did receive. Many answering machines now use matchbook-sized microcassettes rather than standard cassette tapes, so they are smaller and lighter than their predecessors. Some answering machines come with a telephone built in.

You can spend as little as $50 for a model that only delivers a message from you, or several hundred dollars for a unit with all the latest electronic wizardry.

Any machine worth considering should have voice activation (VOX) and remote control, two highly desirable features. VOX means that callers won't be cut off in mid-message; whenever a caller speaks, the machine records. Remote control allows you to collect your messages when you're away from home.

What an answering machine can do

An answering machine does these basic things: It answers the phone with your recorded greeting, records callers' messages (usually on a separate tape), and plays the messages on demand.

Recording your greeting. Many machines answer the phone with a greeting that's as long or as short as you like. A variable message length is much more convenient than a fixed length.

The outgoing message tape should be long enough. Five minutes seems quite adequate. (Of course, you can record a much shorter greeting.) The maximum length on some models ranges from only 20 to 150 seconds. The low end of that range may be too brief.

Some machines lock you into using the full length of their twenty-second tape. If you take less than the allotted time, callers will get dead silence before the beep. A longer greeting won't be recorded in full.

Some use an "endless loop" tape that gives you up to twenty to thirty seconds for your greeting message. You can buy longer tapes, but using one may force callers to wait through several rings while the tape "recycles" after a previous call.

Other machines use a single microcassette for both outgoing and incoming messages. That creates a delay. After the greeting, a caller must wait through a long beep while the tape fast-forwards to the end of the last message taken. As messages pile up, the beep gets annoyingly longer.

Most machines have a built-in microphone for recording the outgoing message; that's more convenient than a separate microphone. After you've recorded your greeting, some units automatically replay your message for a final review. Others allow you to do that manually. The beep is inserted automatically when you've finished recording your greeting.

It's a good idea to use some care in how you record your greeting. Your message should be loud enough and with a minimum number of pauses. That will make things easy on callers' ears, and at the same time conform with some machines' vagaries.

Setting the unit to answer. Some machines are set to record calls as soon as you turn them on. Others require you to push a button or slide a lever after turning them on. That's not difficult, but it's something you have to remember to do.

Nearly all answering machines let you select the number of rings before the call is answered. If you'll be away from home, you might set the machine to pick up quickly; if you're home, you can leave it in the Answer mode and set it to pick up after four rings or so. That gives you a chance to pick up a call if it's convenient.

Leaving the answering machine on when you're at home permits you the luxury of screening calls: You can hear the caller through the built-in loudspeaker. Models that make call-screening easiest automatically stop and reset to the Answer mode when you pick up an extension. (If you pick up the phone too quickly, the machine won't automatically stop; answering machines have a built-in two-second delay so the phone company's equipment can establish that the call went through. You can stop the machine by momentarily depressing the phone's disconnect switch.)

Call-monitoring on other machines isn't quite as simple. If you choose to pick up the call, you have to stop the machine or at least turn down the volume to prevent screeching acoustic feedback noise. That also means that you have to use a phone extension close to the answering machine (assuming your answering machine doesn't have a built-in telephone).

VOX quality. Old-fashioned answering machines without VOX force callers to limit their message to twenty or thirty seconds. Units with VOX continue recording as long as they are activated by a voice. When the voice stops, they usually continue recording for a few seconds before sensing silence or the disconnect signal. A number of models also rewind back over a portion of that dead time, so there's less empty air between messages.

Some VOX circuits can be overly sensitive. A very noisy telephone line might trick a unit into continuing to record after a caller has hung up. As a safeguard against this, some models override the VOX to limit incoming calls to two or three minutes. Most machines give you the option of limiting time on incoming calls if you suspect that your phone line may bollix the VOX.

Call counting. All answering machines visually indicate that calls have come in. Most do it with a light. Some even blink their light in Morse-code fashion to indicate the number of calls. But the best call-counter by far is a digital readout of the number of calls.

The counter on some models counts hang-ups as calls, but many machines ignore callers who hang up during the greeting. Harder to screen out are hang-ups that occur right after the greeting. Some machines are smart enough not to count them, either.

Listening to messages. A typical answering machine is automated to allow playback at the touch of one control. Others require that you press Rewind, then Playback.

All answering machines let you rewind the incoming-message tape to catch a garbled name or number. Nearly all also allow you to fast-forward so you can bypass a message. Some allow you to skip forward or back exactly one message.

Many machines don't know when to stop playing back. They continue past the last current message and inflict you either with silence or, worse, with the tail end of old, unerased messages.

Machines usually let you save messages if you wish. If you choose to save them, new messages get recorded after them. The Save command may require an extra step or be automatic. If you decide not to save messages, new messages will be recorded over them. Taped messages can generally be erased when rewinding.

Sound quality

An answering machine should reproduce your voice and callers' voices so they are recognizable, intelligible, and fairly easy to listen to. Even though you shouldn't have a problem understanding recorded speech on just about any model, there are differences in reproduction fidelity. So, try to listen to a machine before you buy it. If you can't, insist on return privileges.

Remote access

Remote control permits you to call home from a distant phone and hear any messages that have come in. Although surveys indicate that this is among the most wanted features, it's also among the least used. The reason could be that the remote operation is, for some machines, neither convenient nor easy. Indeed, the operation of some remote controls is nearly incomprehensible.

The actual details of operation vary greatly from one model to another. Generally, it works like this. You call your answering machine, and while your greeting plays, you signal the machine with a special sound code that identifies you as the owner. The machine switches to a special mode to allow you to hear messages and, on some machines, operate other functions as well. When you hang up, the unit resets itself to the Answer mode.

Beeper remote. A number of units require the use of a beeper. You "talk" with your answering machine by holding the beeper against the telephone's mouthpiece and playing a set of tones over the phone line. The beepers, which are about the size of a pocket calculator, are a nuisance to carry around, and they are easily lost or forgotten.

Tone-dialing remote. A much better method of remote control is tone-dialing. You signal your answering machine by punching numbers on the keypad of a tone-dialing phone. Although clearly more convenient than using a beeper, it does require you to use a true tone-dialing phone to reach the machine. With a beeper, you can use any kind of telephone.

Security. Remote control of your answering machine means others may also have access to your messages if they have the signal or code.

The least secure remote control is a single-tone beeper. It's easy to duplicate that single tone. A beeper that emits a series of tones is somewhat more secure against eavesdroppers. Units with tone-dialing remote control provide security with a one- to three-digit code number that you punch in. The code number can be assigned by the manufacturer or, better, selected by you.

Some units use what might be called "false" Touch-Tone signals. They detect only one of the two simultaneous tones generated by each key on a tone-dialing phone. That allows the keys in the same row as the code digit to work as well as the original code, and makes the code easier to crack.

Models that offer the most security are the ones that use "true" Touch-Tone signals and that let you change your own code at will.

Toll-saver. This worthwhile feature saves you phone charges. More than one or two rings means that no messages are waiting. You can hang up before the machine answers on the fourth ring.

Listening to messages. Once you reach your machine, it should play messages without a lot of bother. Most do. A typical machine lets you review, replay, and save messages in the same call. Some also let you leave a "marker" on the tape to remind yourself which messages you've already heard.

Many machines also allow you to change your outgoing message remotely. That would be a help if you wanted to refer callers to another number.

Most machines stop after playing the last message and reset to Record. But a few don't sense the end of the messages. If you hang up or get disconnected without commanding those machines to stop, they will tie up your home phone line for whatever time it takes to play the rest of the incoming-message tape. That could mean a wait of nearly thirty minutes before you can call back and reset the tape.

And a few others may not let you use the remote if the incoming-message tape is full. With those, it's important to call and reset the machine often to avoid a full tape.

Additional features

Announce only. Many answering machines give you the option of playing only an outgoing message without taking incoming messages. That's ideal for announcing club meetings or a change in a scheduled event.

Voice-prompting. A machine with voice-prompting is simple to operate. The only thing it doesn't explain is how to set the built-in clock. After you play back the last message, the voice may say, "That was your last message." If you use the remote control to retrieve messages, it might say, "Hello, you have two messages," and then gently prompts you through all the steps of remote operation.

Time-stamping. It's often important to know when a call came in, but callers usually don't say when they've called.

Remote answer-mode setting. Unless you leave the answering machine on all the time, you could easily forget to turn it on when you're leaving the house.

Dual greetings. A few models break the monotony of a recorded greeting by playing one of two messages.

Tape-full indicator. Anyone who gets lots of messages might appreciate a unit that either hangs up if the message tape is full or indicates to callers that the tape is full. Some models also play a special message asking callers to call back.

Two-way recording. Some answering machines can work like an ordinary tape recorder, so you can record a conversation you're having with someone else. Although most machines do this silently, some indicate that the tape is rolling with a periodic beep.

Memo recording. It's possible to use some answering machines as a family message center. Instead of writing out a note, you can leave a brief message on the machine. Other family members can play back the tape when they come home.

Dictation. A few machines can be used for dictation. They let you stop and start the tape at will and may offer an optional external microphone for dictation.

Calling-party control (CPC). In certain areas of the country, the phone company sends a CPC signal when the caller hangs up. At that signal, some machines stop recording immediately without waiting for the VOX circuit to detect the hang-up. That leaves less dead space between messages on the tape.

Ratings of telephone answering machines

Listed in order of estimated quality. Models judged equal in quality are bracketed and listed alphabetically. Ratings should be used in conjunction with Features tables on pages 335–37. Prices are as published in a **May 1986** report.

⦿ ◓ ○ ◒ ●

Better ⟵ ⟶ *Worse*

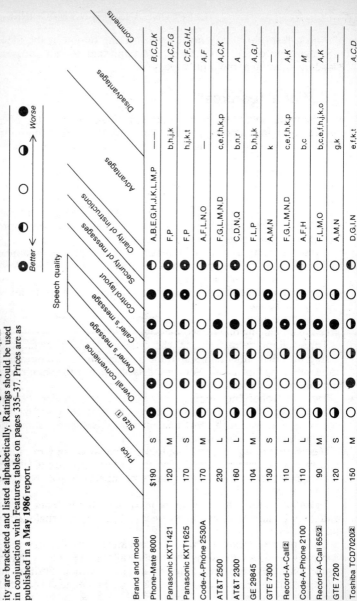

Brand and model	Price	Size [1]	Overall convenience	Speech quality: Owner's message	Speech quality: Caller's message	Control layout	Security of messages	Clarity of instructions	Advantages	Disadvantages	Comments
Phone-Mate 8000	$190	S	⦿	⦿	⦿	⦿	●	◓	A,B,E,G,H,J,K,L,M,P	——	*B,C,D,K*
Panasonic KXT1421	120	M	○	○	⦿	○	⦿	⦿	F,P	b,h,j,k	*A,C,F,G*
Panasonic KXT1625	170	S	○	◓	◓	◓	⦿	⦿	F,P	h,j,k,t	*C,F,G,H,L*
Code-A-Phone 2530A	170	M	◓	◓	○	○	○	◒	A,F,L,N,O	—	*A,F*
AT&T 2500	230	L	○	○	◓	●	○	◓	F,G,L,M,N,D	c,e,f,h,k,p	*A,C,K*
AT&T 2300	160	L	◒	◓	◓	●	◒	⦿	C,D,N,Q	b,n,r	*A*
GE 29845	104	M	◒	◓	◓	◓	○	○	F,L,P	b,h,j,k	*A,G,I*
GTE 7300	130	S	○	○	○	●	⦿	○	A,M,N	k	—
Record-A-Call [2]	110	L	○	○	◒	⦿	○	○	F,G,L,M,N,D	c,e,f,h,k,p	*A,K*
Code-A-Phone 2100	110	L	○	○	◒	⦿	◒	◓	A,F,H	b,c	*M*
Record-A-Call 655 [2]	90	M	◒	◓	◓	⦿	○	○	F,L,M,O	b,c,e,f,h,j,k,o	*A,K*
GTE 7200	120	S	◒	○	○	●	◒	○	A,M,N	g,k	—
Toshiba TCD7020 [2]	150	M	○	●	◒	◒	○	◓	D,G,I,N	e,f,k,t	*A,C,D*

Cobra AN8400	130	L	●	○	◒	●	●	◒	—	b,c,d,k	*A*
Message-Minder MM1850	157	L	●	○	◉	◓	◒	○	—	b,c,d,e,f,g,k,s	*A,K*
Toshiba TCD7010 [2]	127	M	◒	●	◒	◒	○	◓	D,G,H	e,f,h,t	*A,C,D*
Radio Shack TAD212	120	M	●	◓	◓	◓	◒	○	O,P	b,c,d,e,f,g,h,k,o	*A,E*
Phone-Mate 5000	120	S	◒	○	●	◒	●	◓	A,B,G,J,M,O	a,i,l,m,n,t	*J,K*
Sanyo TAS3000 [2]	100	L	●	◓	○	●	◒	●	M,N	b,d,e,f,k,r	*A,C,D*
Radio Shack TAD214	180	L	◒	◒	●	◓	◒	○	D,H	b,c,d,e,f,g,h,o	*A*

[1] *S = small (approx. 6 × 8 × 2 in.); M = medium (approx. 8 × 9 × 2 in.); L = large (approx. 9 × 11 × 13 in.).*

[2] *Model discontinued at the time the article was originally published in* Consumer Reports. The information has been retained here, however, for use as a guide to buying.

[3] *Price CU paid.*

Specifications and Features

All have: • Voice-controlled message length (VOX). • Phone-in remote playback of calls. • Telephone cord with modular plug, plus ac power cord. • Tapes that can be replaced. • Call-screening with adjustable volume.

Except as noted, all: • Use standard cassettes. • Have momentary Rewind and Fast-Forward for caller's message. • Have built-in microphone for recording owner's message. • Have built-in jack for telephone connection. • Have variable-length owner's message. • Lose count of callers' messages when power is interrupted. • Require only one action to play owner's message and to reset for answering. • Allow erasing of callers' messages from tape. • Automatically stop or reset after last caller's message is played back. • Allow remote playback even if callers' message-tape becomes full. • Have ringer equivalence number (REN) of 1.0 or less.

Key to Advantages

A–Automatically enters Answer mode when no other function is selected.
B–Automatically resets for new messages after playback.
C–Rewind backs up by one message when pressed during playback.
D–Fast-forward skips forward one message when pressed during playback.
E–Batteries preserve incoming-message count in case of momentary power loss.
F–Rewinds back over a portion of VOX delay; reduces time between callers' messages.
G–Answer mode cancels automatically if any phone is picked up.
H–Day and time are announced by voice after each message; voice judged to be clear.
I–Time is announced by voice before each message; voice judged not to be very clear.
J–Can be set to Answer mode remotely by letting phone ring 10 or more times.
K–Prompts user with voice for most functions; judged a significant aid to use.
L–Hang-ups after the beep are not counted or recorded.
M–Automatically plays back owner's message for review after recordings.
N–VOX can be overridden to limit length of callers' messages; can prevent tie-ups on noisy phone lines.
O–Hang-ups during greeting not counted.
P–Answer cycle cancels automatically if connected phone is picked up.

Ratings of telephone answering machines (continued)

Key to Disadvantages
a–Uses single tape for both owner's and callers' messages.
b–Requires endless-loop cassette for owner's message; replacements may be harder to find than standard cassettes.
c–Requires leaderless cassette for caller's message; replacements may be harder to find than standard cassettes.
d–Recorder's heads do not retract between uses; may cause premature tape wear.
e–Requires more than 1 action to play back callers' messages.
f–Callers'-message tape must be rewound after playback before new messages can be recorded.
g–Requires more than 1 action to reset machine after remote playback of callers' messages.
h–Rewind control locks when pressed; slightly less convenient than control that is activated only when you hold it down.
i–No Fast-Forward.
j–Fast-Forward control locks when pressed; slightly less convenient than control that is activated only when you hold it down.
k–Does not automatically stop or reset after last message has played back.
l–Can't erase previous messages; new messages merely recorded over old ones.
m–Momentary power loss erases greeting.
n–Momentary power loss loses callers' messages up to point of power loss.
o–Lacks built-in jack for telephone connection; two-way adapter must be used if answering machine and phone share a jack.
p–Requires external microphone (supplied) for recording owner's message.
q–Ties up phone line for up to ½ hr. if remote playback of callers' messages is interrupted.
r–Remote playback of callers' messages not possible after tape becomes full.
s–Length of owner's message fixed by tape length; 20 sec. tape supplied.
t–VOX circuit may cause problems on noisy phone lines.

Key to Comments
A–Controls arranged so that telephone can rest on top of unit.
B–Batteries must be replaced yearly.
C–Can be used with multikey office phone system using A-A1 contacts.
D–Can't be used on phone jacks supplying lighted-dial voltage or two-line phones.
E–Ringer equivalence number (REN) greater than 1.0.
F–Has headphone jack.
G–Caller's-message time can be fixed at 60 sec. (30 sec. for *Uniden*).
H–Has jack for remote-control microphone.
I–Line held briefly for call pick-up if machine is manually stopped during caller's message.
J–Supplied caller tape will hold only 15 min. of calls. Longer tapes are readily available.
K–Remote needs long tones—may not work on phones with short tones.
L–Accepts external microphone with multiple functions for recording dictation.

Telephone answering-machine features

Listed below are specifications and features for the tested answering machines and 9 essentially similar models.

						Owner's message		Caller's message				Remote control							
Brand and model	Suggested retail price	Tape size [1]	Maximum message length, sec. [2]	Built-in microphone	VOX	Maximum time limit, min. [3]	Calling party control	Call counter [4]	Playback stops after end of current messages	Activation [5]	Change security code?	Change message?	Toll-saver	Ring-number selection	Two-way record	Memo	Dictation	Includes telephone	Comments
AT&T 2300	$160	C	20	✓	✓	Long	—	B	—	B	—	—	✓	2,4	✓	—	✓	—	C,F,K,L,M,R,S,T
2500	230	C	Long	—	✓	Long	—	D	—	T[7]	—	✓	✓	1-6	✓	—	✓	—	C,D,F,H,K,M,R,S,T
Cobra AN8400	130	C	20	✓	✓	2	—	B	—	B	—	—	✓	2,4	—	✓	—	—	B,T
Code-A-Phone 2100	110	C	20	✓	✓	Long	—	L	✓	T	—	—	✓	2,4	—	✓	—	—	H,M,R,T
2300	120	C	20	✓	✓	Long	—	D	✓	T	✓	✓	✓	106	—	✓	—	—	C,S,T
2150	150	C	20	✓	✓	Long	—	L	✓	T	—	—	✓	2.4	—	✓	—	✓	H,M,P,R,T
2530A	170	M	150	✓	✓	Long	—	L	✓	T	—	✓	✓	1,4	✓	✓	—	—	C,D,E,F,H,M,R,S,T
2350	160	C	20	✓	✓	Long	—	D	✓	T	✓	—	✓	1-6	—	✓	—	✓	C,Q,R,S,T
2555[6]	180	M	150	✓	✓	Long	—	L	✓	T	—	✓	✓	1,4	✓	✓	—	✓	C,D,E,F,H,M,P, R,S,T
General Electric 29845	104	C	30	✓	✓	Long	✓	B	—	B	—	—	✓	2,4	—	—	—	—	H,T
GTE 7200	120	M	Long	✓	✓	Long	—	B	—	B	—	—	✓	2,4	✓	✓	—	—	G,J,R,T
7300	130	M	Long	✓	✓	Long	—	D	—	T	✓	✓	✓	2,4	✓	✓	—	—	G,J,R

Telephone answering-machine features (continued)

Listed below are specifications and features for the tested answering machines and 9 essentially similar models.

Brand and model	Suggested retail price	Tape size [1]	Maximum message length, sec. [2]	Built-in microphone	VOX	Maximum time limit, min. [3]	Calling party control	Call counter [4]	Playback stops after end of current messages	Owner's message: Activation [5]	Owner's message: Change security code?	Caller's message: Change message?	Caller's message: Toll-saver	Caller's message: Ring-number selection	Remote control: Two-way record	Remote control: Memo	Remote control: Dictation	Includes telephone	Comments
Message-Minder MM1800	157	C	20[9]	✓	✓	3.5	✓	L	—	B	—	—	—	1–4	✓	—	—	—	C,F,M,R,T
MM1850	157	C	20[9]	✓	✓	3.5	✓	L	—	T	—	—	—	1–4	✓	—	—	—	C,F,H,M,R,T
Panasonic KXT1421	120	C	30	✓	✓	Long	✓	B	—	T	✓	—	✓	2,4	—	✓	✓	—	H,K,R,T
KXT1625	170	M	Long	✓	✓	Long	✓	B	—	T	✓	✓	✓	2,4	—	✓	✓	—	C,D,F,H,K,R,S
KXT2421	220	C	30	✓	✓	Long	✓	B	—	T	✓	—	✓	2,4	✓	—	✓	✓	G,H,K,R,T
KXT2425	290	C	30	✓	✓	Long	✓	D	—	T	✓	✓	✓	2,4	✓	✓	✓	✓	G,H,K,R,T
Phone-Mate Minimate 5000	120	M	60	✓	✓	3	—	B	✓	T	—	—	✓	1,4	—	✓	—	—	A,H,T
Minimate 5050	160	M	60	✓	✓	3	—	B	✓	T	—	—	✓	1,4	—	✓	—	✓	A,H,J,P,T
8000	190	M	Long	✓	✓	Long	—	D	✓	T	—	✓	✓	1,4	✓	✓	—	—	F,R,T,U,V
8050	230	M	Long	✓	✓	Long	—	D	✓	T	—	✓	✓	1,4	✓	✓	—	✓	F,P,R,T,U,V
9700	240	M	Long	✓	✓	Long	—	D	✓	T	—	✓	✓	1,4	✓	✓	✓	—	N,Q,R,T,U,V
9750	280	M	Long	✓	✓	Long	—	D	✓	T	—	✓	✓	1,4	✓	✓	✓	✓	N,P,Q,R,T,U,V,
Radio Shack TAD210	80	C	20[8]	✓	—	3	—	L	—	—	—	—	—	1,4	—	—	—	—	C,M
TAD212	120	C	20[8]	✓	✓	3	✓	L	—	B	—	—	✓	1,4	—	—	—	—	C,H,M
TAD214	180	C	30	✓	✓	3	✓	L	✓	B	—	✓	✓	1,4	✓	—	—	—	C,F,H,I,M,R,S,T

Record-A-Call 655[6]	90[7]	C	20	✓	✓	3	—	L	—	T	—	—	—	2,4	—	—	—	—	C,M,R
675[6]	110[7]	C	Long	—	✓	Long	✓	B	—	T	—	✓	✓	4	✓	—	✓	—	C,D,F,H,K,M,R,T,U
Sanyo TAS1100	70	C	30	✓	✓	Long	✓	L	—	—	—	—	—	1	—	✓	—	—	M,R
TAS3000[6]	100[7]	C	30	✓	✓	Long	✓	L	—	B	—	—	—	1,3,5	—	✓	✓	—	C,K,L,M,R,T
Sony ITA500	220	M	60	✓	✓	Long	✓	L	✓	T	✓	✓	—	1	✓	—	✓	✓	G,K,N,O,T
ITA600	260	M	60	✓	✓	Long	✓	L	—	T	✓	✓	—	1	✓	—	—	✓	G,K,N,O,T
Toshiba TCD7010	127	C	90	✓	✓	Long	✓	B	—	T	—	—	✓	3,4,5	—	—	✓	—	M,R,S,T
TCD7020	150	C	90	✓	✓	Long	✓	B	—	T	—	✓	✓	3,4,5	✓	—	✓	—	K,M,R,S,T,U,V

[1] *M = microcassette; C = cassette.*

[2] *Long = 5 min. or more.*

[3] *Long = as long as rest of tape.*

[4] *L = light, no counter; B = blinking light indicates count; D = digital display of count.*

[5] *B = beeper; T = Touch-Tone.*

[6] *Model discontinued at the time the article was originally published in* Consumer Reports. *The information has been retained here, however, for use as a guide to buying.*

[7] *Price CU paid.*

[8] *Fixed message length.*

Key to Comments
A–Uses single cassette.
B–Phone available; hearing-aid compatible.
C–Has "announce-only" mode.
D–Announce-only message can be longer than 1 min.
E–Announce-only message can be interrupted with code to use remote features.
F–Records conversations silently.
G–Records conversations with beep.
H–User can leave "marker" on tape.
I–Tape counter.
J–Can fit on standard wall-telephone base.
K–Can be used for long dictation.
L–Will not answer if caller-message tape full.
M–Has automatic Save function.
N–Dual greeting.
O–Call-forwarding.
P–Phone is hearing-aid compatible.
Q–Answers 2 phone lines.
R–Remote erase of message tape.
S–Remote Skip and/or Fast-Forward.
T–Remote backspace and/or repeat message.
U–Tape-full message.
V–Has time-stamping for callers' messages.

Multiple line capability. The most common machines can handle calls on only one phone line. Most models can also be used on a two-line phone system to answer calls on one line without interfering with the other. Some models can't be connected directly to a two-line phone jack, but you can buy a multiline isolating adapter to go between the unit and the phone line.

Synthesized voice. Some of the budget-priced models on the market cut corners by including an electronically synthesized voice on a microchip for the outgoing message. That eliminates the need and the cost of a second tape drive.

Curiously, some higher-priced units also use synthesized speech to greet callers. On these units, however, the preprogrammed electronic greeting is optional, and you have a choice of a "male" or "female" voice.

Some basics

Instructions. The instruction book should be well-translated (into English) and laid out logically.

Controls. Some machines have rather old-fashioned buttons for Rewind or Fast-Forward—once pressed, they remain on. You have to press Stop to halt the tape. It's better if a control is activated only when you hold it down.

REN number. Answering machines, like telephones, have a ringer equivalence number (REN), an indication of the amount of power required to make the phone ring. The RENs for all the phones and answering machines you have shouldn't add up to more than 5; otherwise, the electrical load may be so great that the phones may not ring. An answering machine should have a REN of 1.0 or less.

Tapes. Some models that use standard cassettes need special tapes that are "leaderless." That means they have no clear section at the beginning, so that the first few seconds of the initial incoming message won't be lost. Some machines require "endless loop" tapes for the outgoing message. These special tapes may be slightly harder to find than standard tapes. It's best if an answering machine doesn't require them.

Most tapes should last several years. "Endless loop" tapes may wear faster than other kinds because of internal friction.

Repairs. You may encounter exorbitant factory-service charges if your answering machine requires a repair after the warranty period. Compare hourly labor charges before you leave your machine anywhere. Most problems with answering machines are mechanical and should take only a short time to fix.

Index